Kanai, Campbell

Film Processing Advances

Polymer Processing Society
Progress in Polymer Processing Series

T. Kanai / G. A. Campbell
Film Processing Advances

T. Kanai / G. A. Campbell
Film Processing

I. Manas-Zloczower
Mixing and Compounding of Polymers, 2nd Edition

M. R. Kamal / A. I. Isayev / S.-J. Liu
Injection Molding

J. L. White / H. Potente
Screw Extrusion

W. E. Baker / C. E. Scott / G.-H. Hu
Reactive Polymer Blending

I. M. Ward / P. D. Coates / M. M. Dumoulin
Solid Phase Processing of Polymers

R. S. Davé / A. C. Loos
Processing of Composites

Toshitaka Kanai
Gregory A. Campbell

Film Processing Advances

With Contributions by:
J. F. Agassant, J. Breil, G. A. Campbell, Y. Demay, A. Funaki, T. Kanai, T. Kikutani, V. Sidiropoulos, M. A. Spalding, W. Takarada, J. Vlachopoulos, K. Xiao, M. Zatloukal

HANSER

Hanser Publishers, Munich   Hanser Publications, Cincinnati

The Editors:
Dr. Toshitaka Kanai
KT Polymer,
5-7-14 Kuranami Dai, Sodegaura, Chiba, 299-0245, Japan

Dr. Gregory A. Campbell
Castle Research Associates
1145 Mason Bay Rd., Jonesport, ME 04649, USA

Distributed in North and South America by:
Hanser Publications
6915 Valley Avenue, Cincinnati, Ohio 45244-3029, USA
Fax: (513) 527-8801
Phone: (513) 527-8977
www.hanserpublications.com

Distributed in all other countries by
Carl Hanser Verlag
Postfach 86 04 20, 81631 München, Germany
Fax: +49 (89) 98 48 09
www.hanser-fachbuch.de

Cataloging-in-Publication Data is on file with the Library of Congress

Bibliografische Information Der Deutschen Bibliothek
Die Deutsche Bibliothek verzeichnet diese Publikation in der Deutschen Nationalbibliografie;
detaillierte bibliografische Daten sind im Internet über <http://dnb.d-nb.de> abrufbar.

ISBN 978-1-56990-529-6
E-Book ISBN 978-1-56990-536-4

Production Management: Steffen Jörg
Coverconcept: Marc Müller-Bremer, www.rebranding.de, München
Coverdesign: Stephan Rönigk

PROGRESS IN POLYMER PROCESSING SERIES

Changyu **Shen**
Dalian University of Technology
CHINA

Lih-Sheng (Tom) **Turng**
University of Wisconsin
U.S.A.

Mark **Smith**
Carl Hanser Verlag GmbH & Co. KG
GERMANY

John **Vlachopoulos**
McMaster University
CANADA

Giuseppe **Titomanlio**
Università degli Studi di Salerno
ITALY

Ica **Manas-Zloczower**
Case Western Reserve University
U.S.A.

Foreword

The polymer and plastics industries have had a profound techno-economic impact on society for almost a century. In fact, it has been suggested that the advent and use of polymers and plastics products have represented a revolutionary technological change. They are used in packaging, furniture, construction materials, automotive, aerospace, sporting goods, biomedical, electronics, communications, and so on. More importantly, they have adapted to the ever changing social and technological demands. Thus, many of the current popular products, such as smart phones, computers, and other technological innovations would be difficult to contemplate in the absence of polymers. It does not seem likely that the foreseeable future will see a reduction in the important role that polymers and plastics will play in future technological development.

Cognizant of the role that polymers played and will continue to play in our lives, a group of polymer scientists and engineers from various countries around the world founded the Polymer Processing Society (PPS) in March 1985 at the University of Akron, Akron, Ohio, USA. According to its constitution, the goal of the PPS is to foster scientific understanding and technical innovation in polymer processing by providing a discussion forum in the field for the worldwide community of engineers and scientists. Thus, PPS has attempted to achieve this goal using the following mechanisms:

1. Organization of annual and regional conferences rotating among the various regions of the world and the dissemination of technical content of the conferences in the form of proceedings.
2. The publication of the International Polymer Processing (IPP) Journal.
3. The publication of the Progress in Polymer Processing (PPP) Series.

So far, these activities have allowed the PPS and its members to exchange information and ideas about the evolution of the principles and methods of polymer science and engineering and their application to the generation of innovative products, processes and applications.

Since the formation of PPS, eleven PPP volumes have been published. Four distinguished leaders in the polymer processing field have served as series editors: Leszek Utracki, Warren Baker, Kun Sup Hyun, and James L. White. Last summer at PPS 29 in Nuremberg, Germany, I was asked by the Executive Committee of PPS to serve as

PPP series editor. It is my hope, that with the help of the Advisory Editorial Board, our colleagues in the polymer processing field, and Hanser Publications, to publish at the rate of about one book every year. We already have two books under preparation. I encourage prospective authors to contact me or any of the Advisory Board members with their ideas and suggestions.

One of my first tasks has been to follow and expedite the completion of *Film Processing Advances*. This has given me the opportunity to refresh and expand my contacts with the editors, Drs. Kanai and Campbell, whom I have known for many years. As I have done some work in the area of film processing, I always benefited from reading their works and meeting them at conferences. Thus, it was easy to work together with them and with the publisher, Hanser, to set up the necessary mechanisms and procedures for a smooth and timely finish for this ambitious project. It is a real pleasure to have *Film Processing Advances* as the first PPP project completed during my first term as a series editor. Obviously, the credit goes to Professors Campbell and Kanai and to the contributors of the chapters for their tireless efforts. We also owe special thanks to the editorial staff at Hanser, especially Ms. Cheryl Hamilton, who handled the details of publication smoothly and efficiently.

As we all know, plastic films represent a major component of the polymer and plastics business. Plastic films are used extensively in packaging products. They have withstood and adapted to various pressures and requirements. Film processing technology continues to advance with the advent of improved extrusion and die design technologies, development of advanced film blowing and casting techniques; temperature, orientation, and crystallization control, and advanced computer simulation, monitoring and control systems. Thus, the publication of *Film Processing Advances*, by the same editors of the successful *Film Processing*, represents a timely technical update on the status of film processing technology.

Finally, on behalf of the Polymer Processing Society and the PPP Editorial Advisory Board, I would like to express our sincerest thanks and appreciation for Professor Gregory Campbell and Professor Toshitaka Kanai for the immense amount of effort, time, and dedication that they have contributed to the editing and preparation of this book. I also wish to thank the other authors for contributing their excellent chapters. Also, we owe a lot of thanks to Ms. Cheryl Hamilton and other Hanser staff for the organization of the copyediting of the book and timely completion of this project.

Musa R. Kamal
Series Editor

Contents

Contributors

Agassant, Prof. J.-F., MINES Paristech, Centre de Mise Forme des Matériaux, UMR 7635 CNRS, 06904 Sophia Antipolis, France, jean-francois.agassant@mines-paristech.fr

Breil, Dr. J., Brückner Maschinenbau GmbH & Co. KG, Königsberger Str. 5–7, 83313 Siegsdorf, Germany, dr.juergen.breil@brueckner.com

Campbell, Dr. G. A., Castle Research Associates, 1145 Mason Bay Rd., Jonesport, ME 04649, USA, swcgac@maineline.net

Demay, Prof. Y., Université de Nice Sophia-Antipolis, Laboratoire Jean Alexandre Dieudonné, UMR CNRS 6621, 06108 Nice Cedex 2, France, yves.demay@unice.fr

Funaki, Dr. A., Performance Materials Laboratories, Idemitsu Kosan Co., Ltd., 1-1 Anesaki-Kaigan, Ichihara-city, Chiba, 299-0193, Japan, akira.funaki@idemitsu.com

Kanai, Dr. T., KT POLYMER, 5-7-14 Kuranami Dai, Sodegaura, Chiba, 299-0245, Japan, toshitaka.kanai@ktpolymer.com

Kikutani, Prof. T., Department of Organic & Polymeric Materials, Graduate School of Science & Engineering, Tokyo Institute of Technology, 2-12-1-S8-32, O-okayama, Meguro-ku, Tokyo, 152-8552, Japan, kikutani.t.aa@m.titech.ac.jp

Sidiropoulos, Dr. V., Department of Chemical Engineering, McMaster University, Hamilton, Ontario, L8S 4L7, Canada, sidir@cogeco.ca

Spalding, Dr. M. A., The Dow Chemical Company, 433 Building, Midland, MI 48667, USA, MASpalding@dow.com

Takarada, Dr. W., Department of Organic & Polymeric Materials, Graduate School of Science & Engineering, Tokyo Institute of Technology, 2-12-1-S8-32, O-okayama, Meguro-ku, Tokyo, 152-8552, Japan, takarada.w.aa@m.titech.ac.jp

Takashige, Dr. M., Idemitsu Unitech Co., Ltd., Idemitsu Nagahori Bldg. 11F, 3-4-26 Minamisennba, Chuo-ku, Osaka-city, Osaka, 542-0081, Japan, masao.takashige@idemitsu.com

Vlachopoulos, Prof. J., Department of Chemical Engineering, McMaster University, Hamilton, Ontario, L8S 4L7, Canada, vlachopj@mcmaster.ca

Xiao, Dr. K., Celgard LLC, Charlotte, NC 28277, USA, karen.xiao@celgard.com

Zatloukal, Prof. M., Polymer Centre, Faculty of Technology, Tomas Bata University in Zlin, TGM 275, Zlin 76272, Czech Republic and Centre of Polymer Systems, University Institute, Tomas Bata University in Zlin, Nad Ovcirnou 3685, Zlin 76001, Czech Republic, mzatloukal@ft.utb.cz

Preface

Film Processing was published in 1999 and the references in that book extended up to 1992. Because of the broad treatment of film processing fundamentals, it became a popular reference book. Many professionals working on polymer films requested that we publish a newer version of *Film Processing* that includes more recent technologies and that addresses the latest literature. The PPS and Hanser agreed that it was time to publish *Film Processing Advances*. This edition is still a technically oriented book to be used by film processing professionals, graduate students, and researchers and it covers recent technologies of film extrusion, extruder screw design, die design, film structure, film temperature, crystallization dynamics, and film properties. Furthermore, *Film Processing Advances* also focuses on multilayered film and recent developments of high performance films. The chapters in this edition are all essentially new and were designed to complement the materials found in *Film Processing*.

The polymeric film industry is going through a major transition, from relying only on commodity resins as their film base, to the use of more well-defined materials and processes. In terms of resins, metallocene catalyzed polymers have been developed, and as a result, the polymer structure can now be tailored to improve or help optimize film processes and film property design. Engineering plastics such as polyethylene terephthalate (PET), polycarbonate, polyamides, and polyphenylene sulfide have become popular film resins that provide high temperature resistance in films when desired. The line speeds of film production have reached more than 500 m/min for oriented polypropylene and oriented PET. In addition, the technology for "crystal" polystyrene film has also been commercially developed and applied extensively to food packaging. These achievements in developing high speed production, very thin film, and alternative starting materials followed the development of new machines, improvements in polymer material characteristics, and optimization of processing techniques. A further demand for more complex, multilayered films for various uses in packaging has led to the development of more complicated die designs and film process control so that the required film thickness and composition can be maintained. All of these advances in film production technology are addressed in this book.

Film extrusion is one of the most popular commercial polymer processing techniques and the resulting film products are widely used in our daily lives. Polymeric film constitutes about 35% of all plastics products at this time. Recently, because of competition from other materials, the economics of film production have moved to higher productivity of film production methods: High speed lines and high quality reproducible film have been required in order to reduce cost and to obtain additional value. A few examples of this economically driven technology are the production of very thin capacitor film, separator film for batteries in mobile devices and electric vehicles, optical films for liquid crystal displays, and multilayered film for the protection of fresh food. To commercially achieve these requirements, one needs to employ a wide range of technology, not only the traditional empirical knowledge that has historically dominated the film industry, but also a more basic understanding of the chemistry, physics, and engineering that influence the film's properties. Topics introduced to the reader in this book are the structure of the polymer, the polymer rheology, theoretical analysis of film processes, structure and morphology of the film product, film temperature profile, process heat transfer, semicrystalline polymer crystallization dynamics, film stability, and finally the physical properties of film.

Unfortunately, it is not possible to present, in one book, all of the science and technology that relates to film extrusion. This book addresses to varying degrees the following topics: extruder screw design, multilayer extrusion die design and analysis, theory and technologies of the blown film process, cast film process, film process stability, multilayered film technology, biaxially oriented film analysis of the tentering and double bubble tubular film processes, and high performance film and sheets.

Film production covers a very wide range of fundamentals. Topics that are discussed in some detail are the theoretical analysis of film extrusion including film temperature dynamics, crystallization dynamics of film blowing, the multilayer die, polymer film structure, film physical properties, as well as how these topics are related.

We appreciate the substantial efforts of all the chapter authors who shared new and recently developed technologies in their specialized technical fields through their contributions. The Polymer Processing Society (PPS) continues to publish a series of books on polymer processing. We wish to thank Professor Musa R. Kamal, the series editor, for all of his efforts during the development of this volume. We hope our contribution to the PPS series will help the reader develop a fundamental introductory understanding of film extrusion. Since *Film Processing* and *Film Processing Advances* were designed to cover complimentary subject matter, the film practitioner is encouraged to consult both *Film Processing* and *Film Processing Advances* when seeking specific information about many aspects of polymeric film extrusion. Both volumes contain extensive references to the original literature that will lead to a more detailed discussion of any of the topics presented and necessarily condensed in a single book.

Chiba, Japan
Jonesport, USA

Toshitaka Kanai
Gregory A. Campbell
March 31st, 2014

1 Extruder and Screw Design for Film Processing

Mark A. Spalding and Gregory A. Campbell

Single-screw extruders are the preferred machines for plasticating and metering resin to downstream film processes. The extruder must provide a homogenous and stable extrudate at high rates and at the target discharge temperature and pressure. Moreover, gels must be at a low and acceptable level. Gels are defined as any particle that creates an optical defect in the film. Because film products are typically very thin and in the range of 15 to 250 µm, very small particles can cause observable defects. In many cases, these particles are created in the extruder, and thus screw design can be used to mitigate gels from the final film product.

This chapter will describe the single-screw extrusion process typically used for film processes, common screw designs, troubleshooting operations, and common gel defects that originate from the extruder. In-depth operation and fundamentals of the process are beyond the scope of this writing. The reader can learn more about the fundamentals of single-screw extrusion and troubleshooting in reference 1.

1.1 The Extrusion Process

All single-screw extruders have several common characteristics. The main sections of the extruder include the barrel, a screw that fits inside the barrel, a motor drive system for rotating the screw, and a control system for the barrel heaters and motor speed. A schematic for an extruder is shown in Fig. 1.1. Many innovations on construction of these components have been developed by machine suppliers over the years. A hopper is attached to the barrel at the entrance end of the screw, and the resin is typically fed by gravity (flood fed) into the feed section of the screw. The resin is typically purchased in pellet form (most polyolefins for example), but powders are common (PVDC resin). Recycle film from edge trim is often chopped and metered into the hopper. The extruder screw must first convey the pellets away from the feed opening, melt the resin, and then pump and pressurize it for a downstream filming process. This type of machine is referred to as a plasticating, single-screw extruder. The barrel is usually heated with a minimum of three temperature zones. These different temperature zones are consistent with the three functions of the screw: solids conveying, melting, and pumping or metering of the resin.

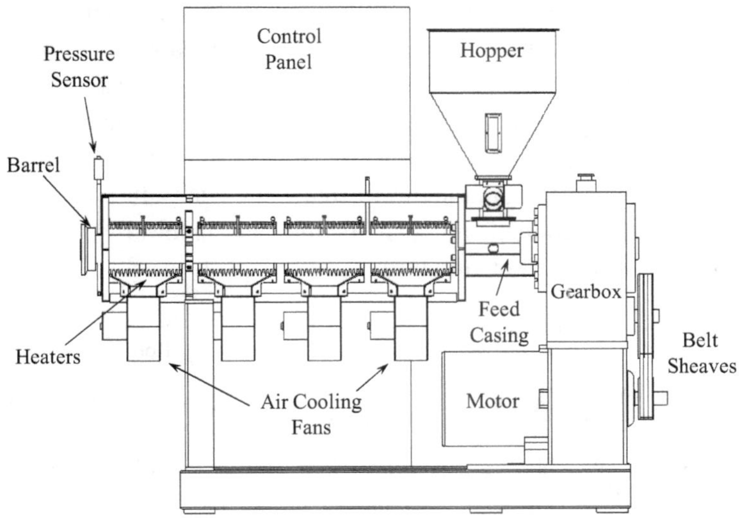

Figure 1.1 Schematic of a typical plasticating, single-screw extruder with a smooth-bore feed section. The extruder is equipped with four barrel heating and cooling zones and a combination belt sheave gearbox speed reduction drivetrain (courtesy of William Kramer of American Kuhne)

The single-screw plasticating process starts with the mixing of the feedstock materials. Typically, several different feedstocks are added to the hopper such as fresh resin pellets, recycle material, additives, and a color concentrate. Often these components need to be blended prior to adding them to the hopper. Next, the feedstock flows via gravity from the hopper through the feed throat of the feed casing and into the solids conveying section of the screw. Typically this feed casing is cooled using water. The feed section of the screw is typically designed with a constant depth and is about 4 to 8 barrel diameters in axial length. Directly after the solids conveying section is a section where the channel depth tapers to a shallow-depth metering section. The tapered depth section is commonly referred to as the transition or melting section. In general, the metering section is also a constant depth, but many variations exist where the channels oscillate in depth. The metering section pumps and pressurizes the material for the downstream unit operations including static mixers, screen filtering devices, and dies. The total length of the extruder screw and barrel is typically measured in barrel diameters or as a length-to-diameter (L/D) ratio. Section lengths are often specified in barrel diameters or simply diameters.

A conventional single-flighted screw is shown in Fig. 1.2. This screw has a single helix wound around the screw root or core. Multiple-flighted screws with two or more helixes started on the core are very common on high-performance screws. Screws with barrier flighted melting sections are very common in film processes because they provide high rates with lower discharge temperatures. Barrier melting sections have a secondary barrier flight that is located a fraction of a turn downstream from the primary flight, creating two flow channels in the transition section:

a solids melting channel and a melt conveying channel. Barrier flighted sections will be discussed in more detail later. Many high-performance screws [1] have two or more flights in the metering section of the screw. The screw is rotated by the shank using either specially designed splines or by keys with rectangular cross sections. The mathematical zero position of the screw is set at the pocket where the screw helix starts. Most extruder manufacturers rotate the screw in a counterclockwise direction for viewers positioned on the shank and looking towards the tip. This rotation convention, however, is not standard.

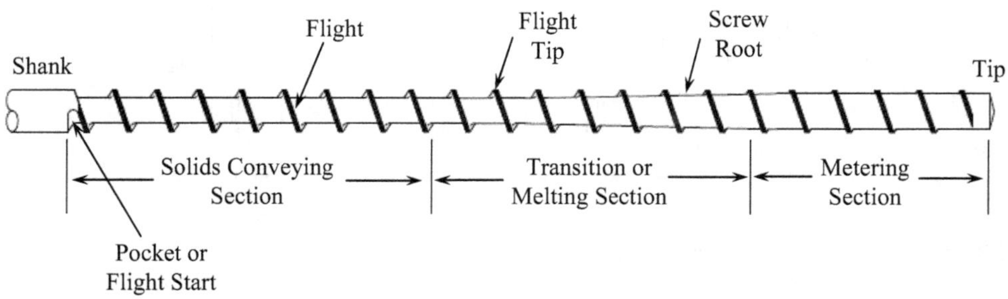

Figure 1.2 Schematic of a typical single-flighted screw (courtesy of Jeff A. Myers of Robert Barr, Inc.)

The flight is a helical structure that is machined into the screw and extends from the flight tip to the screw core or root. The flight has a width at the flight tip called the flight land. The small clearance between the flight land and the barrel wall minimizes the flow of resin over the land. The polymer that does flow between the clearances supports the screw and centers it in the barrel. The radial distance between the flight tip and the screw root is referred to as the local flight height or channel depth. The feed section usually has the largest channel depth and provides the largest cross-sectional volume in the screw. The deep channel conveys the relatively low bulk density feedstock pellets into the machine via the motion of the helix. The feedstock is conveyed forward into the transition section or melting section of the screw. The transition section increases in root diameter in the downstream direction, and thus the channel depth is decreasing. Here, the feedstock is subjected to higher pressures and temperatures, causing the feedstock to compact and melt. As the material compacts, its bulk density can increase by a factor of nearly two or more. As the feedstock compacts, the entrained air between the pellets is forced backwards and out through the hopper. For example, a pellet feedstock such as low density polyethylene (LDPE) resin can have a bulk density at ambient conditions of 0.58 g/cm^3, while as a fully compacted solid bed in the transition section the density will approach 0.92 g/cm^3 before melting starts. Thus for every unit volume of resin that enters the extruder, about 0.4 unit volumes of air must be expelled through the voids in the solid bed and then discharged through the hopper. The transition section is where most of the polymer is converted from a solid to a fluid. The fluid is then conveyed to the metering section where the molten resin is pumped to the discharge opening of the extruder. In general, the metering section of a conventional screw has a constant root diameter, and it has a much smaller channel depth than the feed section. The ratio of the channel depth in the feed section to the channel depth in the metering section is often referred to as the compression ratio of the screw.

The transition section shown in Fig. 1.2 is a conventional single-flighted design. These designs are still used for film operations, but barrier flighted melting sections are much more common. Barrier flighted melting sections will typically provide higher rates, lower discharge temperatures, a more stable discharge pressure, and extrudates that have fewer gels due to poor mixing. Barrier flighted melting sections are constructed by positioning a second flight (or barrier flight) in the transition section such that the solids are maintained on the trailing side and the molten resin on the pushing side. A schematic of a barrier flighted screw with an Egan (or spiral Maddock-style) mixer is shown by Fig. 1.3. A schematic of a cross section of a barrier melting section is shown in Fig. 1.4. The resin that is melted near the barrel wall is conveyed across the barrier flight and collected in the melt conveying channel. The key design parameters include the position of the barrier flight, the depths of the channels, and the undercut clearance on the barrier flight. The undercut clearance is measured by positioning a segment of straight bar stock across the two main flights and then measuring the gap between the bar and the barrier flight land. For most designs, the barrier flight undercut is constant for the entire length of the section. As a very general rule, the undercut is typically about 0.01 times the diameter of the screw. Undercuts that are smaller than this rule, however, are often used. The position of the barrier flight sets the width of both channels. Many styles of barrier melting screws are commercially available, and many different variations of the channel widths and depths are used commercially.

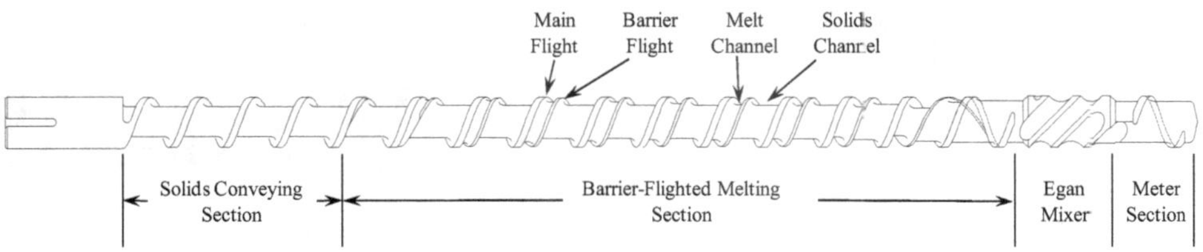

Figure 1.3 Schematic of a Steward barrier flighted screw with a downstream dispersive (Egan) mixer (courtesy of William Kramer of American Kuhne)

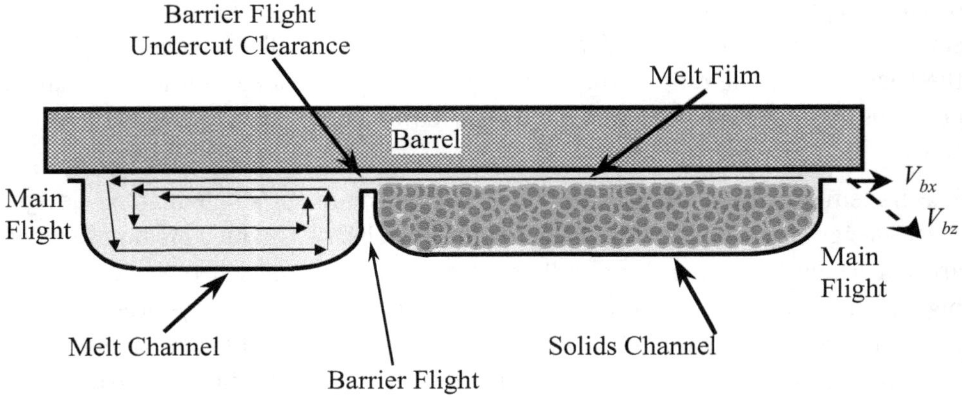

Figure 1.4 Cross-sectional view of a barrier melting section

Solid polymer fragments often exit the solids channel of barrier sections, or they can be discharged from conventional melting sections, especially at high rates and screw speeds. These solids need to be trapped and dispersed before the extrudate is shaped into film. Maddock-style mixers are typical dispersive mixers that are used for this application, but other mixers or high-performance sections are used. An Egan mixer (spiral Maddock-style mixer) is shown in Fig. 1.3 while other Maddock mixer styles are shown in Fig. 1.5.

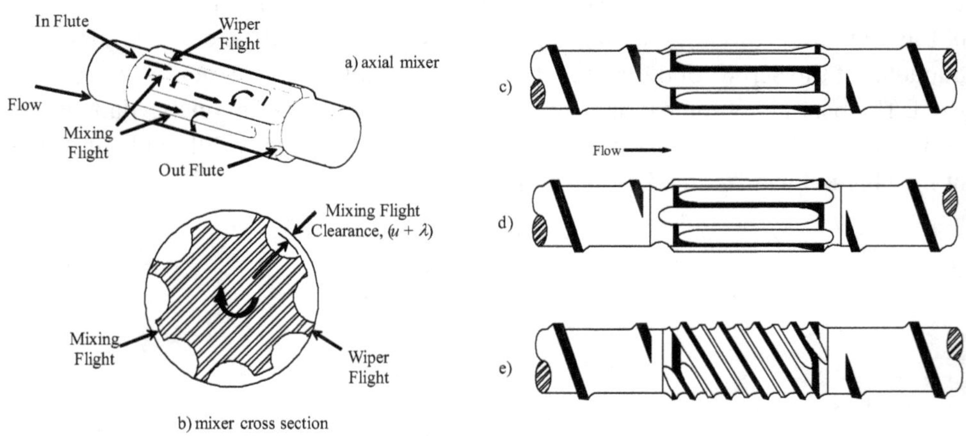

Figure 1.5 Schematic for Maddock-style mixers. (a) A mixer with the flutes aligned in the axial direction; (b) a cross-sectional view perpendicular to the screw axis showing the clearance for the mixing flight; (c) a mixer with the flutes aligned in the axial direction; (d) an axial mixer with pressure relief zones at the entry and exits; (e) an Egan mixer with spiral flutes (courtesy of Jeff A. Myers of Robert Barr, Inc.)

Maddock-style mixers [2] are very commonly used due to their low cost to manufacture and their ability to disperse solid fragments, trap and melt polymer solids, and mitigate color and compositional gradients. Many styles are on the market under two basic types: (1) flutes parallel to the screw axis, and (2) flutes in a spiral pattern in the same direction as the flights. Schematics for these devices are shown by Fig. 1.5. For small-diameter screws, the mixer is generally constructed with four inflow flutes (or channels) and four outflow flutes. Larger diameter screws will have more paired flutes due to the larger available area at the screw circumference. For a Maddock mixer with the flutes parallel to the axis of the screw, molten polymer flows into the inflow flutes via a pressure gradient and then either continues to flow downstream in the flute or is passed through a small clearance between the mixing flight and the barrel wall. This small clearance is responsible for providing the dispersive mixing characteristics of the device. Screw manufacturers typical specify the mixer flight height position relative to the main flight as an undercut. The undercut u for a 63.5 mm diameter screw is typically about 0.5 to 1.2 mm, although for some applications and designs the clearance can be smaller. For this size screw with an undercut of 0.50 mm and a flight clearance of 0.07 mm, the clearance between the mixing flight and the barrel wall is 0.57 mm. The material that flowed across the mixing flight is accumulated in the outflow flute and is then flowed via pressure to the discharge end of the mixer. The wiper flight shown in Fig. 1.5 is set at the same height as the flight in the metering section. For mixers with the flutes in a spiral pattern, some of the forwarding flow in the flutes is due to the rotational movement of the flute relative to the barrel wall. Performance and simulation details can be found in the references [3, 4].

The specification of the undercut on the mixing flight for Maddock-style mixers is critical to its performance. As previously stated, all material must flow through the clearance provided by the sum of the undercut and flight clearance. If the clearance is too large, some medium- and small-size solid polymer fragments will not be trapped and melted by the device. If the clearance is too small, then a high-pressure gradient can occur, and there exists the possibility of increasing the temperature of the resin beyond its thermal capabilities, that is, causing degradation. The shear stress that the material experiences for flow across the mixing flight of the mixer can be estimated using Eqs. 1.1 and 1.2. The shear stress level is responsible for breaking up agglomerates and dispersing solid polymer fragments. A higher shear stress level will improve the ability of the mixer to disperse smaller size fragments. This shear stress calculation is based on screw rotation physics and is as follows:

$$\dot{\gamma}_M = \frac{\pi(D_b - 2u - 2\lambda)N}{(u+\lambda)} \tag{1.1}$$

$$\tau_M = \eta \dot{\gamma}_M \tag{1.2}$$

where $\dot{\gamma}_M$ is the average shear rate for flow over the mixing flight in 1/s, u is the undercut clearance on the mixing flight, λ is the mechanical clearance of the flights, N is the screw rotation rate in revolutions/s, η is the shear viscosity at the temperature of the mixing process and at shear rate $\dot{\gamma}_M$, and τ_M is the shear stress that the material will experience for flow over the mixing flight. The stress level for flow across the mixing flight is typically between 50 and 200 kPa.

Several other design factors are important for the correct operation of Maddock-style mixers. These include the positioning of the mixer downstream from the melting section, the distance between where the meter flight ends and the mixer starts, and the elimination of resin stagnation regions. The mixer must be positioned on the screw downstream far enough such that only low levels of solid polymer fragments exist. If the level of solids is too high in the stream, then the fragments may be melted and dispersed at a rate slower than the rate of the entering solids, causing the mixer to become plugged with solids and reducing the rate of the machine. As shown in Fig. 1.5, the mixer should be positioned about 0.3 to 0.5 diameters away from the end of the upstream metering section flight. This creates an annular gap where the material is allowed to flow evenly into all inflow flutes of the mixer. The annular gap is often undercut as shown by Fig. 1.5(d). If the flights extend close to the mixer entry, then it is possible that the inflow flute near the trailing side of the flight will not operate completely full of resin and thus may cause the resin to stagnate and degrade. Moreover, flute depths should be streamlined and shallower at the entry end of the outflow flute and the exit end of the inflow flute. A common design error is to make these regions too deep, creating stagnation regions and causing resin degradation.

As an example of process design, linear low density polyethylene (LLDPE) resin is commonly used for blown film, cast film, and extrusion coating processes. Even though the resin grades are similar (melt indices or MIs do vary) for these three processes, the extrusion equipment is significantly different due to the requirements of the die and downstream equipment. The blown film process requires an extrudate that is relatively low in temperature and typically in the range of 200 to 220°C. In order to plasticate and produce an extrudate in this temperature range the metering channel is relatively deep and the screw would be designed to rotate at speeds less than 100 rpm. For a 150 mm diameter screw with a square-pitch lead length ($L = D_b$), the metering channel depth (H) would be between 9 and 12 mm. The cast film process requires an extrudate that is slightly higher in temperature and typically in the range of 240 to 260°C. The same 150 mm diameter screw would have a metering channel depth of 6 to 9 mm and the screw would rotate at higher speeds. The extrusion coating process requires an extrudate that is very high in temperature and often approaching 310°C. Here the metering channel depth would typically be about 3 to 4 mm for the 150 mm diameter screw, and the screw would be designed to rotate at very high speeds up to 230 rpm. These examples clearly show that the extrudate temperatures are set in part by the geometry of the metering section channel, the conditions of the process, and the MI of the resin.

The specific rate is often a good measure of the relative discharge temperature for a process. The specific rate is simply the rate divided by screw speed. For the examples above, the specific rate for the metering channels would be the highest for the blown film screw with the deep metering channel and the lowest for the extrusion coating process screw with the very shallow channel. Thus as a general guideline, the extrudate temperature decreases when the specific rate of the screw increases at constant barrel diameter. The specific rates for the screws in these examples increased because the channel depth increased. The specific rate could also be changed by adjusting the lead length. The calculated specific rotation rate can be also used as a similar guideline for discharge temperature. This guideline, however, can be violated if the channel is extremely deep and a large positive pressure gradient exists [1].

1.2 Rate Calculation

For smooth-bore extruders, the rate of the extruder is controlled by the metering section of the screw. The expected rate for the process can be calculated based on the geometry of the metering channel, screw speed, pressure gradient, and the melt density and shear viscosity for the resin. The basic screw geometry for a single-flighted (p = 1) metering channel is shown in Fig. 1.6.

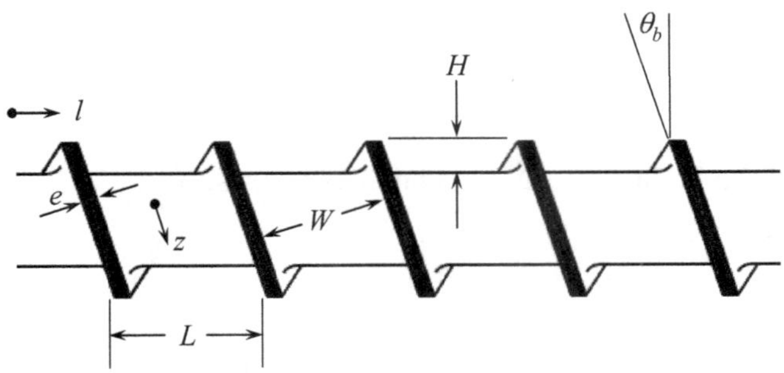

Figure 1.6 Geometric parameters for a single-flighted screw in the wound state

Two driving forces for flow exist in the metering section of the screw. The first flow is due just to the rotation of the screw and is referred to as the rotational flow component. The second component of flow is due to pressure gradients that exist in the z direction, and it is referred to as pressure flow. The sum of the two flows must be equal to the overall flow rate. The overall flow rate, Q, the rotational flow, Q_d, and the pressure flow, Q_p, for a constant depth metering channel are related as shown in Eq. 1.3. The subscript d is maintained in the nomenclature for historical consistency even though the term is for screw rotational flow rather than the historical drag flow concept. The method described here was developed by Rowell and Finlayson [5, 6] and later modified by Tadmor and Klein [7].

$$Q = Q_d - Q_p \tag{1.3}$$

The volumetric rotational flow term (Q_d) depends on the several geometric parameters and rotation speed. Since most extruder rates are measured in mass per unit time, the term Q_{md} is defined as the mass rotational flow:

$$Q_d = \frac{pV_{bz}WHF_d}{2} \tag{1.4}$$

$$Q_{md} = \frac{p\rho_m V_{bz}WHF_d}{2} \tag{1.5}$$

where ρ_m is the melt density at the average fluid temperature of the resin, V_{bz} is the z component of the screw velocity at the flight tip, W is the average width of the channel, p is the number of flight starts, H is the depth of the channel, and F_d is the shape factor for plane couette flow. The analysis using plane couette flow does not take into account the effect of the flights (channel helix) on the flow rate. For an infinitely wide channel, no flights, F_d would be equal to 1. As the channel width approaches the height, F_d is about 0.5. It is important to include the shape factors when evaluating commercial screw channels. This becomes extremely important for deep channels where H/W does not approach zero. An additional correction factor can be used to improve the calculation of the rotational flow term [1]. The shape factors are calculated as follows:

$$F_d = \frac{16W}{\pi^3 H} \sum_{i=1,3,5,\ldots}^{\infty} \frac{1}{i^3} \tanh\left(\frac{i\pi H}{2W}\right) \tag{1.6}$$

The velocity of the flight tip V_{bz} is calculated as follows:

$$V_{bz} = \pi N D_b \cos\theta_b \tag{1.7}$$

$$\tan\theta_b = \frac{L}{\pi D_b} \quad \text{thus} \quad \theta_b = \arctan\frac{L}{\pi D_b} \tag{1.8}$$

where N is the screw rotation rate in revolutions per second, D_b is the diameter of the inside barrel wall, L is the lead length, and θ_b is the helix angle at the barrel wall.

Because of the helical nature of the screw, the width of the channel is narrower at the core of the screw as compared to that at the barrel wall. The calculation of the rotational flow rate, however, requires the average width of the channel. The average width of the channel is calculated as follows:

$$W_b = \frac{L}{p}\cos\theta_b - e \tag{1.9}$$

$$W_c = \frac{L}{p}\cos\theta_c - e \tag{1.10}$$

$$\tan\theta_c = \frac{L}{\pi(D_b - 2H)} \quad \text{thus} \quad \theta_c = \arctan\frac{L}{\pi D_c} \tag{1.11}$$

$$W = \frac{W_b + W_c}{2} \tag{1.12}$$

where e is the width of the flight perpendicular to the edge, D_c is the diameter at the screw core, W_c is the channel width at the screw core, and θ_c is the helix angle at the screw core.

The volumetric pressure flow term, Q_p, and the mass flow pressure flow term, Q_{mp}, are computed as follows:

$$Q_p = \frac{pWH^3F_p}{12\eta}\left[\frac{\partial P}{\partial z}\right] \tag{1.13}$$

$$Q_{mp} = \frac{p\rho_m WH^3F_p}{12\eta}\left[\frac{\partial P}{\partial z}\right] \tag{1.14}$$

$$F_p = 1 - \frac{192H}{\pi^5 W}\sum_{i=1,3,5,\ldots}^{\infty}\frac{1}{i^5}\tanh\left(\frac{i\pi W}{2H}\right) \tag{1.15}$$

where F_p is the shape factor for pressure flow, $\partial P/\partial z$ is the pressure gradient in the channel in the z direction, and η is the shear viscosity of the molten polymer at the average channel temperature and at an average shear rate, $\dot{\gamma}$:

$$\dot{\gamma} = \frac{\pi D_c N}{H} \tag{1.16}$$

The shear rate in the channel contains contributions from the rotational motion of the screw and the pressure-driven flow. The calculation of the shear rate, $\dot{\gamma}$, using Eq. 1.16, is based on the rotational component only and ignores the smaller contribution due to pressure flow.

The relationship between the pressure gradient in the z direction to the axial direction, l, is as follows:

$$\frac{\partial P}{\partial z} = \frac{\partial P}{\partial l}\sin\theta_b \tag{1.17}$$

The pressure gradient is generally unknown, but the maximum that it can be for a single-stage extruder screw is simply the discharge pressure, P_{dis}, divided by the helical length of the metering section. This maximum gradient assumes that the pressure at the start of the metering section is zero. For a properly designed process, the actual gradient will be less than this maximum, and the pressure at the start of the metering section will not be zero.

$$\frac{\partial P}{\partial z} = \frac{P_{dis} \sin \theta_b}{l_m} \tag{1.18}$$

where l_m is the axial length of the metering section.

The total mass flow rate, Q_m, is calculated by combining the flow components as provided in Eq. 1.19 for the total mass flow rate. An additional correction factor can be used to improve the calculation of the rotational flow term [1].

$$Q_m = \frac{p\rho_m V_{bz} WHF_d}{2} - \frac{p\rho_m WH^3 F_p}{12\eta}\left[\frac{\partial P}{\partial z}\right] \tag{1.19}$$

Estimation of the rate and pressure gradient using Eq. 1.19 should be performed for each troubleshooting operation. Examples for its use are available elsewhere [1].

For grooved-bore extruders, the design of the feed section including the grooves and screw section control the rate of the process. This calculation of the rate is considerably more complex and is out of the scope of this chapter. Grooved bore extruders are discussed in detail elsewhere [8–10].

1.3 Gels

A common contaminant in polyolefin film products is gels. The term gel is commonly used to refer to any small defect that distorts a film product, creating an optical distortion. There are many types of gels [11, 12], and the most common include (1) gels that are crosslinked via an oxidative process, (2) highly entangled resin gels (undispersed but not crosslinked), (3) unmelted resin, and (4) a different type of resin or contaminant such as wood, cloth fibers, or dirt. A crosslinked resin gel is typically formed during an oxidation process, resulting in the crosslinking of the resin chains and the generation of color bodies. These gels will not melt fully during analysis using a hot-stage microscope. Highly entangled gels are typically high molecular weight polymer chains that are highly entangled and thus difficult to disperse during the extrusion process. When analyzed using a hot-stage microscope, this gel type will melt as the stage temperature is increased. When the stage temperature is then decreased, the gel will crystallize before the surrounding material, creating the appearance of a gel. Since these gels are not oxidized they are not associated with color. They are commonly referred to as undispersed or unmixed gels. Unmelted resin exiting with the discharge can sometimes occur, especially at high extrusion rates. These gels will melt during heating with a hot-stage microscope, and typically they will not re-form during the cooling phase. Numerous sophisticated methods are available for analyzing gels, including epi-fluorescence microscopy, polarized light microscopy, and electron microscopy with X-ray analysis.

Gels can be generated from many different sources and include (1) the resin manufacturer, (2) the converting process, (3) pellet blending of resins with significantly different shear viscosities, (4) pellet blending of different resin types, and (5) direct contamination. Modern resin manufacturing processes exclude oxygen from the system and are very streamlined such that process areas with long residence times do not exist. As such crosslinked and oxidative gels are likely not generated by the manufacturer. Improperly designed extrusion equipment and processes, however, are common, leading to the oxidative degradation of resins and crosslinked gels. Several case studies in the next sections show how poorly designed processing equipment can lead to crosslinked and unmixed gel contamination of products.

Established protocols for gel analysis in polymer films are well documented in the literature [11–14]. Typically a film with defects is visually inspected using a low power dissecting microscope. The gels can be classified based on size, color, and shape and isolated using a razor blade or scissors. Cross sections of the gels ranging from 5 μm to 10 μm thick are collected at temperatures below the glass transition (T_g) temperature of the film using a cryogenic microtome, about −80°C to −120°C. For optical examination, a thin section containing the gels is placed on a glass microscope slide with a drop of silicone oil and covered with a glass cover slip. Additional sections are collected for examination via hot-stage microscopy and for compositional identification if needed.

After collecting the sections, the remaining polished block-face containing the remainder of the gel is retained. In many instances, gels arise from inorganic contaminants such as the metals from pellet handling equipment, extruders, or components from masterbatches. Examination of these inorganic components are best performed with the block-face sample using a scanning electron microscope (SEM) equipped with an energy dispersive X-ray detector (EDX) [15, 16]. In some cases, additives or inorganic residues are present in low concentrations within the gels. A method to enrich the concentration of these materials is to expose the block-face containing the gel to oxygen plasma. Etching will preferentially remove the polymer at a much faster rate than the inorganic materials, enriching these components for elemental analyses. It must be noted that prior to SEM and EDX analyses, a thin conductive coating like carbon is typically evaporated onto the sample to render it conductive under the electron beam.

The most common type of gel is caused by oxidative processes that crosslink the PE chains. The best way to identify this gel type is by observation with polarized light and ultraviolet (UV) light sources. Transmitted polarized light microscopy represents an effective technique [17] that can be used to investigate structures in crystalline films. For example, black speck gels were contaminating a multilayer film product. The gels were relatively brittle when cut for analysis. The source was unknown. The detail of the gel is clearly visible using transmitted polarized light, as shown in Fig. 1.7(a). Close examination of this gel using epi-fluorescence with an ultraviolet light source caused an intense fluorescence emission, as shown in Fig. 1.7(b). This type of emission suggests thermal oxidation and crosslinking of the polymer. Microinfrared analysis of the gel indicated that it contained oxidized PE and maleic anhydride [1, 12]. This material likely formed on the metal surfaces of the extruder and then flaked off during a minor process instability. The material then flowed downstream and contaminated the film as a gel.

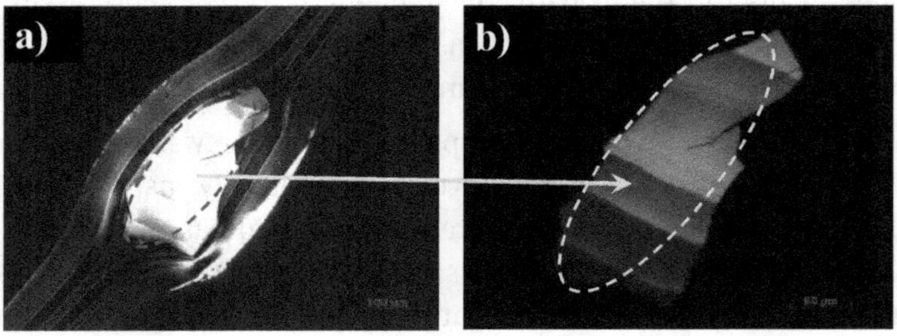

Figure 1.7 Transmitted polarized light images of a thermally oxidized and crosslinked gel in a multilayer film: (a) photograph in polarized light, and (b) the gel fluorescing under UV light [12]. Photographs were provided by E. Garcia-Meitin of The Dow Chemical Company

Crosslinked gels are oxidized gels, but the level of oxidation is not enough to cause them to fluoresce under UV light. The gels may have a level of crystallinity and thus be birefringent under polarized light. For example, the slightly birefringent gel shown in Fig. 1.8(a) was studied using a temperature-programmable, hot-stage, polarizing light microscope [16]. The optical melting temperature (T_m) of the gel was measured at 128 °C and consistent with the PE used to make the product, as shown in Fig. 1.8(b). To determine if the gel was unmixed (highly entangled but not crosslinked), the gel was held above the melting temperature (135 °C) and then stressed. A dental tool was used to stress the top of the glass slip cover. Crosslinked gels will appear birefringent, as shown in Fig. 1.8(c), in response to the anisotropy of stress distribution in the gel to polarized light. The gel dimensions and shape remained after cooling, verifying crosslinking, as shown in Fig. 1.8(d). If the gel was highly entangled and not crosslinked (unmixed gel), the gel would have disappeared after stress and cooling were applied.

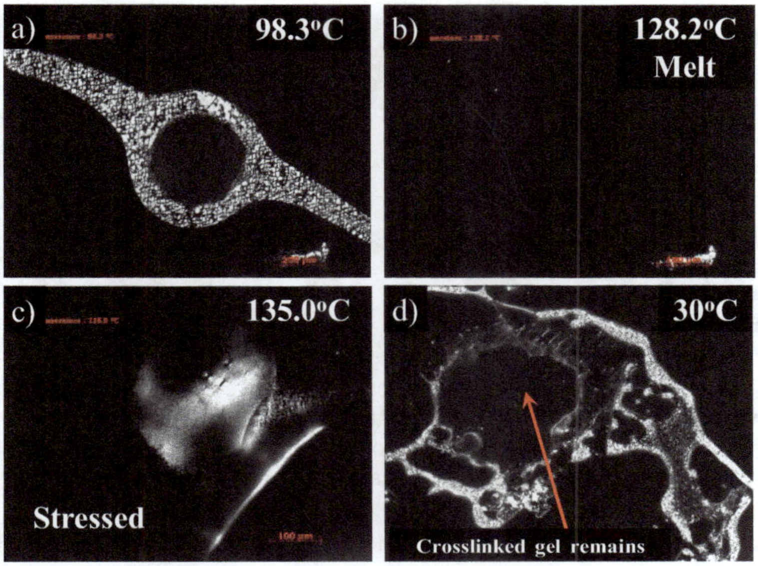

Figure 1.8 Hot-stage microscopy of a crosslinked gel in a crystalline monolayer film: (a) below the melting temperature, (b) optical melting point at 128°C, (c) appearance of birefringence after stressing at 135°C, and (d) intact crosslinked gel after cooling to 30°C [12]. Photographs were provided by E. Garcia-Meitin of The Dow Chemical Company

The origin of defects causing discoloration in polyolefin pellets can be identified using light and electron microscopy. For example, PE pellets from an in-plant recycle repelletizing process contained pellets that were off-color and had black specks, as shown in Fig. 1.9(a). One of these defects was isolated using the cross-sectioning technique, as shown in Fig. 1.9(b). The cross section revealed an intense reddish particle that caused the discoloration of the pellet. SEM and EDX microanalyses were used to determine that the defects contained primarily iron and oxygen, and it likely was iron oxide. A backscatter electron image (BEI) of the pellet block-face sample showed the defect causing the discoloration, and the elemental spectrum was shown to be iron oxide [1]. Metal-based defects can originate from process equipment, railcars used for shipment, pellet transfer lines, and poor housekeeping. The origin of the iron oxide was likely from a storage bin.

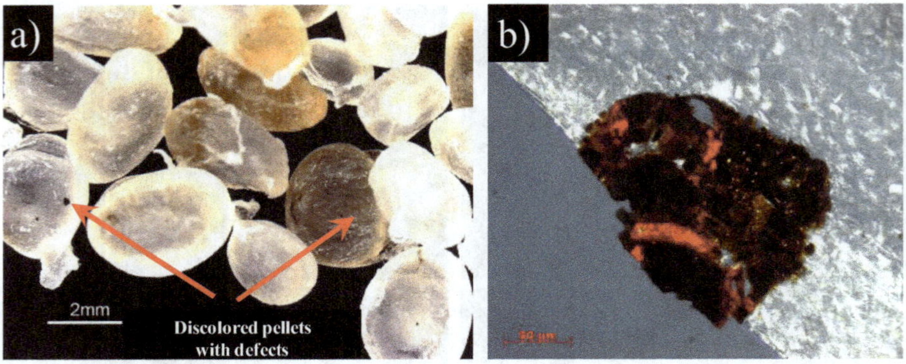

Figure 1.9 Photographs of foreign contamination in pellets of a repelletized reclaim stream: (a) photomicrograph of discolored PE pellets containing dark defects, and (b) transmitted polarized light micrograph of a pellet cross section containing a defect [12]. Photographs were provided by E. Garcia-Meitin of The Dow Chemical Company

In another example, a multilayer film product was experiencing occasional gels. The gels were isolated and the cross sections were collected as shown in Fig. 1.10(a). These gels contained highly birefringent particles that resided in the core layer. The outer film layers appeared amorphous, and the core layer was slightly birefringent. The optical melting temperature of the core layer was determined to be 123°C while the birefringent gels melted at 265°C. The melting temperature of 123°C was consistent with the polyethylene (PE) resin used to produce the core layer. The higher melting temperature of the material and microinfrared analyses of the defects indicate that they were foreign contaminants, and they were identified as a polyester resin. The polyester resin was used in another process in the converting plant, and it inadvertently contaminated the PE feedstock.

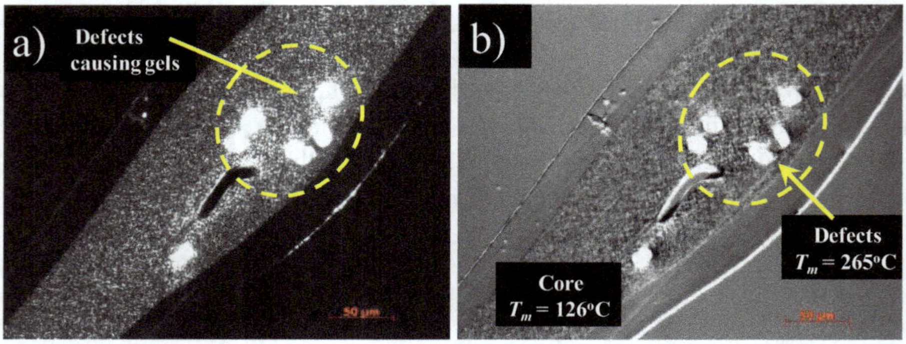

Figure 1.10 Photographs of gels in the core layer of a three-layer film: (a) transmitted polarized light, and (b) hot-stage microscopy was used to determine the melting temperatures of the core resin and defects [12]. Photographs were provided by E. Garcia-Meitin of The Dow Chemical Company

Another common contaminant that produces gels is fiber, as shown in Fig. 1.11. In many cases, these contaminants are cotton fibers from clothing and gloves or cellulosic fibers from packaging materials. Fourier transform infrared (FTIR) spectroscopy is one of the best techniques for determining the chemical functionality of organic-based defects in PE films.

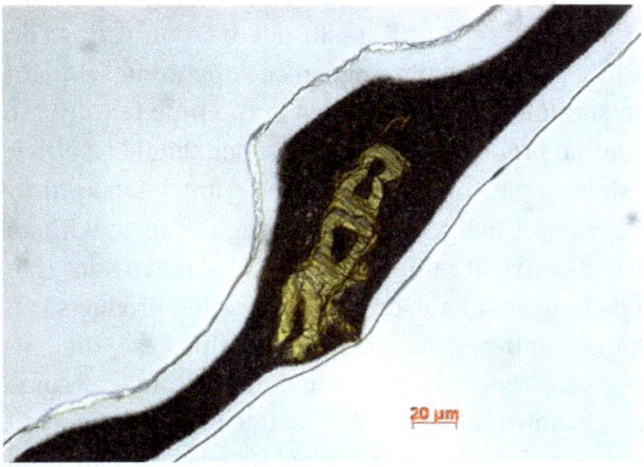

Figure 1.11 Transmitted bright-field image of a PE film containing a fibrous gel [12]

Once the contaminant is identified, the troubleshooter must determine how the material entered the feedstock stream. Process controls must be identified and implemented to mitigate the contaminant source.

■ 1.4 Troubleshooting Extrusion Processes

Eventually every single-screw extrusion process will experience periods when the machine is operating at a performance level that is less than designed. During these periods, the cost of manufacturing will increase due to the production of off-specification products, loss of production rates, high levels of recycle, higher labor costs, and lower daily production of prime product. In extreme cases, the problem can be so severe that the line must be shut down. Obviously the plant needs to restore the operation of the line to the original performance level as soon as possible to maximize profitability. Many things can cause an extruder to malfunction, including mechanical and electrical failures, installation of new equipment, process changes, and resin changes. A complete process for troubleshooting an extrusion process can be obtained in reference 1. Several of the most common problems associated with film production processes are presented in this section.

1.4.1 Improper Shutdown of Processing Equipment

Shutting down an extrusion line occurs for many reasons, including planned shutdowns for maintenance, shift changes, changing filtering screens, product changes, and many unplanned events. A shutdown period is defined here as a period when the screw is not rotating and thus resin is not discharging from the line. If the shutdown period is relatively short such that very little resin degradation can occur, then the extruder barrel temperatures can be maintained at the operating set point temperatures. If the length of the shutdown is long relative to the time required to create a significant level of degradation products, then the extruder should be either purged with a more thermally stable resin or the barrel set point temperatures should be decreased to considerably lower temperatures. Purging the resin with an inert gas to exclude oxygen is also effective at mitigating gels [1]. An extruder that is maintained at process conditions long enough to create degradation products can be very difficult to bring back online running prime product. In this case, the surfaces of the screw and all metal surfaces in contact with molten resin may become coated with degradation products, as shown in Fig. 1.12. The time to purge them out can be extremely long and very expensive. For example, LLDPE resins can form crosslinked gels and black specks after 30 minutes of being off-line at process temperatures. If the shutdown is under 30 minutes, the barrel can be held at the process temperature. If the shutdown is longer but the line will be brought back online soon, the screw could be rotated at a low speed of 5 rpm to keep resin flowing, mitigating the formation of degradation products. For longer shutdown periods, the extruder should be purged using an LDPE resin and then cooled to ambient temperature.

Figure 1.12 Photograph of a screw that had numerous shutdowns where the extruder was maintained at operating temperatures for an extended time. The extruder was purged prior to removing the screw, yet dark degraded resin covers most areas of the screw [1]

Antioxidant chemicals are typically added at levels that stabilize the resin during normal melt processing. They are not meant to protect the resin from degradation during an extended shutdown period. Antioxidants are slowly consumed during the extrusion process, and thus they can be fully consumed during an extended shutdown period. When they are fully consumed, the resin system is not protected from degradation, and thus degradation reactions will occur at accelerated rates.

1.4.2 Gel Showers in a Cast Film Process

Crosslinked gels can form in stagnant regions of screw channels, transfer lines, and dies. The time required for these gels to form range from about 30 minutes for linear low density polyethylene (LLDPE) resin up to 12 days for low density polyethylene (LDPE) resin. Stagnant regions can occur at entries and exits of mixers [1] and barrier sections, and they can occur when the metering channel of smooth-bore extruders is not controlling the rate. In these cases, a section upstream of the metering section is rate limiting, causing portions of the metering section to operate partially filled [18, 19]. When these channels operate partially filled, the main flow is on the pushing side of the channel while the trailing side operates void at first. After a period of time, clean resin gets into the void regions and rotates with the screw for long durations. Eventually the resin will degrade, forming crosslinked gels. Slight process upsets can dislodge this material, allowing the material to flow downstream, creating a gel shower in the film.

A film plant was extruding an LDPE resin into a specialty product using a cast film process [18, 19]. Due to high demand, a new 88.9 mm diameter, 33 *L*/*D* extruder was installed in the plant. Soon after start-up the product was acceptable and high quality. After 12 days, the line began to experience intermittent discharges of crosslinked material (gel showers) and carbon specks. Photographs of these gels are shown in Fig. 1.13. In some cases, the gel showers were observed two to three times per day and would last from 1 to 5 minutes. The gels were clearly crosslinked and were brown in color. The extrudate temperature was higher than expected for the process. The intermittent gels resulted in production downtime due to purging and in numerous customer complaints. A high and costly level of quality control was required to remove the gel-contaminated product from the prime product. Due to the high amount of downtime and the high levels of quality control needed, the operation of the new line was considerably more expensive than planned.

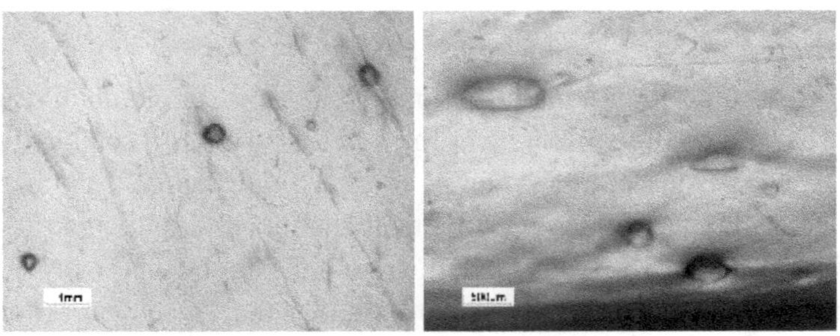

Figure 1.13 Photographs of crosslinked gels in an LDPE film

It was hypothesized that the extruder was operating partially full due to the low specific rate during operation. To determine if partially filled channels were the root cause of the reduced rate, high discharge temperature, and degraded material, screw rotation was stopped and the screw was removed while hot from the extruder. Examination of the polymer on the screw indicated that in the meter section about half of the channel width on the trailing sides of the flights for all but the last diameter were filled with a dark-colored, partially carbonized LDPE resin, indicating that these regions were stagnant. The reduced flow rate caused these regions to be partially filled, creating void regions on the trailing side of the channel. Some of the resin adhered to the trailing side of the screw in the void regions and stayed there for extended time periods, as shown in Figure 1.14. The resin adhering in the void regions eventually degraded into the dark-colored, crosslinked material. Small process variations dislodged some of this material and caused the intermittent gel showers that contaminated the product. Moreover, compacted solids were found wedged in the channel at the entrance to the barrier section. The wedged material was caused by the relatively large width of the entering solid bed being forced into the continually decreasing width of the solids channel of the barrier section.

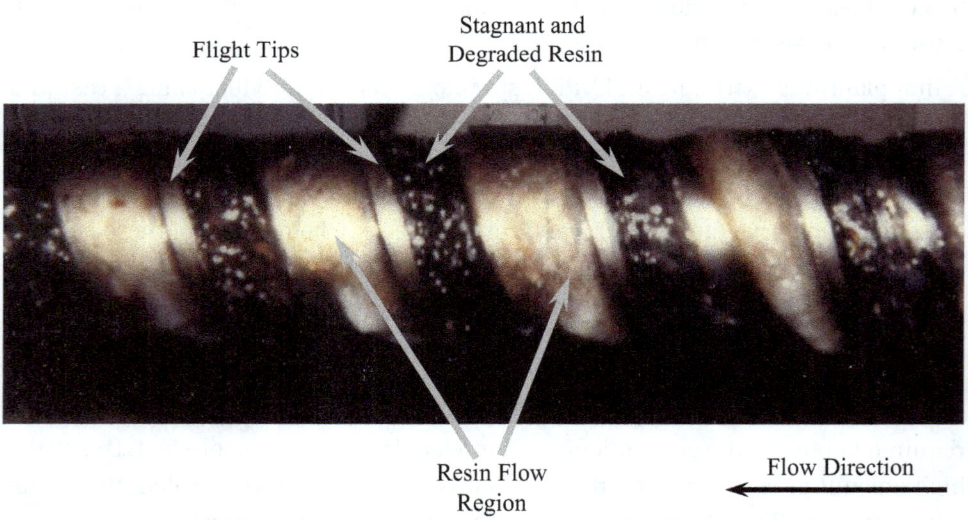

Figure 1.14 Photograph of a removed screw showing the resin flow and degraded resin due to stagnant regions [1]

The technical solution to eliminate this problem was a simple modification to the entry of the barrier melting section. For this modification [18], some of the metal in the melt conveying channel was removed along with a portion of the barrier flight, allowing some solid material to enter the melt channel and reducing the restriction at the entry. By reducing the restriction, the rate-limiting step of the process changed from the entry region of the barrier section to the metering section. After the modification was made, the gels were eliminated from the process.

1.4.3 Unmixed Gels

As stated previously, unmixed gels are highly entangled species that are molten when they are discharged from the die but solidify first upon cooling to produce a gel that appears as a solid polymer fragment. These types of gels are easily removed from an extrusion process by subjecting all molten resin to a one-time high level of stress near the discharge of the extrusion screw. This stress is easily applied using a Maddock-style mixer with a relatively tight clearance on the mixing flight.

A film process was producing a monolayer film that had a low level of gels [1]. The gels were tested using hot-stage microscopy and identified as highly entangled species (unmixed gels). These gels melted and then disappeared when heated and stressed via pressure smearing using a dental tool, as shown in Figure 1.15.

The unmixed gels were removed by increasing the stress level in the Maddock mixer. The stress level was increased by decreasing the clearance on the mixing flight. The stress level required to disperse unmixed gels depends on the resin and the level of chain entanglement. In past experiences, the stress level required to disperse PE unmixed gels is about 100 to 200 kPa. The shear stress that the material experiences for flow across the mixing flight of the Maddock mixer can be estimated using Eqs. 1.1 and 1.2.

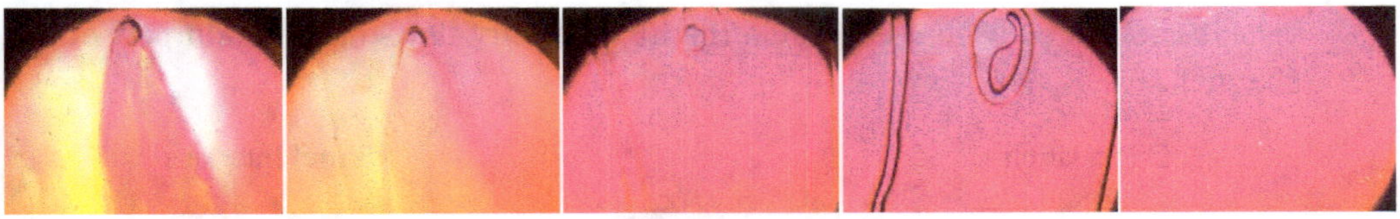

Figure 1.15 Photographs of an unmixed gel at select temperatures using a hot-stage microscope. The unmixed gel melted at about 135 °C. When the gel was smeared by moving the glass cover slip, the stress was enough to disentangle the polymer chains such that the gel would not reappear upon cooling [1]

A similar problem with solid polymer fragments occurred for a thermoplastic polyurethane (TPU) resin [20] in the production of sheet. For this case, a combination of a lower compression ratio, a longer barrier section with a very small barrier flight clearance, a Maddock mixer with a small mixing flight clearance, and deeper metering channels allowed the TPU resins to extrude at twice the rate and provide high quality extrudates that were free of solid polymer fragments.

1.4.4 Carbon Specks in a Film Product

Carbon specks can be generated in the extruder channels and in downstream transfer lines and dies if stagnant regions are present. In general, these regions are not very large like those in Fig. 1.14. Instead, they are thin coverings that occur at the flight radii or at entries and exits of mixing devices [1]. In general, the region will first create small crosslinked-type materials that adhere to metal surfaces. With additional residence time, the crosslinked material will form a thin carbon layer of highly oxidized material. When the layer breaks away from the metal, it is discharged as black specks in the PE film. These specks will fluoresce under UV light.

An LLDPE blown film line was experiencing black specks in the product. In order to locate the source, a Maddock solidification experiment [21] was performed where a small amount of a red color concentrate was added to the feedstock resin. After the red color appeared in the extrudate screw, rotation was stopped, and the resin was solidified in the channels. A photograph of the experimental sample [22] is shown in Fig. 1.16. Here a thin layer of carbonaceous material was formed at the pushing flight due to the formation of Moffat eddies [23]. Moffat eddies are recirculation or vortices that occur at sharp corners as shown in Fig. 1.17. When fluid is put in motion with top-driven cavity flow, the main circulation is shown in Fig. 1.17. A secondary circulation is set up in the stationary corners of the channel, creating a low-velocity helical eddy that is outside the high-velocity flows of the main part of the channel.

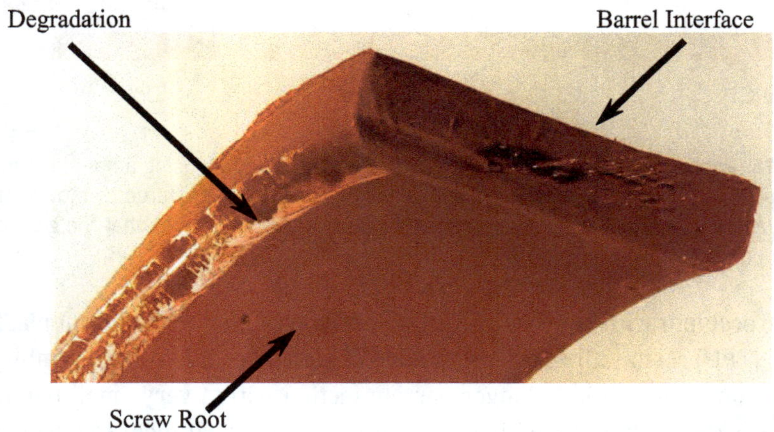

Figure 1.16 Photograph of degradation at the pushing flight for a screw running LLDPE resin [22]

The Moffat eddies that created the degraded resin occurred because the flight radii were too small for the depth of the channel. If the flight radii had been larger, the Moffat eddies would not have occurred, and thus carbon deposits would not have formed.

The Society of the Plastics Industry, Inc. (SPI) guidelines state that [24] "unless otherwise specified the root radius will not be less than one-half of the flight depth up to 25 mm radius." Many screws are often designed, however, with flight radii that are very small and approach values that are between 10 and 20% of the channel depth. Previous research [22] has indicated that the SPI guideline as a minimum is appropriate for many resins. But for thermally sensitive resins, radii up to 2.5 times the depth are optimal. Flight radii sizes are shown in Fig. 1.18. When a new screw with radii equal to the depth of the channel was built and installed into the blown film line, the black specks were essentially eliminated.

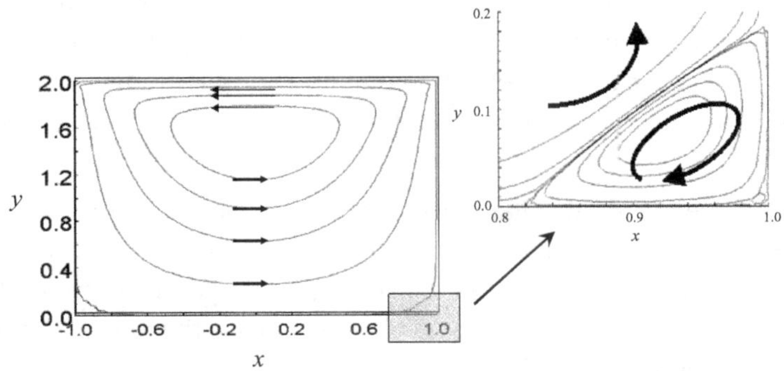

Figure 1.17 Two-dimensional flows in a screw channel with an $H/W = 1$ (channel depth/channel width). The arrows show the recirculation flows. The shaded area in the lower right corner is expanded to show the Moffat eddy [1]

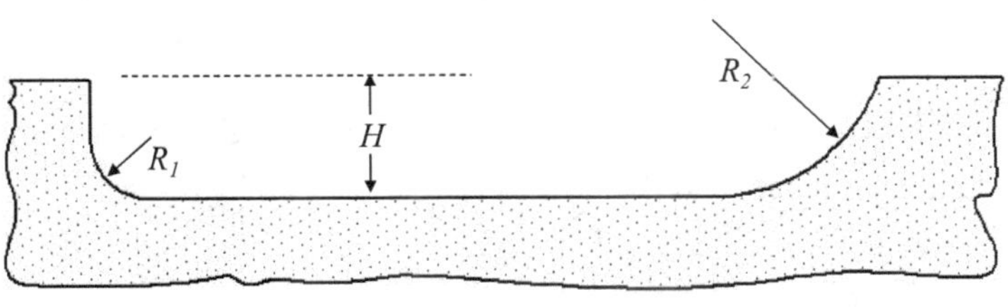

Figure 1.18 Schematic of small (R_1) and large (R_2) flight radii

1.4.5 Rate Limitation Due to a Worn Screw

A specialty blown film product line [1] was rate limited by the discharge temperature of the extrudate at 130 kg/h. At this rate, the discharge temperature from the 88.9 mm diameter extruder was 245°C. At higher rates, the discharge temperature increased to even higher levels such that the bubble process could not be maintained. The operators decreased the barrel temperature setting over the metering zone to 190°C, but the discharge temperature would not decrease. The proprietary screw design appeared acceptable for this process. When the process was numerically simulated, the discharge temperature of the process was estimated at 220°C. Here the process based on the screw schematic could not be simulated to provide the measured discharge temperature of 245°C.

The screw was removed from the extruder and inspected to determine if the screw was in specification as defined by the schematic. The measured geometries were consistent with the schematic except that the diameter at the flights was measured at 88.15 mm, indicating that the flight clearance was 0.38 mm. The original flight clearance was specified at 0.09 mm, and thus the screw had worn such that the flight clearance was more than four times the specified value. The screw was placed back into the extruder, and production was continued at the reduced rate.

An identical screw was ordered with the correct flight clearance. As soon as the screw was delivered, the screw was installed in the extruder. Soon after start-up, the extruder was discharging at a temperature near 220°C at the base rate of 130 kg/h. Next the rate was slowly increased to 180 kg/h. The discharge temperature increased to about 224°C. Higher rates are likely possible. The worn screw was sent back to the screw manufacturer for refurbishment. This screw was then stored as the spare screw.

This case study clearly shows that a worn screw will cause the discharge temperature to increase due to material flow over the flights, and it will decrease the heat transfer coefficient, limiting the transfer of energy through the barrel wall. With the restored flight clearance, the process was able to increase the rate by about 35%, and the quality of the product was improved.

Nomenclature

D_b inner diameter of the barrel

D_c diameter of the screw core

e flight width of the screw and perpendicular to the flight edge

F_d shape factor for rotational flow

F_p shape factor for pressure flow

H local distance from the screw root to the barrel

l axial distance

l_m axial distance for the metering section

L lead length

N screw rotation speed in revolutions/s

p number of flight starts

P_{dis} discharge pressure

P pressure in the channel

Q volumetric flow rate

Q_d volumetric rotational flow rate

Q_m mass flow rate

Q_{md} mass rotational flow rate

Q_{mp} pressure-induced mass flow rate

Q_p volumetric pressure flow rate

u undercut clearance on the mixing flight of a Maddock mixer

V_{bz} z component of velocity of the screw flight at the barrel wall

W average channel width

W_b channel width perpendicular to flight at the barrel wall

W_c channel width perpendicular to flight at the screw core

x independent variable for the cross-channel direction perpendicular to the flight edge

y independent variable for the direction normal to the barrel surface (channel depth direction)

z independent variable in the down-channel direction (or helical direction)

ρ_m melt density of the fluid

$\dot{\gamma}$ average shear rate in the channel

$\dot{\gamma}_M$ average shear rate for flow over the mixing flight of a Maddock mixer

λ mechanical clearance between the top of the screw flight and the barrel wall

θ_b helix angle at the barrel

θ_c helix angle at the screw core

η shear viscosity of the polymer at the average channel temperature and average shear rate, $\dot{\gamma}$

τ_M shear stress that the material will experience for flow over the mixing flight of a Maddock mixer

References

1. Campbell, G. A. and Spalding, M. A., *Analysis and Troubleshooting of Single-Screw Extruders* (2013) Hanser Publications, Munich
2. Maddock, B. H., "An Improved Mixing-Screw Design," *SPE J.* (1967) 23, p. 23
3. Tadmor, Z. and Klein, I., "Design of Certain Fluted Mixing Sections of Extruder Screws," *Polym. Eng. Sci.* (1973) 13, p. 382
4. Zitzenbacher, G., Karlbauer, R., Thiel, H., "A New Calculation Model and Optimization Method for Maddock Mixers in Single-Screw Plasticating Technology," *Int. Polym. Process.* (2007) 22, p. 73
5. Rowell, H. S. and Finlayson, D., "Screw Viscosity Pumps," *Engineering* (1922) 114, p. 606
6. Rowell, H. S. and Finlayson, D., "Screw Viscosity Pumps," *Engineering* (1928) 126, p. 249

7. Tadmor, Z. and Klein, I., *Engineering Principles of Plasticating Extrusion* (1970) Van Nostrand Reinhold Co., New York
8. Potente, H., "Single Screw Extruder Analysis and Design," Chapter 5 in *Screw Extrusion*, White, J. L. and Potente, H. (Eds.) (2002) Hanser Publications, Munich
9. Potente, H., *Kunststoffe* (1985) 75, p. 439
10. Potente, H. and Schöppner, V., "A Throughput Model for Grooved Bush Extruders," *Int. Polym. Process.* (1995) 10, p. 289
11. Butler, T. I., "Gel Troubleshooting," Chapter 19 in *Film Extrusion Manual*, Butler, T.I. (Ed.) (2005) TAPPI Press, Atlanta, GA
12. Spalding, M.A., Garcia-Meitin, E., Kodjie, S.L., Campbell, G.A., "Troubleshooting and Mitigating Gels in Polyolefin Film Products," *Annu. Tech. Conf. - Soc. Plast. Eng. (ANTEC)* (2013) 59, p. 1205
13. Scheirs, J., *Compositional and Failure Analysis of Polymers* (2000) Wiley, New York
14. Kodjie, S., Guerra, S., Savargaonkar, N., *2012 Microscopy and Microanalysis Proceedings* (2012) p. 715
15. Garcia-Meitin, E.I., Bar, G., Blackson, J., Reuschle, D., *Microsc. Microanal.* (2008) 14 (Suppl. 2), p. 1380
16. Blackson, J., Garcia-Meitin, E.I., Darus, M., *Microsc. Microanal.* (2007) 13 (Suppl. 02), p. 1062
17. Hemsley, D.A., *Applied Polymer Light Microscopy* (1989) Elsevier Appl. Science, London
18. Hyun, K.S., Spalding, M.A., Powers, J.R., "Elimination of a Restriction at the Entrance of Barrier Flighted Extruder Screw Sections," *Annu. Tech. Conf. - Soc. Plast. Eng. (ANTEC)*, (1995) 41, p. 293
19. Spalding, M.A., "Metering Channel Flows and Troubleshooting Single-Screw Extruders," *Annu. Tech. Conf. - Soc. Plast. Eng. (ANTEC)* (2004) 50, p. 329
20. Hyun, K.S. and Spalding, M.A., "Troubleshooting TPU Resin Extrusion Rate Limitations Due to Solids in the Discharge," *Annu. Tech. Conf. - Soc. Plast. Eng. (ANTEC)* (2012) 58
21. Maddock, B.H., "A Visual Analysis of Flow and Mixing in Extruder Screws," *SPE J.* (1959) 15, p. 383
22. Spalding, M.A., Dooley, J., Hyun, K.S., "The Effect of Flight Radii Size on the Performance of Single-Screw Extruders," *Annu. Tech. Conf. - Soc. Plast. Eng. (ANTEC)* (1999) 45, p. 190
23. Moffat, H.K., "Viscous and Resistive Eddies Near a Sharp Corner," *J. Fluid Mech.* (1964) 18, p. 1
24. "Recommended Dimensional Guideline for Single Screws," The Society of the Plastics Industry, Inc.

2 Kinematics, Dynamics, Crystallization, and Thermal Characteristics and Their Relationship to Physical Properties of Blown Film

G. A. Campbell

2.1 Abstract

Understanding the physics relating to the development of blown film physical properties is potentially very important to both the industrial and academic practitioner. This chapter focuses on many of the processes that affect the development of physical properties during production of semicrystalline polymer such as polyethylene (PE) film. The physical properties of PE are frozen into the film as a result of physical crosslinks developed during crystallization. The crystallization of low-density polyethylene is investigated as received from the polymer producer and relative to the influence of a small amount (less than 6%) of high-density PE added to the as-received resin as a nucleation agent. The process was evaluated by the use of simultaneous online SALS (small angle light scattering) and IR (infrared) temperature measurements to study structure development as influenced by nucleation during tubular blown film extrusion of LLDPE. As the concentration of the nucleation aid HDPE was increased, the observed scattering intensity decreased. The effect of melt temperature was also investigated, and the scattering intensity increased as the temperature increased. The later processes may reflect secondary crystallization and orientation processes. By allowing the processor to track the crystallization dynamics, this online SALS system is potentially a useful tool for monitoring process-property interactions. These observations of the interaction of the crystallization dynamics and the cooling dynamics of the film led to the desire to model the process using a finite difference temperature model with the incorporation of crystallization dynamics. The crystallization is dominated by the removal of energy from the film, which is of course a function of the heat transfer coefficient. In order to check the utility of these models, a technique was developed that allows the practitioner to measure simultaneously both the surface temperature and the average bulk temperature of the film. The film temperatures measured were always higher than predicted by the original models above the frost-freeze line, so an investigation of bubble heat transfer was carried out. These several aspects of the blown film process will be developed in the following sections of this chapter. The crystallization dynamics further indicate that models used to simulate the blown film process should be based on rheological models that incorporate plasticity and a yield stress since the crystallization dynamics occur starting at or near the freeze line and continue well above the freeze line.

■ 2.2 Introduction

The tubular blown film extrusion process, Fig. 2.1, is used extensively to make thin polymer films. In this process, polymer melt is extruded through an annular die to form a continuous tube. The tube is drawn upwards between two rotating rolls that collapse the tube into a flat two-ply sheet. Usually air is introduced through the center of the die to inflate the tube into a bubble that is several times the original tube diameter. Air is introduced above the die through an air ring on the outside of the bubble to cool the film to room temperature. In many commercial systems the line will also have internal bubble cooling. For a semicrystalline polymer, the molten polymer crystallizes as the film cools. The dynamics of the cooling above the air ring is complex, and the heat transfer coefficient that controls the film cooling is an important consideration when simulating the cooling characteristics of the bubble. The cooling of the bubble provides the driving force for the crystallization of semicrystalline polymer. This process of crystallization leads to the formation of the physical crosslinks that freeze in the structure developed during the expansion and elongation of the molten polymer as it moves from the die to the freeze line. Depending on the relaxation time, some of this potential structure development will be lost if the relaxation time is small compared to the time of flight of the polymer as it is stretched between the die and the freeze-frost line.

Since the film is of finite thickness, the temperature in the film will not be a constant; it will be a minimum at the air-polymer interface and a maximum at some position across the film thickness. If the film is being cooled at the same rate from both surfaces, the maximum temperature will be in the center of the film. Determining this gradient is necessary if one is to develop models and then certify the models to show that they actually predict the cooling characteristics of the film. This would be necessary to accurately predict the crystallization within the film since the crystallization process is exothermic and will only proceed as the heat of crystallization is removed from the film. It has historically been assumed that the majority of the crystallization in semicrystalline polymers occurred during or before the constant temperature plateau that was associated with the freeze-frost line. The crystallization that produced the frozen-in structure was thought to produce machine direction (MD) and transverse direction (TD) properties that historically were primarily related and correlated to the drawdown and blow-up ratio of the bubble. The physical crosslinks developed during crystallization strongly suggest that modeling the blown film process requires a visco-plastic or viscoelastic-plastic rheological model for the film in order to accurately simulate the process during the crystallization phase.

In the remainder of this chapter, a discussion will be presented that develops and demonstrates several techniques that can be used to obtain the fundamental data needed to understand the relationship between physical properties, the kinematics, and the process dynamics dominated by the heat transfer from the bubble. It will be demonstrated that, using the process stretching dynamics, MD and TD properties can be presented on a single surface and correlated to the frozen-in structure of the film.

2.3 Real-Time Crystallization of the Blown Film Process

In this section, experiments are described utilizing online simultaneous SALS (small angle light scattering) and IR (infrared) temperature measurements to study crystallization during tubular blown film processing of an LLDPE. The technique is of course applicable to any polymer that produces a spherulitic crystalline structure.

The tubular blown film extrusion process is used extensively to make thin polymer films. A schematic of the process is shown in Fig. 2.1.

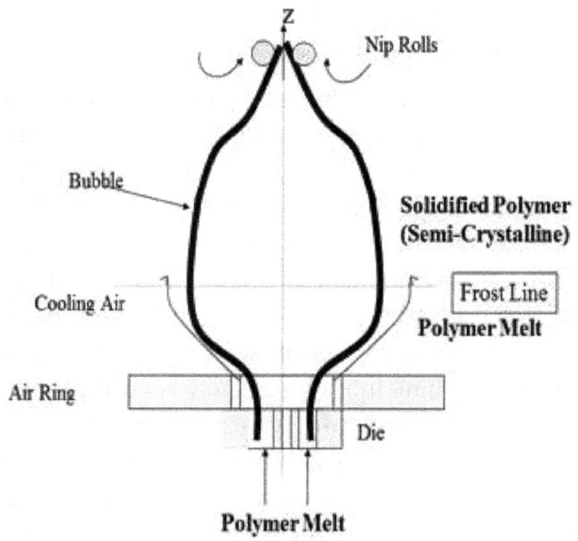

Figure 2.1 Schematic of the blown film process

In this process, a polymer melt is extruded through an annular die to form a continuous tube. The tube is drawn upwards between two rotating rolls that collapse the tube into a flat two-ply sheet. Air is injected into the center of the die to inflate the tube into a bubble that is usually several times the original tube diameter. Air is introduced above the die that impinges the outside of the bubble through a ring on the outside of the bubble to cool the film, and the film continues to cool as it goes through the collapsing system, ultimately being reduced to room temperature. In many commercial film lines there is internal bubble cooling where air flow inside the bubble and the pressure is maintained using process control to keep the bubble expanded at a constant diameter or to keep the layflat constant. This can lead to a substantial increase in extrusion rate. For a semicrystalline polymer, the molten polymer crystallizes as the film cools.

The final physical properties of semicrystalline blown films are extremely dependent on the crystalline morphology, extent of crystallization, and orientation of polymer chains within the film. Orientation of the polymer chains within the film results from the stretching in the vertical machine direction due to the nip rolls and the stretching in the cross-machine direction due to bubble inflation. As the crystallization occurs while the film cools, physical crosslinks are developed that freeze in the molecular stretching to form the film structure. Several techniques including DSC, SAXS, WAXS, birefringence, IR dichroism, SALS, SEM, and TEM [1–3] are typically used to characterize polymer morphology and orientation. In general, these methods have only been applied after processing, although, as mentioned below, online systems for some of these techniques have been developed. There have been several efforts to try to understand the relationship between polymer material properties, processing conditions, and final properties. One approach is to examine deformation history that occurs during the transition from the amorphous to semicrystalline state in light of orientation and deformation levels that are obtained by comparing the relaxation time(s) of the polymer with the elapsed time of flight of the polymer as it moves up the bubble. Various critical quantities during processing at which the final properties become set have been suggested, such as plastic strain near the freeze line [4] and the stress at the freeze line [5].

Polyethylene crystallizes by forming lamellae of folded chains. Different superstructure organizations of lamellae result from different process conditions during crystallization. Under quiescent conditions, the superstructure is spherulitic with radial fibrils of lamellae. Keller and Machin [6] proposed a row nucleated structure ("shish kebab") model for crystalline morphology development during stretching of a melt in which the lamellae are twisted at relatively low stress levels. At higher stress levels the lamellae are not twisted. Choi et al. [2] used this model to explain results of SAXS and orientation measurements for HDPE blown films for different stress levels. They proposed a superimposed row structure for the case of biaxially oriented films. Dees and Spruiell [7] found that a spherulitic structure is formed when very low stress levels occur during melt spinning of LDPE. In addition, Kwack et al. [8] found that a spherulitic structure was formed for LDPE blown films as long as the stress level was not too high. Nagasawa et al. [9] proposed that HDPE crystallized from the melt under stress is composed of an assembly of fine rods. The thickness of rods depends on stress and cooling rates.

In order to understand these factors better, many workers are developing systems to follow the dynamics of polymer crystallization and orientation, either during processing or under conditions that approximate those experienced during processing. Polarized light microscopy [10, 11] has been implemented using a hot stage to obtain high cooling rates that more closely simulate conditions experienced during processing. In situ SAXS and WAXS [12] have been applied during cold drawing to study lamellar deformation. Raman spectroscopy and spectrographic birefringence techniques [13, 14] have been used to track orientation and structural development during crystallization. Recently, an online birefringence [15] has been used to track orientation development during blown film processing.

SALS analysis techniques that are applicable to spherulitic structures have typically been developed and utilized for postprocessed film samples. Stein and Rhodes published the pioneering work using SALS for polymer films in the early 1960s [16, 17]. This technique has been extended to treat situations where the sample is subject to deformation as well as isothermal and constant rate crystallization. Stein discusses some of the considerations necessary for an accurate analysis [18]. Studies of crystallization kinetics have utilized SALS and a hot stage [19–21] at isothermal conditions with constant cooling rates. In order to perform quantitative SALS analysis, which is based on absolute intensities, various corrections and calibrations to the measured intensities need to be made for reflection, refraction, multiple scattering, spherulite truncation, interspherulitic effects, and internal disorder [20].

There are potentially many more modifications to the existing theory that need to be included in the analysis to accurately take into account the conditions for SALS experiments during tubular blown film processing. These include the fact that two films are scattering instead of the usual single film, refraction effects at a non-normal interfaces, rough interfaces, variable temperatures, oscillations in the film to screen distances due to small amount of bubble instabilities, and film thickness effects. For this preliminary analysis, it was assumed that the light scattered from the first film impacted by the coherent light can be neglected because it has spread and become more diffuse than the light scattered from the second film. The data collected indicates that enough coherent light reaches the second film to produce accurate data regarding the crystallization process for PE blown film. In addition, the process data indicate that the film thickness is essentially constant for most of the important SALS data.

While Nagasawa et al. [9] briefly described online application of SALS to the film-forming process, the process described in the following section can be found in detail in Bullwinkle et al. [22]. This online SALS technique proved to be robust and can potentially be used for process development and control. It can lead to optimization of film properties by determining where and when the crystallization structure develops. The following sections detail the procedures that were undertaken for this set of experiments. More details of the results for these resins can be found in [22].

2.4 Experiments

LLDPE (Exxon Chemical, Escorene, Type LL 1001.59, Lot M93040171C) films were processed on a blown film process line consisting of a Killion extruder with a 3.81 cm 24/1 screw, 76.2 cm layflat nip roll section, and a single-lip air ring. The annular die gap was 0.71 mm and had an initial radius of one inch (2.54 cm). As shown in Fig. 2.2(a), a motorized table containing an IR surface noncontact temperature probe and SALS equipment (polarized He-Ne laser, analyzer, screen, and CCD camera) translated in the axial direction along the outside of the bubble between axial positions of 21 and 74 cm above the die. Other details about the experimental setup of the IR surface temperature probe are described in Cao et al. [23] and Campbell et al. [24]. SALS images were recorded on a VCR for later analysis while the IR data were recorded on a computer. Figure 2.2(b) shows the traditional SALS experimental geometry. All SALS experiments were made under the *Hv* polarization condition, which denotes that the analyzer was set perpendicular to the polarization direction of the laser. As indicated in the above text, the image from the first side of the bubble would be inside the bubble; thus one might expect that this process would not provide useful data. However, as will be shown, enough coherent light reaches the second side of the film to provide reproducible and useful data regarding the crystallization as a function of position above the die.

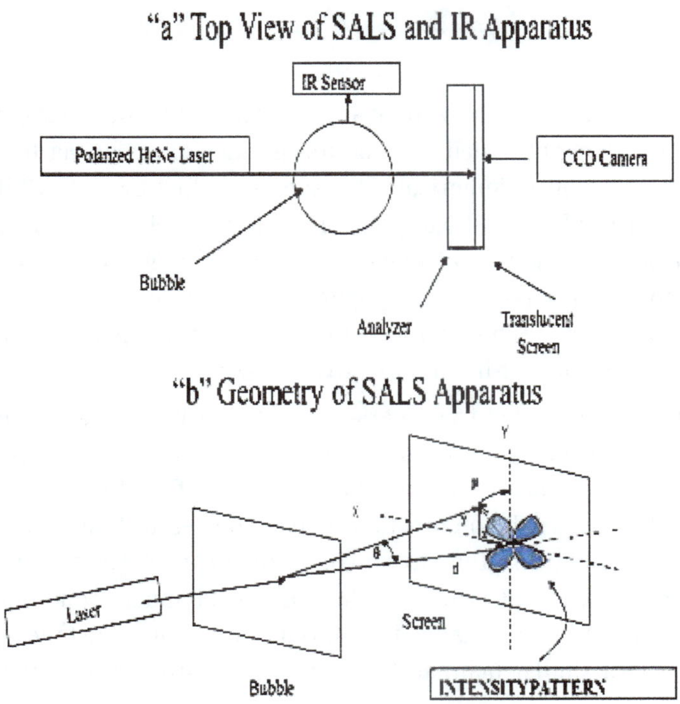

Figure 2.2(a–b) SALS equipment for temperature and crystallization

The cooling air flow rate and extruder temperature were adjusted to give different deformation histories relative to the relaxation time of the polymer. Frost line heights of 11.5, 21.5, and 31.5 cm above the die were run. These relatively high frost line heights were chosen to allow SALS and IR probes to view the area below the frost line due to the physical size of the motorized table. Extrusion temperatures of 190 °C and 221 °C were used. Process conditions for the complete series of experiments may be found in [22]. The nip speed, extruder speed, and inflation pressure were adjusted to give nominally constant values of film gauge, BUR (blow-up ratio), and DR (draw ratio). As a result, the same magnitude of deformation and integrated deformation rate were experienced for each run.

After steady state was achieved, images of the bubble shape were recorded on videotape. To determine bubble velocity and deformation rates, markers were placed on the bubble surface near the die exit. The position of the marker as a function of time as the bubble moved upward was also recorded on video [25]. Mass flow rates were measured by taking samples of the film. The infrared and SALS equipment were mounted on a platform that was computer controlled as a function of height above the die exit. To check for steady state, SALS and IR temperature data were recorded for sequential up and down direction scans along the bubble.

2.5 Process Data Analysis

Radius, velocity, time to crystallization, angle, strain, and strain rates as a function of axial position were calculated from digitized video data. Fitting functions were used to smooth data noise. Numerical cubic spline routines CSSMH, CSDEC, CSVAL, and CSDER from IMSL (Visual Numerics, Inc.) were implemented in a FORTRAN program to make the function calculations. CSSMH provides piecewise coefficients to a fitting polynomial for smoothing input data based on a user-input smoothing parameter. CSDEC provides piecewise coefficients to a fitting polynomial for smoothing input data based on user-input derivative end conditions.

Velocity data were calculated from marker vertical position versus time. This data was smoothed using the spline routine to give two interpolation functions of position versus time. The velocity was calculated from this fit using the derivative routine and the values interpolation functions of position versus time data. The velocity data was supplemented with the known $v(0)$ at the die exit. The final velocity functions were calculated using CSDEC and choosing the boundary condition parameter to give the smoothest fit. The change in radius as a function of distance from the die was calculated in a similar manner. Other quantities are based on the radius and velocity functions.

The film thickness was calculated from:

$$2\pi\rho r \cdot h \cdot v = \dot{m} \tag{2.1}$$

where ρ is the polymer density, r is the bubble radius, h is the film thickness, v is the bubble velocity, and the mass flow rate $\dot{m}$ (g/s) was measured for each trial by weighing a sample collected for one minute. The angle θ was calculated from the radius function using:

$$dr / dz = \tan\theta \tag{2.2}$$

Henky strains ε_{11} and ε_{33} in the machine and cross-machine directions were calculated as were the respective Henky strain rates $\dot{\varepsilon}_{11}$ and $\dot{\varepsilon}_{33}$, which are the time-dependent derivatives of the strains.

First the velocity and the radius are fitted with nonlinear functions from the data produced by the video analysis:

$$v = \beta_1 - \left(\beta_2 + \beta_3 x + \beta_4 x^2 + \beta_5 x^3\right) e^{\beta_6 x^{\beta_7}} \tag{2.3}$$

$$r = \alpha_1 - \left(\alpha_2 + \alpha_3 x + \alpha_4 x^2 + \alpha_5 x^3\right) e^{\alpha_6 x^{\alpha_7}} \tag{2.4}$$

The stain rates can be obtained from the deformation rate tensor:

$$D = v\cos\theta = \begin{pmatrix} \frac{1}{v}\frac{dv}{dz} & 0 & 0 \\ 0 & \frac{1}{h}\frac{dh}{dz} & 0 \\ 0 & 0 & \frac{1}{r}\frac{dr}{dz} \end{pmatrix} \tag{2.5}$$

Henky strain rate in the machine direction:

$$\dot{\varepsilon}_{11} = \cos\theta \cdot \frac{dv}{dz} \tag{2.6}$$

Henky strain rate in the cross-machine direction:

$$\dot{\varepsilon}_{33} = v \cdot \cos\theta \cdot \frac{1}{r}\frac{dr}{dz} \tag{2.7}$$

And the strains are developed by integrations of the strain rates:

$$\varepsilon_{11} = \ln\frac{v(t)}{v(t_0)} \tag{2.6}$$

$$\varepsilon_{33} = \ln\frac{r(t)}{r(t_0)} \tag{2.7}$$

At each axial position z, the time elapsed for the element after it left the die was calculated:

$$t(z) = t(z - dz) + \frac{dz}{v \cdot \cos\theta} \tag{2.8}$$

These functions will be used to correlate both the crystallization and physical property data.

It is observed in Fig. 2.3 (top) that the radius increased nonlinearly and the thickness and radius reach a plateau at some axial position, about 30 cm for this run. A two-dimensional heat transfer analysis was developed, and a typical set of curves are presented also in Fig. 2.3 (bottom). It is observed that, using the classical literature assumption that the crystallization occurs during the temperature plateau at about 30 cm above the die, the model and data agree on a several degree temperature rise in that region (see the insert). However, in almost every case the predicted temperature above the temperature plateau was always lower than the measured temperature. This led to the hypothesis that energy was being generated above the temperature plateau as the film continued to cool or the heat transfer coefficient was acting in a strange way. Both of these potential causes will be discussed in the further development of this chapter. It was these discrepancies between the predicted temperature and the data that led to the investigation using online SALS to investigate if the PE was continuing to crystallize above the freeze line. Also a new analysis of the heat transfer coefficient for the blown film process was developed that is bubble shape dependent.

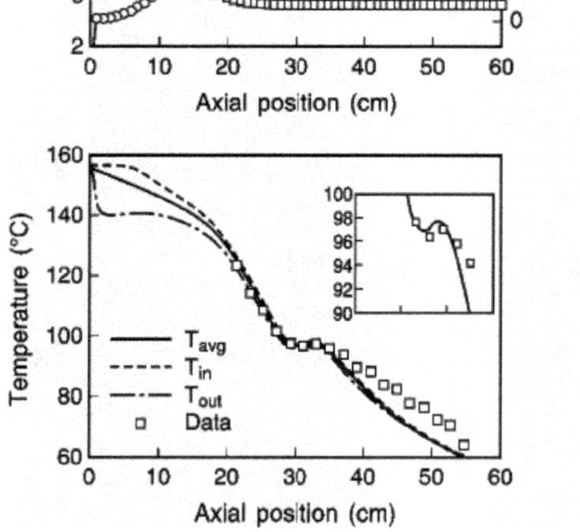

Figure 2.3 Typical blown film process data

■ 2.6 SALS Image Analysis

Images of the SALS patterns at different axial positions were captured from the videotape using a DT3152 board (Data Translation) and HL Image software (Western Digital). A total of 96 frames were digitized for each up or down direction scan so that the axial spacing between each image was about 0.56 cm for the 53 cm total scan length. This spacing proved to be short enough to get detailed results. After the images were stored, image processing was done using functions provided with the Sigma Scan Software (Jandel Scientific Software) as described below.

Figure 2.4 shows representative SALS images obtained at various axial positions along the bubble. Similar patterns were obtained for all of the process conditions studied. The four-lobed pattern indicates the presence of a spherulitic crystalline morphology. The first distinct appearance of the lobed pattern occurs about 35 cm above the die, which corresponds to the end of the temperature plateau (Fig. 2.3). While there is an obvious increase in the intensity of the patterns with increasing axial positions, the shape and orientation of the lobes do not change. These first SALS data demonstrated that classical understanding of structure development in PE resin on a blown film line was not complete. Although the crystallization did initiate at the temperature plateau in the vicinity of the freeze line, most of the crystallization occurred as the film was transported from the freeze line to the nips as indicated by the large increase in intensity in the lobes. Therefore, when simulating the process it was shown in the previous edition of this book that it is necessary to use a visco-plastic- or viscoelastic-plastic-type rheology model to obtain the proper bubble shape above the freeze line.

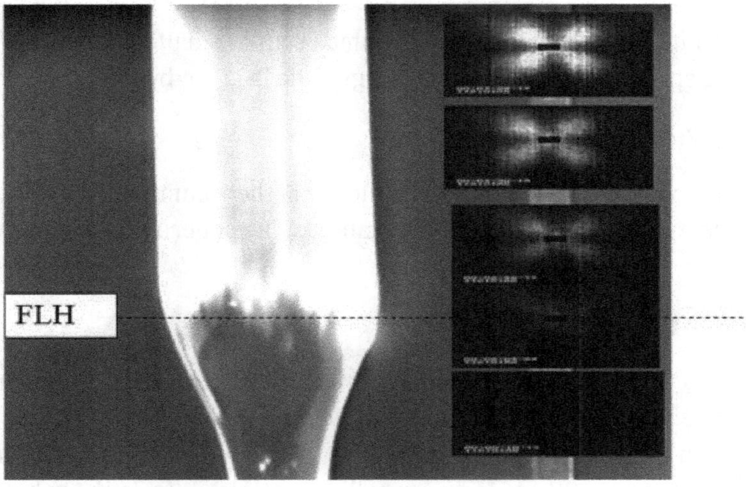

Figure 2.4 Typical SALS patterns observed at different axial positions

The spatial intensity distribution of the SALS patterns may be approximated by the theories developed by Stein and coworkers [16, 19, 20, 26] for isolated three-dimensional spherulites under *Hv* polarization conditions. With the geometry shown in Fig. 2.2 (b), the two-dimensional intensity distribution $I_{Hv}(\mu,\theta)$ can be written:

$$I_{Hv}(\mu,\theta) = I(z)\cdot F(U)(z) \tag{2.9}$$

where z is the axial distance along the bubble and:

$$U(z) = \frac{4\pi R_s(z)}{\lambda}\cdot \sin\left(\frac{\theta(z)}{2}\right),$$

$$F(U) = \left[\left(\frac{3}{U^3}\frac{\cos^2(\theta/2)}{\cos(\theta)}\right)(\sin\mu\cos\mu\cos\rho_2)(4\sin U - U\cos U - 3SiU)\right]^2$$

$$SiU = \int_0^U \frac{\sin x}{x}dx \tag{2.10}$$

where λ is the wavelength of light and R_s is the spherulite radius.

The function $F(U)$ gives rise to the four-lobed spatial distribution of intensity while $I(z)$ gives the magnitude of the intensity.

Along a given scattering vector $\frac{4\pi}{\lambda}\sin\left(\frac{\theta}{2}\right)$, U depends only on the spherulite radius. For spherulites grown under quiescent conditions, the intensity maximum on each lobe is found at odd multiples of $\mu = 45°$ and at a scattering angle of θ_{max}. For three-dimensional spherulites U_{max} is 4.1. This can be used to calculate the spherulite radius:

$$U_{max} = \frac{4\pi R_s}{\lambda}\cdot \sin\left(\frac{\theta_{max}}{2}\right) = 4.1 \tag{2.11}$$

As the spherulite grows, the intensity maxima occur at smaller angles. For a system with growing spherulites, the intensity magnitude is given by:

$$I(z) = \varphi_s V(\alpha_\tau - \alpha_t)^2 \tag{2.12}$$

where φ_s is the volume fraction of spherulites, V is the volume of each spherulite α_τ, and α_t are the radial and tangential polarizability. The spherulite anisotropy is given by:

$$(\alpha_\tau - \alpha_t) = (\alpha_\tau - \alpha_t)_C^0 \varphi_{Cs} \tag{2.13}$$

where $(\alpha_\tau - \alpha_t)_C^0$ is the (constant) intrinsic anisotropy of the pure crystal and where φ_{Cs} is the volume fraction crystallinity within the spherulites. This assumes that spherulites can be modeled as having perfect alignment of crystals and negligible form anisotropy and that there is no crystalline material outside the spherulite.

The equation can then be rewritten as:

$$I(z) = K''R_s^3(z)\varphi_{Cs}^2(z)\varphi_s(z) \tag{2.14}$$

Once the film contains volume-filling spherulites of radius R_s, $\varphi_s = 1$, and the intensity magnitude is given by:

$$I(z) = K''R_s^3\varphi_c^2(z) \tag{2.15}$$

where R_s is constant. φ_{Cs}^2 was replaced with φ_c^2 to emphasize that the overall crystallinity is completely represented by crystallization within the spherulites since the spherulites are volume-filling.

To compare the experimental intensity distribution with that predicted from Equations 2.9 to 2.15, $I_{Hv}(\mu,\theta)$ intensity maps were calculated from the experimental SALS image patterns. Figure 2.5 shows a typical SALS pattern. Superimposed on the pattern are concentric rays in the (θ) direction. The rays are spaced at five-degree increments in the μ direction [22]. An automated image processing software routine extracted the x,y rectangular coordinate position (Fig. 2.2(b)) and pixel intensity data along each ray from the intensity images. Intensities were recorded at between 283 to 400 positions for each ray, depending on the ray orientation. The procedure was repeated for 96 axial positions for each scan. A computer program calculated the x,y position of maximum intensity and total average scattered intensity from the intensity map data. Deviation of the maxima from 45° indicates bulk deformation of the spherulites [22].

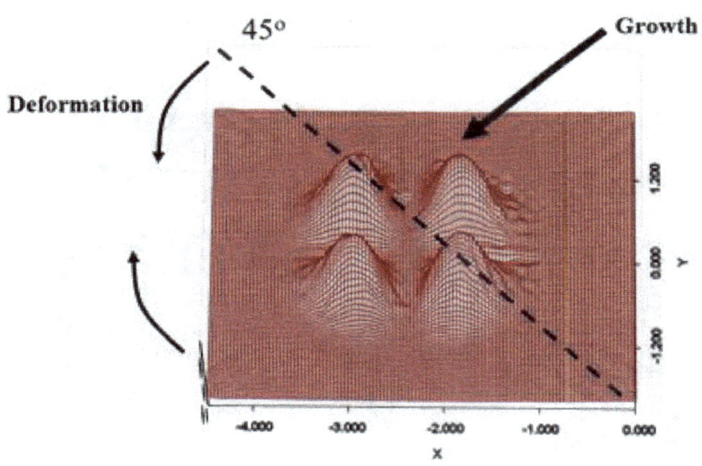

Figure 2.5 Lobed pattern shape indicates that spherulite morphology intensity maxima move inward as spherulite grows

Figure 2.6 shows a typical plot of the locations of the intensity maxima of the four lobes at different axial positions above the frost line. For all the process conditions studied, the lobes were oriented along lines of $\mu = 45°$. The same orientation is observed for quiescent cooling from the melt, that is, in the absence of bulk deformation [21]. In a separate off line experiment, it was shown in that the maxima were distorted away from 45° if a cooled film of the same resin was stretched at room temperature [22].

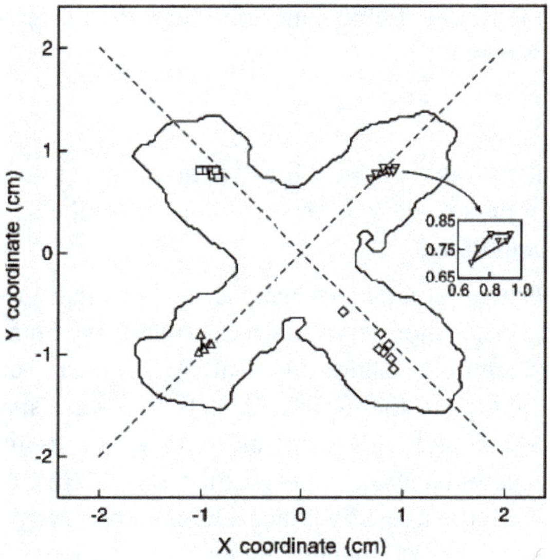

Figure 2.6 Position of intensity maxima for each lobe of the SALS pattern at various axial positions along the bubble

Figure 2.7 shows representative average intensity versus axial position calculated from the SALS images. For comparison, average intensities and temperature data for two consecutive up and down scans are shown to demonstrate the reproducibility of the data.

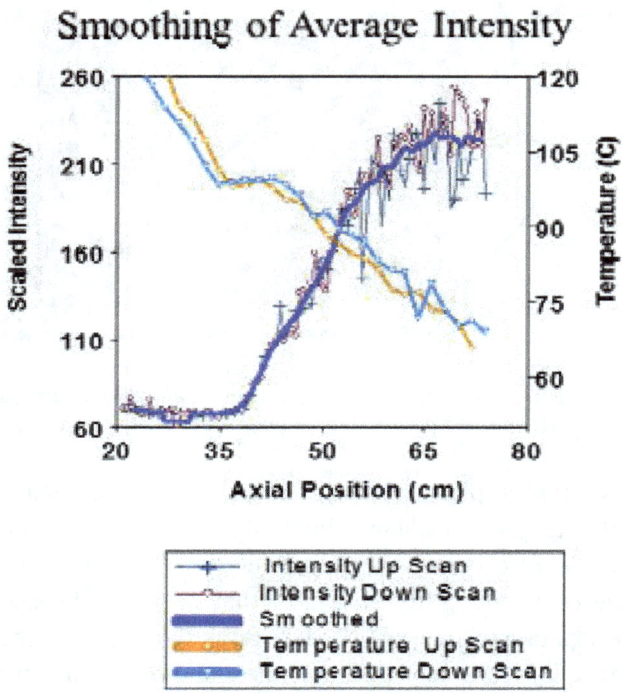

Figure 2.7 Temperature correlation with raw and smoothed SALS intensities

It is observed that there are large periodic oscillations in intensity that were found to be primarily due to movement of the bubble. The solid blue line shows a smoothed average intensity. The smoothed average at each position was calculated by averaging the intensities of the ten nearest axial positions. The data for both the up and down scan data were used in calculating the smoothed average.

The average intensity increases along a roughly sigmoidal curve with increasing axial position. There is an inflection point on the curve corresponding to the end of the temperature plateau, which suggests that there is a change in the mechanism for the SALS intensity increase. The intensity curve can be described using two Avrami kinetic processes in parallel:

$$I(z) = I1\left(1-\exp\left(k1\left(z-z1\right)^{n1}\right)\right) + I2\left(1-\exp\left(k2\left(z-z2\right)^{n2}\right)\right) \tag{2.16}$$

where $I1$ and $I2$ are intensity levels at the end of each process, $z1$ and $z2$ are starting positions for each process, and $k1$, $k2$, $n1$, and $n2$ are constants. For spherulitic growth, the Avrami exponent $n1$ was found to be near 3.0 [10]. The fit to the smoothed intensity data with $z1$ equal to the location of the start of the crystallization plateau and $z2$ equal to the location of the inflection point had the following constants: $k1 = 0.16$, $n1 = 2.5$, $k2 = 0.05$, and $n2 = 2.5$.

Equation 2.16 is consistent with the implications from Equations 2.9, 2.14, and 2.15, which predict that there will be a two-stage mechanism for the change in intensity with axial position along the bubble. The initial rise in intensity (Equation 2.14) occurs during the nucleation and growth and is due to changes in the radius of the spherulite, the volume fraction of spherulites, and the volume fraction crystallinity within the spherulite.

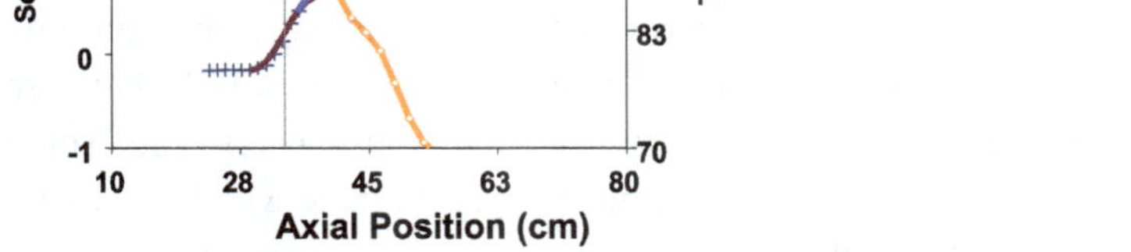

Figure 2.8 Avrami fit to smoothed average intensity. Second process starts at end of temperature plateau

This is followed by an additional increase in intensity (Equation 2.13) due to crystallization within or between the spherulites. Ree et al. [19] also observed multistep increases in intensity in LLDPE, LDPE, and their blends. A close observation of the data indicates that a third process may occur starting at about 45 cm.

The majority of the total intensity increase occurs in the second stage. The precise physical mechanism for the transition to the second process of Equation 2.16 and its relation to Equation 2.15 is not clear at this point. Equation 2.15 predicts that the rise in intensity is due to secondary crystallization within the volume-filling spherulites. An inflection point in polarized light transmission has also been observed at the end of the temperature plateau during quiescent cooling of LLDPE melts [11]. This has been attributed to the crystallization of lower molecular weight material excluded from the spherulite growth fronts. In off-line SALS studies, φ_c has been separately determined using DSC. It was found [19] that for LLDPE, φ_c tracks the radius growth with a rapid increase in weight fraction that quickly reaches a constant value. The fact that the predicted temperature drop (Fig. 2.3) after crystallization is faster than observed is consistent with the intraspherulitic crystallization mechanism because an additional enthalpic term from crystallization would give rise to a higher film temperature. Kamal et al. [27] and Verhoyen et al. [28] have both applied modified Avrami kinetics to include a secondary crystallization process. Another possibility is that orientation processes accompany secondary crystallization. For example, online birefringence measurements for LLDPE [15] indicate that the in-plane birefringence increases with axial position well beyond the crystallization plateau.

2.7 Nucleation with High-Density Polyethylene

Linear and branched polyethylene blends continue to be of great interest in the polymer industry. Blending polyethylenes can improve the properties of the final products such as toughness and processability. Many studies have been devoted to the morphology of melt and solid states of polyethylene blends [29]. For a semicrystalline polymer, the molten polymer crystallizes as the film cools. The properties of the film are dependent on the size and number of crystallites, which are instrumental in forming the physical crosslinks that freeze in the structure in the film. A previous paper describes our initial results in developing the online SALS tool [30]. Further details of experiments are available elsewhere [22]. Wilkes et al. used SALS and other techniques to study the optical properties of the PE film [31].

2.8 Experiments

LLDPE resin with 0, 1, and 4 weight percent HDPE added were processed on a blown film process line consisting of a Killion extruder with a 1.5 inch (3.81 cm) 24/1 screw, 30 inch (76.2 cm) layflat nip roll section, and a single-lip air ring [32]. The annular die gap was of 28 mil (0.71 mm) and had an initial radius of one inch (2.54 cm). A motorized table containing an IR surface temperature probe and SALS equipment (polarized He-Ne laser, analyzer, screen, and CCD camera) translated in the axial direction along the outside of the bubble between axial positions of 16 and 82.5 cm above the die. Other details about the experimental setup of the IR surface temperature probe are described in Cao et al. [33] and Campbell et al. [34]. SALS images were recorded on a VCR for later analysis while the IR data were recorded on a computer that recorded the temperature as a function of position above the die. Figure 2.2 shows the traditional SALS experimental geometry. All SALS experiments were made under *Hv* polarization conditions, which denotes that the analyzer was set perpendicular to the polarization direction of the laser.

The same equipment for developing the SALS images was used in this investigation as previously described [22]. The same analysis as described in Equations 2.9 through 2.15 were used to quantify the image structure and intensity. A typical intensity map is presented in Fig. 2.9.

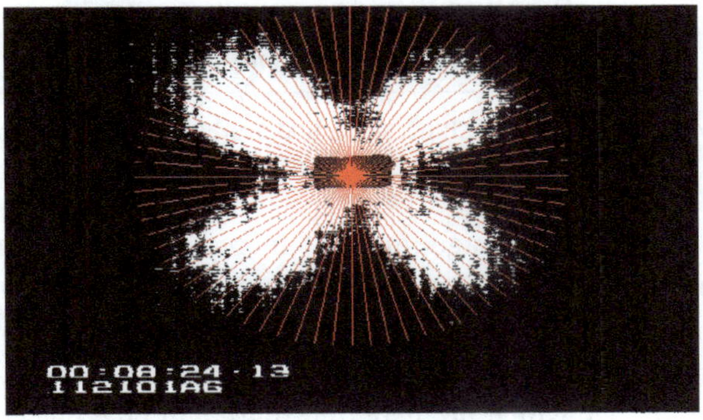

Figure 2.9 Scanning procedure to develop image density map

2.9 Results

Data from these experiments (Fig. 2.10) shows the changes in film velocity, thickness, and temperature as well as the scattering intensity as a function of the axial distance from the die for the LLDPE used in these experiments with 380°F melt temperature. The temperature drop is initially nearly linear. This is followed by an abrupt transition to a crystallization plateau at the subcooled crystallization temperature, with about 5 °C due to the difference between heat generated by crystallization and heat removed by cooling [33, 34]. After the plateau, the primary crystallization process is complete and the temperature again drops along a smooth curve. Moreover, it shows that the increase in intensity starts after the film thickness and velocity become fairly constant, that is above the frost line height (FLH, also called the freeze line), which led to the assumption that one may exclude the thickness effect on the scattering intensity for the majority of the process.

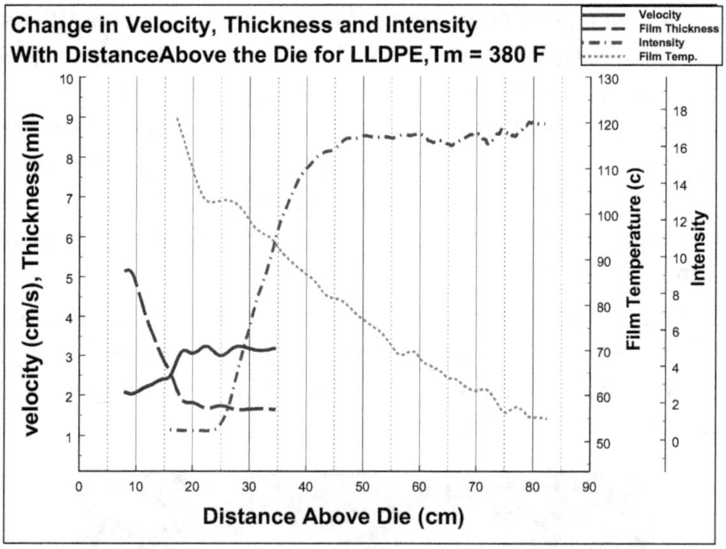

Figure 2.10 Film velocity, thickness, temperature, and average total intensity vs. distance above the die

As with the previous resin, average intensity increases along a roughly sigmoidal curve with increasing axial position. There is an inflection point on the curve corresponding to the end of the temperature plateau, which suggests that there is a change in the mechanism for intensity increase. This is consistent with the implications from Equation 2.16, which predicts that there will be a two-stage mechanism for the change in intensity with axial position along the bubble. The initial rise in intensity occurs during the nucleation and growth and is due to changes in the radius of the spherulite and the volume fraction of spherulites. It is currently thought that this is followed by an additional increase in intensity due to crystallization within or between the spherulites. In some cases, a decrease in the intensity in the end of the process was observed. That can be due to impingement and truncation effects.

In Fig. 2.11, the intensity curves of pure LLDPE and of 1% HDPE/LLDPE and 4% HDPE/LLDPE blends were plotted for two different melt temperatures.

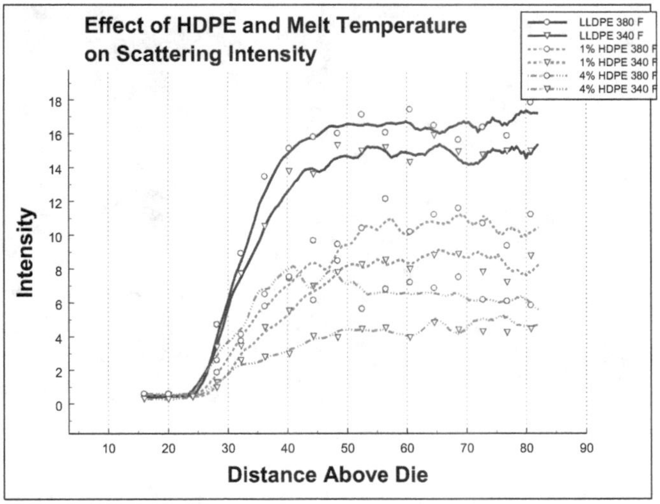

Figure 2.11 Effect of HDPE concentration and melt temperature on the scattering intensity

Increasing the melt temperature causes an increase in the scattering intensity and more distinct spherulitic patterns, which could suggest a lower degree of inhomogeneities, that is internal disorder, in the morphology of spherulites. Figures 2.11 and 2.12 show SALS images of LLDPE film and 4% HDPE/LLDPE film at two different temperatures.

Although it is known that higher cooling rate generally results in higher crystallinity [3] and, consequently, higher scattering intensity, the cooling rate change in this study is not so high that it produced such a significant difference in the intensity. In addition, it is shown that adding a small percentage of HDPE considerably affects the spherulitic pattern of pure LLDPE and causes a decrease in scattering intensity.

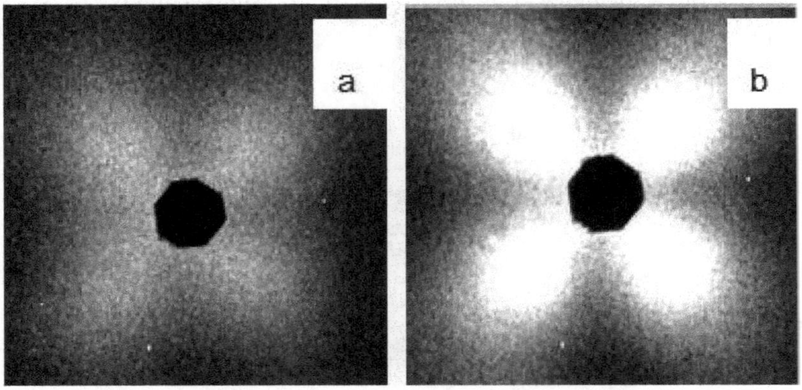

Figure 2.12(a) LLDPE film with melt temperatures (a) 300°F and (b) 380°F

It could be that the large difference between the melting temperature of HDPE and LLDPE allows the HDPE to play a role as nucleation agent and increase the crystallization sites.

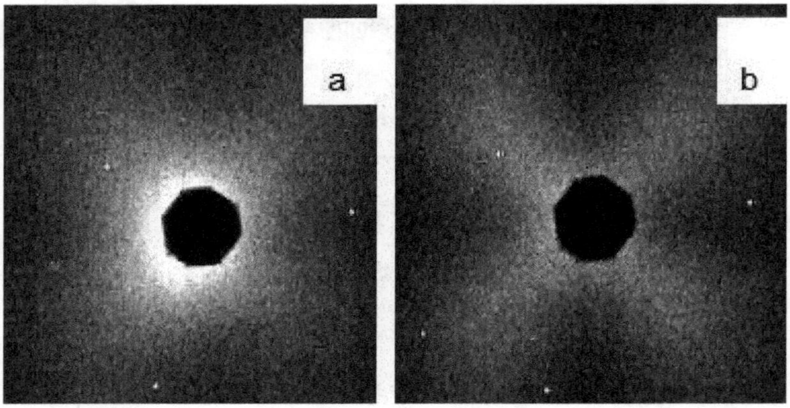

Figure 2.12(b) 4% HDPE film with melt temperature (a) 300°F and (b) 380°F

Consequently, this will lead to a larger number of spherulites and early impingement causing a drastic decrease in both scattering intensity and the distinct four-lobe patterns. However, it was reported that LLDPE/HDPE blends are miscible and cocrystallizable [35]. No considerable change in the spherulite size was observed, which ensures cooling rate insignificance.

Summarizing, an online SALS system was used to study the effect of HDPE as a nucleation aid for crystallization during tubular blown film extrusion. For LLDPE and LLDPE/HDPE blends, effects of both melt temperature and HDPE composition have been studied. Increasing melt temperature leads to an increase in the scattering intensity and well-defined patterns. However, increasing the HDPE composition results in distortion in the spherulite pattern, which can be caused by impingement and truncation of the spherulites. The measured change in average scattered intensity with axial position was consistent with temperature data, and a crystallization process consisting of primary spherulite growth followed secondary crystallization and orientation processes.

2.10 Temperature Measurement and Heat Transfer from the Blown Film Bubble

As pointed out previously, the tubular blown film extrusion process is used extensively to make thin polymer films and may have a complex temperature profile in the film and a temperature coefficient dependent on bubble shape.

The dynamics of the cooling above the air ring is complex, and the heat transfer coefficient that controls the film cooling is an important consideration when simulating the cooling characteristics of the bubble. Early on in the cooling process the film is hot enough such that radiation can account for up to about 30 percent of the heat loss. As the film cools, the heat transfer transitions from an almost equal part radiation and convective heat transfer to almost exclusively convective heat transfer. This can be even more complicated if the bubble has internal bubble cooling. As discussed previously, the cooling of the bubble provides the driving force for the crystallization of semicrystalline polymer. This process of crystallization leads to the formation of the physical crosslinks that freeze in the structures developed during the expansion and elongation of the molten polymer as it moves from the die to the freeze line. Above the freeze line, the film acts as a plastic-elastic solid and undergoes very little additional deformation.

Since the film is of finite thickness, the temperature in the film will not be a constant; it will be a minimum at the air–polymer interface and a maximum at some position across the film thickness. If the film is being cooled at the same rate from both surfaces, then the maximum temperature in the film will be in the center of the film. If there is more cooling from one side of the film, then the maximum will be nearer to the surface that has the lowest heat transfer rate. Determining this gradient is necessary if one is to test developed models that actually predict the cooling characteristics of the film. This would be necessary to accurately predict the crystallization within the film since the crystallization process is exothermic and will only proceed as the heat of crystallization is removed from the film. It has historically been assumed that the majority of the crystallization in semicrystalline polymer occurred at the constant temperature plateau that was associated with the freeze line. The discussion in the earlier sections of this chapter demonstrate that this assumption in really not valid. Rapid crystallization at the freeze line was thought to produce MD and TD properties that were primarily related to the drawdown and blow-up ratio of the bubble. It would now appear that the heat transfer that controls the crystallization rate affects the physical property development.

In the remainder of this chapter, a discussion will be presented on the development of several techniques that can be used to obtain the fundamental data needed to understand the relationship between physical properties, the kinematics, and the process dynamics [36]. It will be demonstrated that, using the process stretching dynamics, MD and TD properties can be presented on a single surface.

2.11 Measuring Film Emissivity and Film Thickness

It is apparent that one cannot place a thermocouple, RTD, or thermistor into these thin films to measure temperature because the devices would not give an accurate measurement for the very thin films generally produced in the blown, cast, or tentered film processes. Therefore, the experimenter is forced to estimate the temperature using a noncontact system. It was reported in 1986 by Campbell and Huang [37] that the surface temperature of a polyethylene film could be monitored using 3.34 micron infrared detectors. This infrared noncontact technique was extended by Cao, Sweeney, and Campbell [38] by developing a technique for characterizing the apparent emissivity of a thin polyethylene film as a function of film thickness and temperature using the radiation spectrum from 1.8 to 3.0 microns. In the experiment a heater was used to heat a metal block with an aluminum tray sitting on the block; see Fig. 2.13.

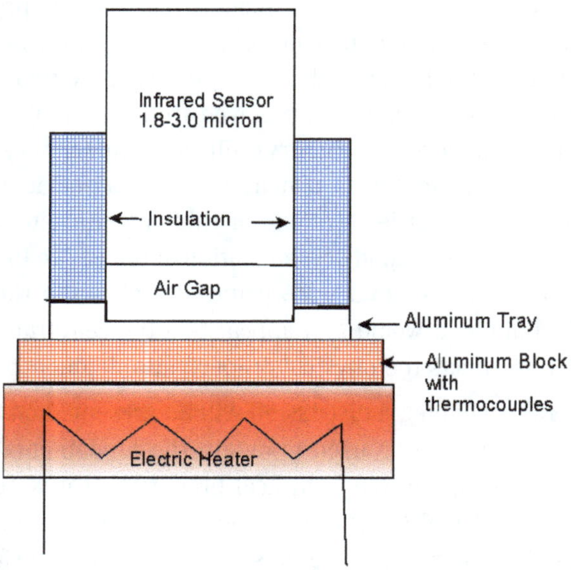

Figure 2.13 Device for measurement of emissivity

The device was insulated to minimize heat loss so that an assumption that the aluminum tray was at the same temperature as the block would be a reasonable assumption. Differing thicknesses of polyethylene film were added to the tray. Data were taken that correlated the infrared response as a function of film temperature and film thickness. A representative sample of the relationship for emissivity as a function of film thickness at several temperatures is found in Fig. 3 in [38]. The functional relationships for the thermal constants are developed according to McMahon [39].

$$\tau^* = \frac{\tau_i (1-\sigma_s)^2}{1-\sigma_s^{\,2}\tau_i^{\,2}} \tag{2.17}$$

$$\sigma^* = \sigma_s \left(1 + \frac{\tau_i^2 (1-\sigma_s)^2}{1-\sigma_s^{\,2}\tau_i^{\,2}} \right) \tag{2.18}$$

$$\varepsilon^* = \left(\frac{(1-\sigma_s)(1-\tau_i)}{1-\sigma_s \tau_i} \right) \tag{2.19}$$

The above three equations represent the apparent transmissivity, reflectivity, and emissivity, respectively. The parameter τ_i is the true or inner transmissivity and σ_s is the true or surface reflectivity. The apparent radiation properties obey the conservation law of radiation:

$$1 = \varepsilon^* + \sigma^* + \tau^* \tag{2.20}$$

The true surface reflectivity for polymers is considered to be in the range of 0.03 to 0.05 for polymers from an IRCON data sheet [40]. The inner transmissivity is a strong function of the material thickness, temperature, and wavelength used in the measurement:

$$\tau_i(h, T, \lambda) = \exp(-\alpha(T, \lambda) h) \tag{2.21}$$

When the film is placed in the aluminum tray, the total radiation entering the sensor is:

$$R = R_f^{(b)} \varepsilon^* + R_e^{(b)} \varepsilon_e \sigma^* + R_b^{(b)} \varepsilon_a \tau^* \tag{2.22}$$

The first term represents the direct radiation from the film, the second term represents the indirect radiation reflected from the film due to the environment, and the third term indirect radiation transmitted through the film from the aluminum tray. It is necessary to know the black body temperature.

$$R^{(b)} = \int_{\lambda_1}^{\lambda_2} \frac{C_1}{\lambda^5} \exp\left(-\frac{C_2}{\lambda T} \right) d\lambda \tag{2.23}$$

where $R^{(b)}$ is the absolute black body radiation, which is a function of temperature according to the Stefan Boltzmann's law.

The data relating apparent emissivity and thickness at three temperatures is presented in Fig. 2.14.

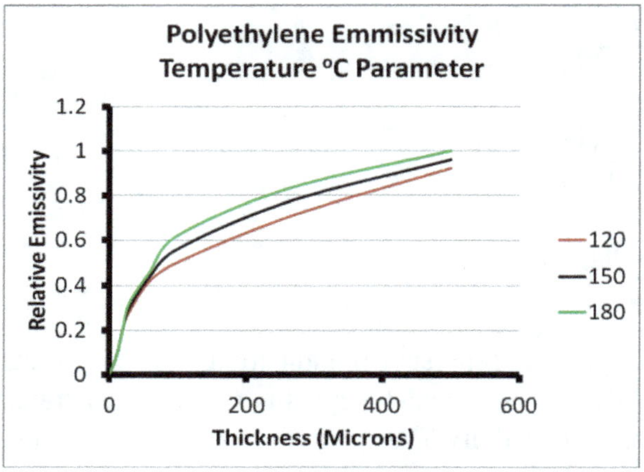

Figure 2.14 Relationship for emissivity, film thickness, and temperature

Using the above analysis, the apparent emissivity for this polyethylene sample as a function of temperature and film thickness was found to be [38]:

$$\varepsilon^* = 2.5*10^{-3}h - 3.12*10^{-6}h^2 + 4.55*10^{-6}hT - 4.13*10^{-4}T + 0.206 \tag{2.24}$$

where h is the film thickness in microns and T is the temperature in centigrade and ε^* is the emissivity. The complete description of the method to obtain the emissivity and temperature for development of the equation may be found in [38]. Once two parameters are known, then the equation reduces to a second-order algebraic equation and h is the positive root.

With this understanding of the interrelationship of emissivity, film thickness, and temperature a technique could then be developed that provided two simultaneous temperatures at the same position on the expanding bubble. The stress in the film as it expands and is drawn is a function of the local temperature in the film since temperature affects both the modulus and the viscosity. The film properties are a strong function of the frozen-in strain that results from the effect of the process kinematics, viscosity, and local modulus. Understanding the temperature gradients in the film is particularly important since the physical crosslinks due to partial crystallization freeze in the local molecular strain and thus affect the film properties. It is important to have a technique that can be used to validate the cooling and crystallization models used in the analysis of the film blowing process.

■ 2.12 Film Average Bulk Temperature and Surface Temperature

The surface temperature of a polymeric film can be deduced from its radiation characteristics, in particular the infrared sorption. If the infrared is absorbed to the extent that it does not penetrate the polymer, then the polymer is essentially a black body at the film thickness being measured. The surface temperature can be obtained using an infrared detector that is tuned to that frequency. A detector operating at 3.43 microns will thus give the average temperature of about the outer 0.001 inch (25.4 micron) of polyethylene [37, 41].

The bulk average temperature is more difficult to determine. A technique was reported in [41] that used 1.8–3.0 micron infrared detection and a variable temperature black body to evaluate average bulk temperature of a blown film from just above the air ring to well above the freeze line. Several devices were placed on a screw-driven platform that could be rotated almost 360° so that the temperatures could be taken over the whole film cooling area; see Fig. 2.15.

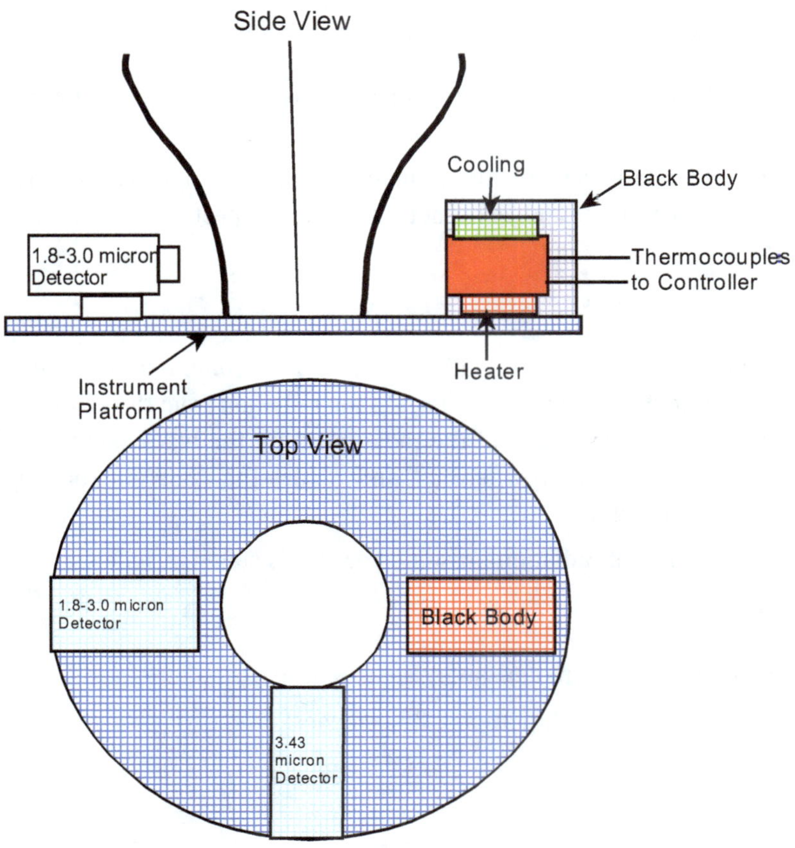

Figure 2.15 Surface and bulk film temperature measurement device

The black body was calibrated to maintain its temperature within 1 K. A typical infrared detector response for several black body temperatures is represented in Fig. 2.16.

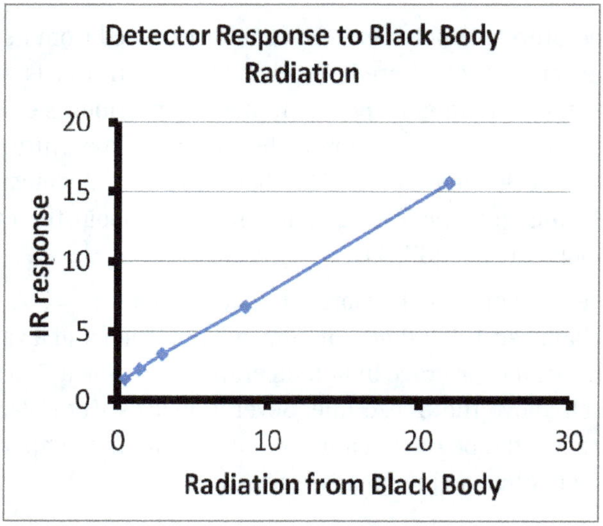

Figure 2.16 Black body radiation vs. infrared detector response

It is apparent that the IR detector's response is a linear function of the black body radiation.

The use of these data is dependent on understanding the concepts that are used to evaluate radiation from any object. There are three aspects of radiation from any "hot" body:

$$R = R_f + R_e\sigma^* + R_b\tau^* \tag{2.25}$$

The total radiation seen by the radiation detector, R, is composed of three contributions, the radiation directly from the film, R_f, the radiation from the environment, R_e, which is reflected by the film, and the radiation from the black body, R_b, which is transmitted through the film, where σ^* and τ^* are the apparent reflectivity and transmissivity of the film.

The conservation law of radiation presented earlier is stated as follows:

$$1 = \varepsilon^* + \sigma^* + \tau^* \tag{2.26}$$

where ε^* is the film apparent emissivity.

It has been shown [41] that Equation 2.25 becomes:

$$R = R_f^{(b)}\varepsilon^* + R_e^{(b)}\sigma^* + R_b^{(b)}\varepsilon_b\tau^* \tag{2.27}$$

where $R^{(b)}$ is the absolute black body radiation, which is a function of temperature according to the Stefan Boltzmann's law:

$$R^{(b)} = \int_{\lambda_1}^{\lambda_2} \frac{C_1}{\lambda^5} \exp\left(-\frac{C_2}{\lambda T}\right) d\lambda \qquad (2.28)$$

The black body emissivity, ε_b, was calculated by the procedure developed by Buckley [42]. The second term on the right-hand side of Equation 2.27 is negligible since the environment temperature is small compared to the film temperature and the apparent reflectivity of polymeric films is generally small. At steady state the response from the radiation detector, R, (Fig. 2.16) is thus a linear function of the black body radiation with the slope τ^* and the intercept of $R_f^{(b)}\varepsilon^*$. Since the apparent reflectivity can be considered to be a constant and small, in this case the reflectivity was considered to be 0.04 from IRCON [40]. Using the slope and Equation 2.26, the apparent emissivity is calculated and from the intercept $R_f^{(e)}$ can be obtained. Then Equation 2.28 is used to evaluate the "average" film temperature. The surface temperature, the outer 0.001 in. (25.4 micron), is obtained directly from the 3.43 micron detector.

This now provides the data to validate a heat transfer model for the cooling and crystallization of the film. Heat transfer models have tended to underpredict the temperature as discussed previously; see Fig. 2.3.

2.13 Experimental Evaluation of the Bubble Heat Transfer Coefficient

The focus of this phase of the investigation was to develop a relatively simple model for the heat transfer coefficient (hc) that could be used in blown film simulation. It was desired to have the model represent the major contributors to heat transfer for blown film because the heat transfer controls the crystallization rate and thus the physical crosslinks that lead to the film's physical properties. The experimental results presented here suggested that the heat transfer coefficient can go through a maximum as the bubble expands for some bubble shapes. A physical model is proposed that is consistent with our experimental observations. The model is based on a 3-D representation of the boundary layer of air around the bubble that changes thickness as it is stretched by the expansion of the bubble radius.

The blown film process is depicted in Fig. 2.1. A polymer melt is extruded through an annular die, and the film is stretched due to the trapped air in the bubble and the nip rolls while being cooled by an air ring. A great deal of effort has been expended on developing models for blown film in an effort to ultimately predict properties of the film. As a predictive tool, the models have a major defect: the bubble temperature profile and frost line height should be specified in order to obtain the bubble shape. Petrie [43] introduced viscoelasticity into the analysis without taking temperature effects into account. The importance of heat transfer was first discussed by Petrie [44]. In his paper he also recognized the importance of cooling air flow around the bubble in predicting the temperature and ultimately the properties of the film. An attempt to model the cooling rate of the film was first attempted by Menges and Predohl [45]. They proposed a new formula for estimating the heat transfer coefficient by determining the heat transfer conditions with real experiments. Similar attempts were made by Ast [46] and Zippenfeld [47]. More recently Kanai and White [48] proposed a different model. These several proposed models for heat transfer coefficient are described below.

Menges and Predohl:

$$h_c = 3.3\,(v_{max})^{1.5}$$

Zippenfeld:

$$h_c = 7.11\,(v_{max})^{0.78}$$

Petrie:

$$h_c = 4\,(v_{max})^{1.5}$$

Kanai and White:

$$h_c = 0.043\,K_{air}\,(Re)^{0.78}/L$$

All of these models predict different heat transfer characteristics, and there were no data available to test which is best. Cao et al. developed a two-dimensional heat transfer model for blown film that incorporated a phase change based on convective heat transfer in the machine direction and diffusive heat transfer in the film thickness direction [41, 49]. Recently, Vlachopoulos and his colleagues have developed numerical models for the blown film heat transfer based on an analysis of turbulence [50]. These models focus on the characteristics of turbulent fluid dynamics and the formation of local persistent macroscopic eddies. In order to experimentally characterize the effect of bubble shape on blown film heat transfer, a method was developed for taking data on structures that mimic the general shape of the blown film bubbles. A first-order model is then proposed that qualitatively fits the data developed.

Two of the experimental devices used in this investigation are shown in Fig. 2.17.

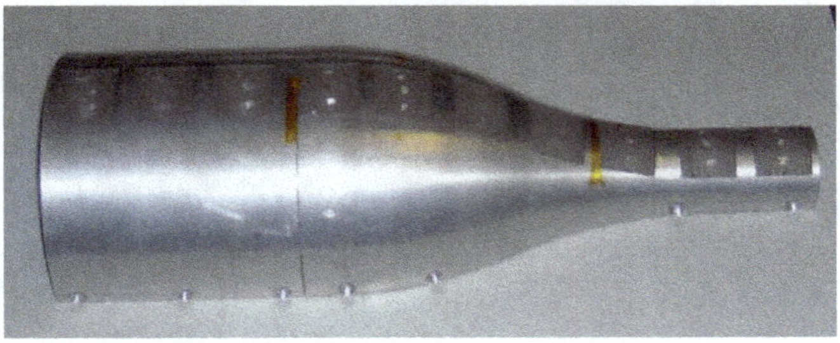

Figure 2.17(a) Slow transition bubble

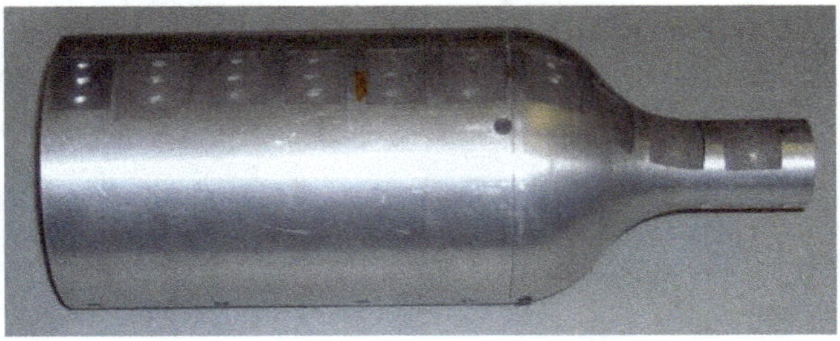

Figure 2.17(b) Rapid transition bubble

Aluminum structures were machined with an external shape of typical blown film bubbles. Figure 2.17(a) depicts a shape typical of low-density blown film, while Fig. 2.17(b) depicts a shape approaching a high-density high stalk transition. Several other bubble shapes were also evaluated [51]. All of the aluminum bubbles had square relief areas machined into the surface.

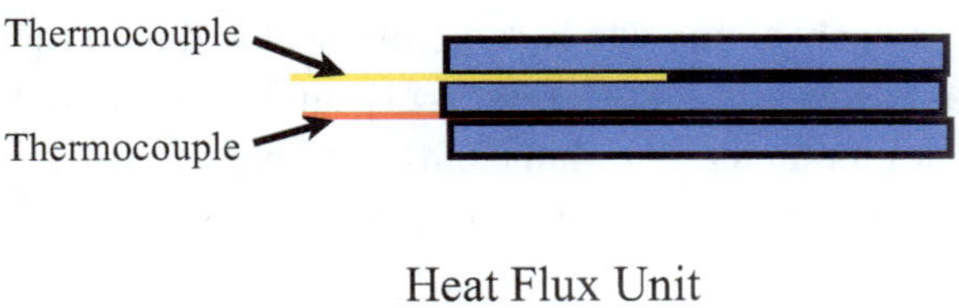

Figure 2.18 Heat flux sensor

These relief areas were filled with heat flux meters (Fig. 2.18), which provide local heat flux normal to the bubble surface and local temperature simultaneously. The heat flux sensors were covered with a very thin film of PTFE to minimize the potential turbulence at the transitions; see Fig. 2.19. The sensor wires obvious in Fig. 2.19 were passed into the center of the bubble through two holes drilled under the sensor so that they did not interfere with the air flow on the bubble surface. A more complete description of the overall process may be found elsewhere [51–53].

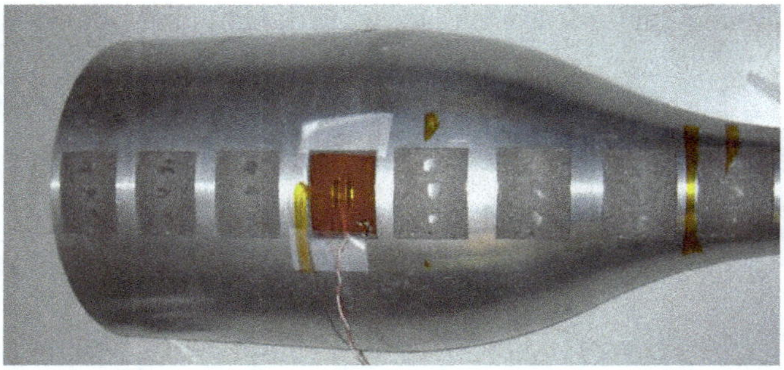

Figure 2.19 Bubble with the sensor attached to the surface

Data was taken in real time. The experiment used in this investigation was a transient heat transfer experiment where the aluminum bubble was heated to a constant temperature, as indicated by the series of thermocouples in the heat flux sensors in Fig. 2.20.

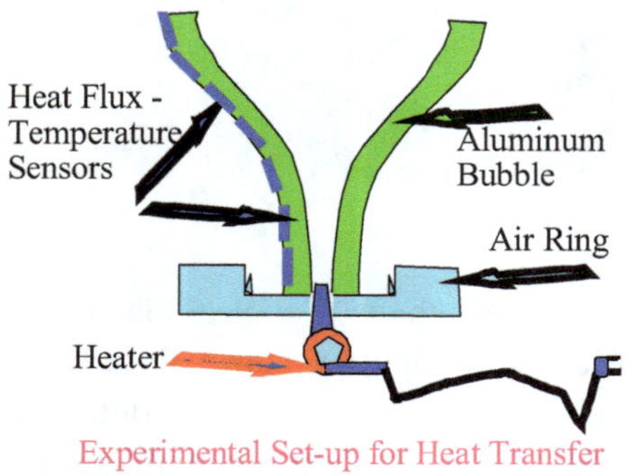

Figure 2.20 Heat gun heater, air ring, and sensors

Then the heater was turned off and the air ring turned on. The heat flux and temperature were then recorded continuously as the whole structure cooled due to the cooling air. The sensors and the data collection system (Fig. 2.21) can be considered as a combined temperature and heat flux meter. The shape of the bubble and the air velocity produced different shapes of the heat flux. Data was taken until the signal-to-noise ratio was high.

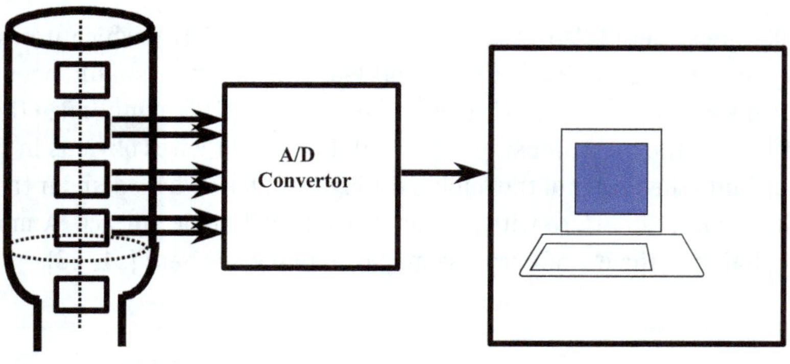

Figure 2.21 Heat flux meter

2.14 Results and Discussion

A typical set of results of a heat transfer experiment with a bubble with a high curvature rate is found in Fig. 2.22.

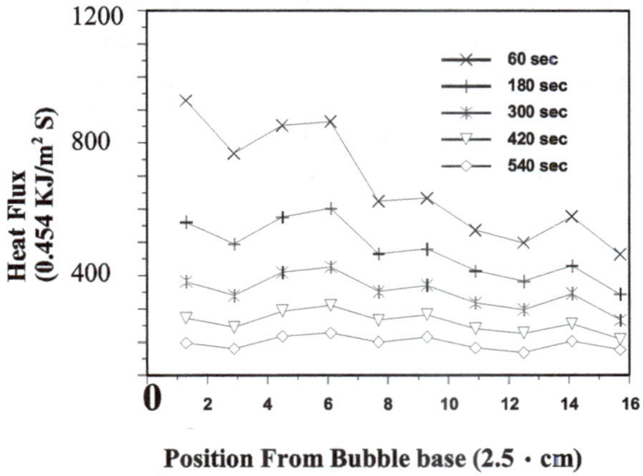

Figure 2.22 Heat flux as a function of height above the die

The most important characteristic is the increase in heat flux from positions 3 to 6 in Fig. 2.22. This increase is present at all times for this bubble shape and is not consistent literature models or with conventional flat film or two-dimensional heat transfer analysis based on boundary layer development on a flat plate due to incorporation of the fluid into the boundary layer as the fluid flows up the surface; see Fig. 2.23.

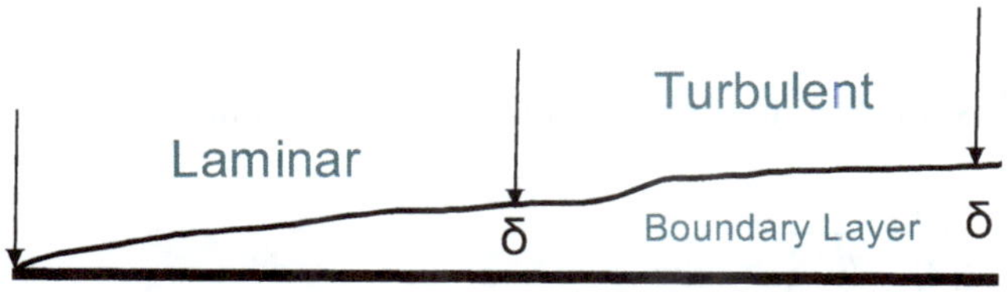

Figure 2.23 Boundary layer in two dimensions

The physics that controls heat transfer is the boundary layer thickness. For simple laminar flow on a flat plate the analysis leads to Equation 2.29 [54] with the Reynolds Number defined as in Equation 2.30. This has a different form for turbulence. However, the boundary layer essentially only grows faster with turbulence in the same monotonic manner, so in this evaluation the boundary layer thickness will be modeled as the laminar function recoganizing that this could lead to an overprediction of the heat transfer coefficient.

$$\delta_{flat} = \frac{5}{\sqrt{\mathrm{Re}_z}} z \tag{2.29}$$

$$\mathrm{Re}_z = \frac{\rho U z}{\mu} \tag{2.30}$$

where U is the free stream viscosity, μ is the viscosity of the air, and ρ is the air density.

Although the boundary layer film thickness can be modeled using several assumptions, Nagaragan [51] found after evaluation of several sets of assumptions that a differential approach led to the best fit to the shape of the data presented in Fig. 2.22.

Assuming the thickness of the boundary layer is small, the volume of air in the boundary layer per unit length can be expressed as:

$$V_{bl(z)} = 2\pi r \delta_{bubble(z)} \tag{2.31}$$

where r is the radius of the bubble at z.

Rearranging:

$$\delta_{bubble(z)} = \frac{V_{bl(z)}}{2\pi r} \tag{2.32}$$

Differentiating and simplifying:

$$\frac{d\delta_{bubble(z)}}{dz} = -\frac{\delta_{bubble(z)}}{r}\frac{dr}{dz} \tag{2.33}$$

Recognizing that this is an alteration of the flat plate boundary layer function:

$$\frac{d\delta_{bubble(z)}}{dz} = \frac{5}{\sqrt{\mathrm{Re}_z}} z - \frac{\delta_{bubble(z)}}{r}\frac{dr}{dz} \tag{2.34}$$

That leads to this function for calculating the bubble boundary layer thickness:

$$\delta_{bubble(z)} = \frac{\delta_{flat(z)}}{1 + \sum_z \frac{1}{r}\frac{dr}{dz}\Delta z} \tag{2.35}$$

A working equation for calculation of the ith bubble boundary layer thickness given these assumptions is:

$$\delta_{bubble(z),i} = \int_0^{z_i} \frac{5}{\mathrm{Re}_z} - \frac{\delta_{bubble(z),i-1}}{r}\frac{dr}{dz} dz \tag{2.36}$$

The heat transfer coefficient that relates to the film thickness $\delta_{bubble(z),i}$ is:

$$Nu_{bubble(z),i} = \frac{2z}{\delta_{bubble(z),i}} \mathrm{Pr}^{\frac{1}{3}} \tag{2.37}$$

$$h_{bubble(z),i} = Nu_{bubble(z),i}\frac{k}{z} \tag{2.38}$$

Where Pr is the Prandtl number and k is the thermal conductivity of the air.

Figure 2.24 shows a bubble with a relatively rapid change in the radius, in red, and the associated derivative in the blue dotted line, as a function of z.

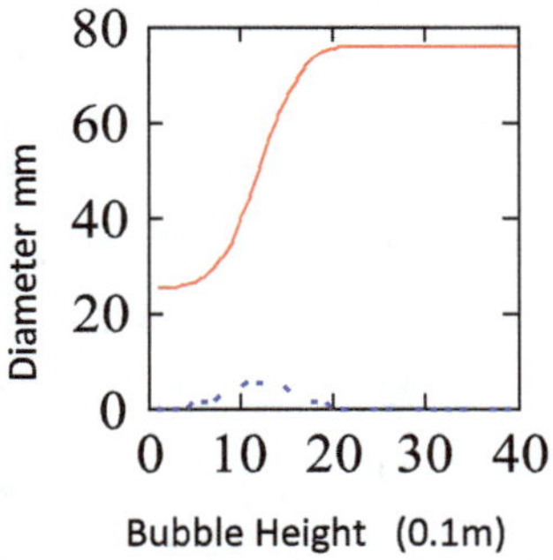

Figure 2.24 Bubble shape (red); derivative (blue)

Figure 2.25 is a graphical representation of the boundary layer thickness for this bubble shape using the differential assumptions presented previously, red using the thickness change predicted in the previous section, compared with the flat plate assumption, blue.

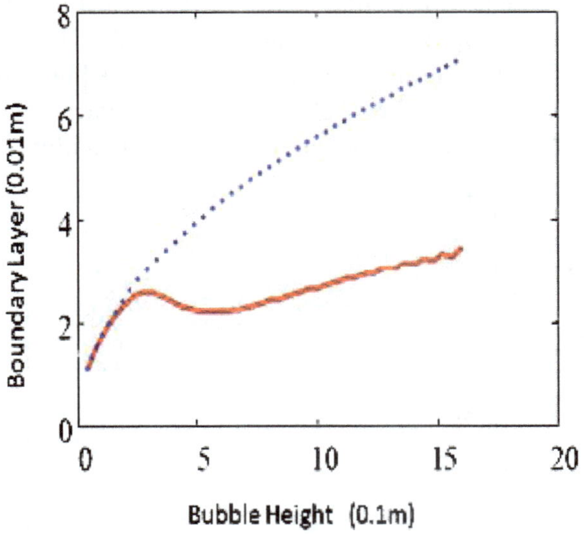

Figure 2.25 Boundary layer thickness

The three-dimensional assumptions lead to the boundary layer thickness decreasing with any large change in curvature dr/dz. A prediction of the heat transfere coefficient from the theoretical analysis is presented in Fig. 2.25. The solid red line is the prediction of the boundary layer thickness for the 3-D differential analysis; the blue dotted line is the prediction for the flat plate assumption. This leads to a prediction for the heat transfer coefficient for this bubble shape.

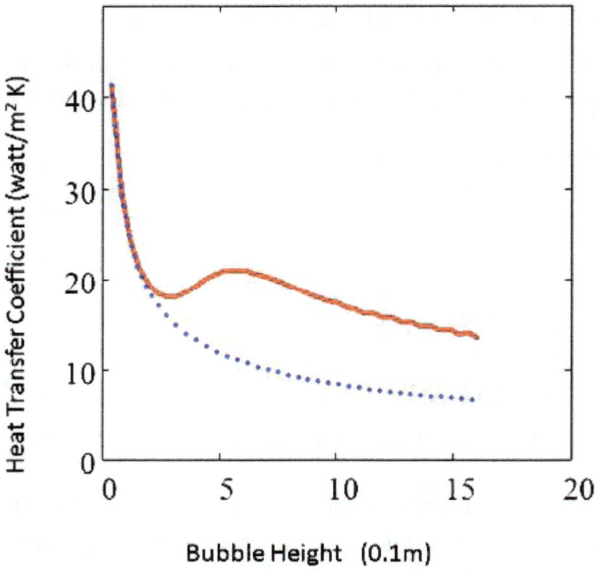

Figure 2.26 Predicted heat transfer coefficients

The 3-D corrected boundary layer bubble heat transfer coefficient (the red line in Fig. 2.26) is consistent with the data presented in Fig. 2.22. The blue dashed line is the prediction from the flat film and does not correlate. It is interesting that the predicted shape of the heat transfer coefficient for this analysis is consistent with Fig. 10 in Chapter 4 in this book although they are based on different assumptions.

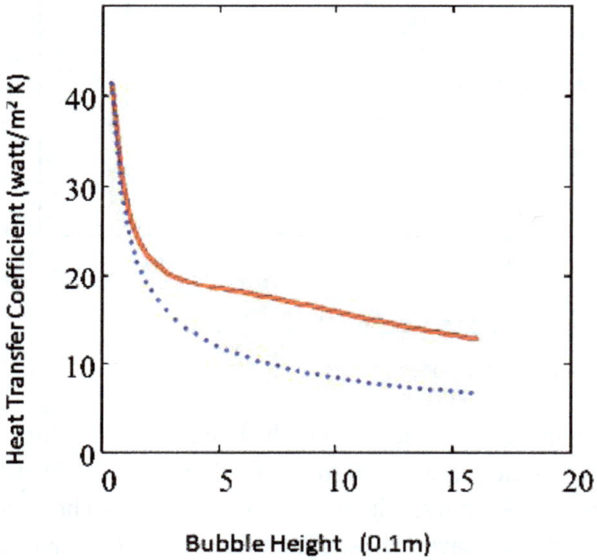

Figure 2.27 Heat transfer coefficient for low-curvature bubble

For a bubble with lower curvature (Fig. 2.27), the heat transfer using the differential boundary layer analysis (red line) is still higher than the flat plate assumption (blue line) due to the thinning of the boundary layer. However, there is no maximum in the heat transfer function.

This thinning characteristic of the boundary layer as a function of distance from the die would look like the width of the red rectangles in Fig. 2.28 as predicted by the differential 3-D boundary layer analysis.

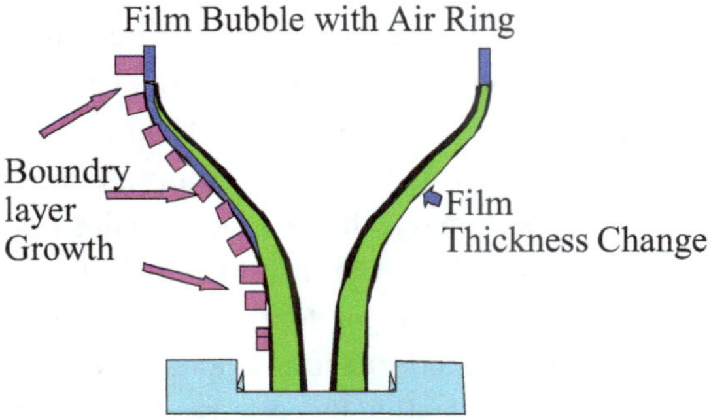

Figure 2.28 2-D representation of the boundary layer thinning

The results from the heat flux measurements and theory were confirmed based on the cooling air velocity measurements [51]. The cooling air velocity distribution around the bubble was measured by means of a hot wire anemometer. It was found that the air velocity goes through a maximum normal to the surface of the bubble. The thickness of the boundary layer near the curvature region decreases because the radius of the bubble grows at a faster rate than the boundary layer. This is consistent with the increase in heat loss from the curvature region (Fig. 2.22).

The heat transfer coefficient for the blown film process was measured and modeled. A corrected boundary layer thickness was calculated by incorporating the effect of the bubble radius growth on the thickness of the cooling air boundary layer. The heat transfer coefficient calculated based on the corrected boundary layer thickness is in excellent agreement with the experimental results. Use of this heat transfer function in blown film modeling will lead to better predictions of cooling and thus crystallization dynamics.

2.15 Correlating of MD and TD Properties of Blown Film

It has been traditional to relate MD and TD film properties with the process parameters of drawdown ratio and blow-up ratio respectively. In this section of the chapter a process will be presented that indicates that the MD and TD properties can be related to the kinematics and dynamics of the film-blowing process. This section presents as a thumbnail review of some of the work presented by Campbell, Cao, and Babel [36]. In order to develop such a correlation, the velocity and bubble radius must be evaluated as the polymer film is stretched from the die to the freeze line. A method for developing this data using video techniques was first published by Campbell and Huang [37]. In summary, the technique relies on following frame by frame a mark placed on the bubble at the die exit. Since the video camera has a frame rate of 1/60 of a second, a time-dependent velocity and radius can be simultaneously recorded. The data presented in the previous section of this chapter indicates that only a fraction of crystallization has occurred by the time that a polyethylene molecule has reached the freeze line. The freeze line is defined as the point in the process where the detection of the film stretching has come to an end. It is now understood that dynamic changes in the film continue to occur as the film continues to cool and crystallize. However, most of the strain and effects of the strain rate occur in the bubble region below the freeze line. The crystallization and the associated physical crosslinks lend credence to the work first published by Campbell and Cao [33] and subsequently expanded by Campbell and others [36] that the blown film dynamics as well as other processes where crystallization is occurring are best simulated with visco-plastic or viscoelastic-plastic rheological models. Unfortunately, most commercial modeling software packages do not have these models and thus are limited to modeling the process only to the freeze line. The physics of this limitation lies in the fact that the dynamic equations used to simulate the blown film process essentially represent a spherical shape, and if conventional viscous-dominated rheology is used to represent the strain-stress relationship for the deformation, then at some point the predicted bubble shape will close as in a spherical shape. By using a plastic transformation as the material crystallizes, the deformation ceases and the bubble shape is consistent with the actual blown film process above the frost-freeze line. In an R&D environment it is advantageous to be able to relate MD and TD properties to the kinematics of the molecular stretching as the film is produced regardless of the commercial interpretation. This technique was fully developed previously elsewhere [36, 51]. To help the reader it is reviewed in the following paragraphs.

The assumption in this correlating analysis is that the physical properties of the semicrystalline film are dependent on the frozen-in strains caused by the molecules being deformed by the plastic deformation and the stress caused by the viscous deformation rate at or near the freeze line. A complete discussion can be found in the Babel dissertation [55]. Babel used Equations 2.2 through 2.8 to obtain the strain rates and strains; the normalized plastic strain and strain rate then are defined in his work as:

$$\varepsilon_N = \frac{|\varepsilon_{11} - \varepsilon_{33}|}{\varepsilon_{11} + \varepsilon_{22}} \tag{2.39}$$

$$\dot{\varepsilon}_N = \frac{|\dot{\varepsilon}_{11} - \dot{\varepsilon}_{33}|}{\dot{\varepsilon}_{11} + \dot{\varepsilon}_{22}} \tag{2.40}$$

These definitions are then used to correlate the MD and TD properties for these resin systems. Figure 2.29(a) is a correlation of impact strength and Fig. 2.29(b) the yield stress. A more complete set of correlations can be found in a previous publication [36]. It is observed that the MD and TD properties fall on a surface with an independent axis of normalized molecular strain and strain rate. In his dissertation [51], Nagarajan found that the tensile strength had more scatter about the surface. This is to be expected because the tensile strength is flaw sensitive. Nagarajan also found that resins with short relaxation time relative to the transition time to complete the crystallization had more scatter than convention LDPE, which typically has a long relaxation time and thus the molecules remember their deformation history during the process of physical crosslinking due to crystallization. Further research is thus indicated to find how to incorporate relaxation time into the correlation. This correlation is still of substantial value when trying to balance the properties of a blown film process for most resins.

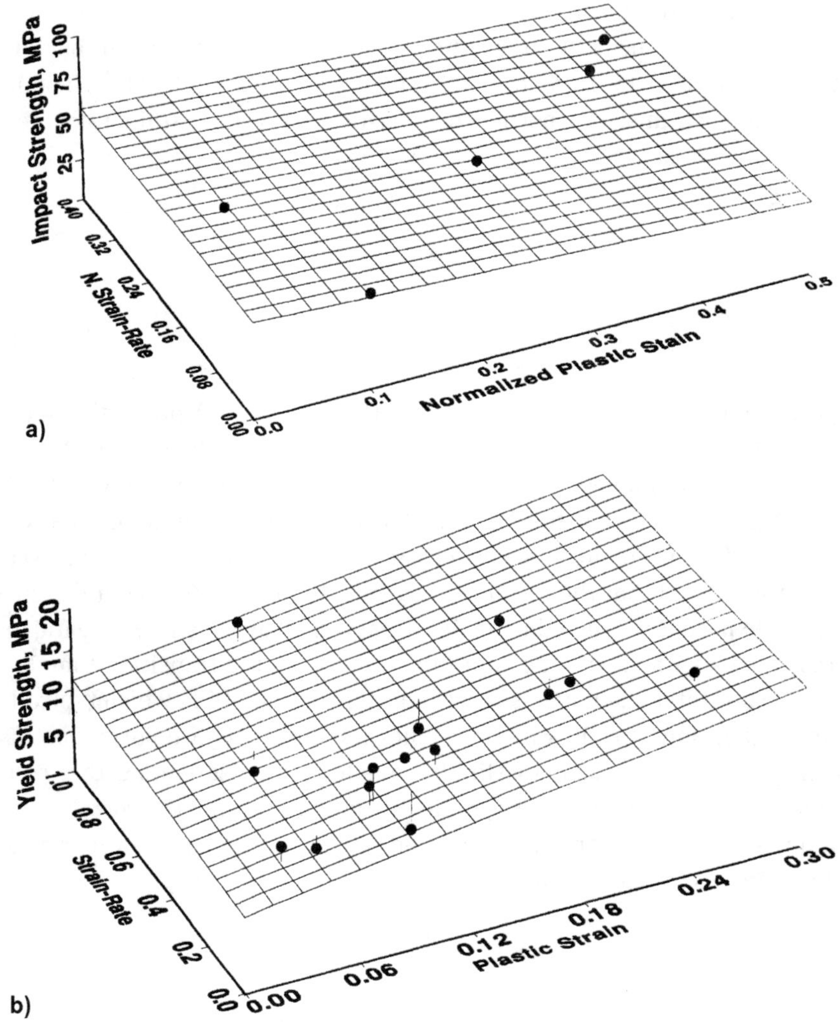

Figure 2.29(a)(b) Physical property correlation (a) impact strength (b) yield stress

References

1. Yu, T.H., Wilkes, G.L., *Polymer* 37, p. 4675 (1996)
2. Choi, K.J., Spruiell, J.E., White, J.L., *J. Appl. Polym. Sci.* 20, p. 27 (1982)
3. Ryan, A.J., Stanford, J.L., Bras, W., Nye, T.M.W., *Polymer* 38, p. 759 (1997)
4. Babel, A.K., Campbell, G.A., *J. Plast. Film Sheeting* 9, p. 246 (1993)
5. Tas, P.P., Ph.D. Thesis, Eindhoven Technical University (1994)
6. Keller, A., Machin, M.J., *J. Macromol. Sci., Part B: Phys.* 1 (1), p. 41 (1967)
7. Dees, J.R., Spruiell, J.E., *J. Appl. Polym. Sci.* 18, p. 1053 (1974)

8. Kwack, T.H., Han, C.D., Vickers, M.E., J. Appled Polym. Sci. 35, p. 363 (1988)
9. Nagasawa, T., Takenobu, M., Hoshino, S., *Appl. Polym. Symp.* 20, p. 295 (1973)
10. Supaphol, P., Spruiell, J.E., *J. Polym. Sci. B: Polym. Phys.* 36, p. 681 (1998)
11. Ding, Z., Spruiell, J.E., *J. Polym. Sci. B: Polym. Phys.* 34, p. 2783 (1998)
12. Butler, M.F, Athene, D.M., Ryan, A.J., *Polymer* 39, p. 39 (1998)
13. Serhatkulu, T.F., Cakmak, M., Graves, M., Galay, J., *Proc. Int. Polym. Proc. Soc.*, Toronto, p. 104 (1998)
14. Serhatkulu, T.F., Cakmak, M., Graves, M., Galay, J., *Proc. Int. Polym. Proc. Soc.*, Toronto, p. 142 (1998)
15. Ghaneh-Fard A., Carreau, P.J., Lafleur, P.G., *Int. Polym. Process.* 12, p. 136 (1997)
16. Stein, R.S., Rhodes, M.B., *J. Appl. Phys.* 31, p. 1873 (1960)
17. Rhodes, M.B., Stein, R.S., *J. Polym. Sci.* 45, p. 519 (1960)
18. Stein, R.S., *Structure and Properties of Polymer Films*, Lenz, R.W, Stein, R.S. (Eds.), Plenum Press, New York, p.156 (1973)
19. Ree, M., Kyu, T., Stein, R.S., *J. Polym. Sci. B, Polym. Phys.* 25, p. 105 (1987)
20. Tabar, R.J., Ph.D. Dissertation, University of Massachusetts, Amherst (1983)
21. Culberson, W.T., Tant, M.R., *J. Appl. Polym. Sci.* 47, p. 395 (1993)
22. Bullwinkel, M.D., Campbell, G.A., Rasmussen, D.H., Krexa, J., Brancewitz, C.J., *Int. Polym. Process.* XVI (1), p. 39 (2001)
23. Cao, B., Sweeney, P., Campbell, G.A., *J. Plast. Film Sheeting* 6, p. 117 (1990)
24. Campbell, G.A., Cao, B., Babel, A.K., in *Film Processing*, Kanai, T., Campbell, G.A. (Eds.), Hanser Publishers, Munich, p. 113 (1999)
25. Campbell, G.A., Huang, T.A., *Adv. Polym. Technol.* 5, p. 181 (1985)
26. van Aarsten, J.J., Stein, R.S., *J. Polym. Sci.* 31, p. 1873 (1971)
27. Kamal, M.R., Chu, E., *Polym. Eng. Sci.* 23 (1), p. 27 (1983)
28. Verhoyen, O., Dupret F, Legras, R., *Polym. Eng. Sci.* 38 (9), p. 1594 (1998)
29. Galante, M.J., Mandelkern, L., and Alamo, R.G., *Polymer* 39, pp. 5105–5119 (1998)
30. Bullwinkel, M.D., Campbell, G.A., Rasmussen, D.H., Brancewitz, C.J, *ANTEC 2000*, p. 92 (2001)
31. Johnson, M.B., Wilkes, G.L., Sukhadia, M., Rohlfing, D.C., *J. Appl. Polym. Sci.* 77, p. 2845 (2000)
32. Alothman, O.Y., Masters Thesis, Clarkson University (2001)
33. Cao, B., Campbell, G.A., *AIChE J.*, 36 (3), pp. 420–430 (1990)
34. Bullwinkel, M.D., Campbell, G.A., *Proc. Int. Polym. Process. Soc.*, Toronto, pp. 130–131 (1998)
35. Norton, D.R., Keller, A., *J. Mater. Sci.* 19, p. 447 (1984)
36. Campbell, G.A., Cao, B., Babel, A.K., Chapter 3.2 in *Film Processing*, Kanai, T., Campbell, G.A. (Eds.), Hanser Publishers, Munich (1999)

37. Campbell, G. A., Huang, T. A., *J. Plast. Film Sheeting* 2, pp. 30–39 (1986)
38. Cao, B., Sweeney, P., Campbell, G. A., *J. Plast. Film Sheeting* 6, p. 153 (1990)
39. McMahon, H. O., *J. Opt. Soc. Am.* 40, p. 376 (1950)
40. IRCON, Inc., Plastic Film Measurement, Technical Notes (1979)
41. Cao, B., Sweeney, P., Campbell, G. A., *J. Plast. Film Sheeting* 6, p. 117 (1990)
42. Buckley, H. *Philos. Mag.* 7, p. 576 (1928)
43. Petrie, C. J. S., *Rheol. Acta* 12, p. 92 (1973)
44. Petrie, C. J. S., *Plast. Polym.* 12, p. 259 (1974)
45. Menges, G., Predohl, W. O., *Polym. Eng. Sci.* 15, p. 394 (1975)
46. Ast, W., *Kunstst.* 63, p. 427 (1973)
47. Zippenfeld, R., *Kunstst.* 61, p. 147 (1971)
48. Kanai, T., White, J. L., *J. Polym. Eng.* 5, p. 135 (1985)
49. Campbell, G. A., Cao, B., *J. Plast. Film Sheeting* 3, p. 158 (1987)
50. Sidiropoulos, V., Vlachopoulos, J. J., *Reinf. Plast. Comp.* 21 (7), p. 629 (2002)
51. Nagarajan, G., Ph. D Dissertation, Clarkson University (1996)
52. Campbell, G. A., Nagarajan, G., *SPE ANTEC '95 Technical Papers* p. 160, Boston, MA, May 7–11 (1995)
53. Campbell, G. A., Nagarajan, G., Campbell, S. W., Burl, L., *SPE ANTEC 2003 Technical Papers* p. 49, Nashville, TN, May 4–8 (2003)
54. White, F. M., *Viscous Fluid Flow* McGraw-Hill (1991)
55. Babel, A. K., Ph. D Dissertation, Clarkson University (1995)

3 Multilayer Die Design and Film Structures

Karen Xiao and Martin Zatloukal

3.1 Background

Flexible packaging is used in our everyday lives, such as in food packaging, medical packaging, and protective films. Multilayer coextrusion of up to 11 layers has become increasingly common over the past 15 years. There are many inherent advantages of coextrusion. First, with the ability to coextrude different materials into one single structure, different applications with various properties can now be achieved. For example, in developed countries such as the United States and Canada where refrigeration is adequate, milk pouches often consist of three layers of polyethylene; in developing countries where refrigeration is lacking, milk pouches made from only polyethylenes are no longer adequate to preserve the freshness of milk. Milk gets spoiled easily and has a shortened shelf-life. With the introduction of five-layer or seven-layer structures, milk can now be transported in a package that consists of both polyethylene and ethylene-vinyl alcohol (EVOH) for an additional gas and aroma barrier. The shelf-life of such milk pouch packaging can be extended up to 90 to 120 days. Without the use of coextrusion, it would be impossible to transport milk while preserving its freshness economically and within a reasonable time frame.

Second, coextrusion can be used to reduce cost by reducing the amounts of expensive polymers used while increasing the percentage of inexpensive resins. Further cost reduction can be realized by using recycled materials in one of the layers or by incorporating different resins so that the film can be down-gauged to reduce the overall thickness of the film while maintaining the same physical properties. In addition, coextrusion can be used to reduce the number of process operations required; for example, some films can be coextruded to eliminate the additional lamination stage. This, needless to say, provides another significant cost savings to the processors. Processors are now working with both equipment manufacturers and resin suppliers to fully explore the possibilities that coextrusion can bring to their operations.

Coextrusion, as the name implies, is a process in which two or more materials are extruded into a single die. Technological advancements in both machinery and resins have allowed processors to obtain innovative and differentiated products through multilayer coextrusion. The greatest advantage of coextrusion comes from the ability to combine the important properties of different materials into a single structure. For example, when designing a film structure to package food, the process engineers may need to take into consideration the following characteristics:

1. Surface characteristics such as the coefficient of friction, gloss, and sealability of the film
2. Physical properties of the film such as tensile strength, elongational properties, impact strength, and puncture strength
3. Some additional requirements such as gas and aroma barrier properties and optical and shrink properties

It is quite evident that it is impossible for one material to meet all of the requirements that a packaging film demands.

To achieve the desirable final properties, processors need to understand the structure-processing-property relationship; in other words, the final properties of the film are a function of the polymer molecular structures, the design of the processing equipment, and the actual processing conditions. This relationship is best described in Fig. 3.1. To get a full understanding of the process, the processors need to work with

1. resin manufacturers to address the polymer properties and quality consistencies, including any additive packages present,
2. the original equipment manufacturer (OEM), to understand the equipment design and its limitations, and
3. the manufacturing engineers, to understand the proper processing window of the materials being processed.

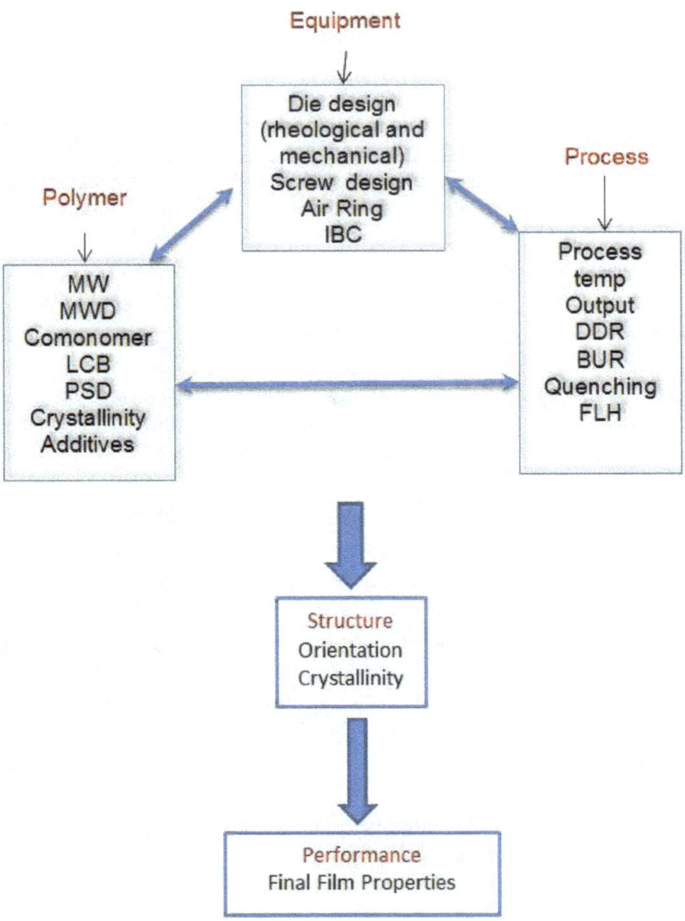

Figure 3.1 Structure-processing-property relationship that affects the final film properties

In this chapter, the emphasis will be on multilayer die design, which is at the heart of any film extrusion process.

3.1.1 Materials

Common materials used in film extrusion and their uses are indicated in Table 3.1. The most common resins used in extrusion are polyolefins, which include low-density polyethylene (LDPE), linear low-density polyethylene (LLDPE), metallocene linear low-density polyethylene (mLLDPE), and polypropylene, both homopolymer and copolymer (PP). These materials are often used for moisture barrier and mechanical properties. As multilayer coextrusion becomes increasingly popular, more engineering resins are now being used in the film structures to improve oxygen barrier properties, heat resistance, and aesthetics among others. Such materials include polyamide (PA), ethylene-vinyl alcohol (EVOH), polyvinylidene chloride (PVdC), polyester (PET), and styrene-butadiene copolymer (SBC) to name a few.

Table 3.1 Materials That Are Commonly Used in Coextrusion

Material	Properties	Common Uses
Low-density polyethylene (LDPE)	Ease of processing, heat sealability, clarity, flexibility	Industrial liners, agricultural films, bags, stretch and shrink films
Linear low-density polyethylene (LLDPE)	Superior strength	Food packaging, stretch/cling film, grocery bags
High-density polyethylene (HDPE)	Vapor barrier, stiffness	Food packaging, grocery bags
Metallocene linear low-density polyethylene (mLLDPE)	Tailored properties for balance of strength, toughness, puncture strength, heat sealability, and clarity	Food packaging, stretch and shrink films
Ethylene vinyl acetate (EVA)	Flexibility, heat sealability, toughness	Often in coextrusion as a heat seal layer
Ethylene acrylic acid (EAA)	Hot tack and adhesion, toughness	Often in coextrusion as a sealant layer
Ionomers	Formability, toughness, clarity, gloss, heat sealability, resistance to oil, adhesion to cardboard and aluminum foil	Coextruded as a heat-sealing layer for fresh fish/meat, skin packaging, thermoforming film
Polypropylene homopolymer (PP)	Heat resistance, stiffness, orientability	Coextruded as a heat-resistant layer, oriented film
Random copolymer polypropylene (RCPP)	Toughness, clarity and gloss, good chemical resistance to acids, alkalis, and alcohol, heat seal	Coextruded as a heat-seal layer, oriented film

Material	Properties	Common Uses
Vinylidene chloride copolymers (PVdC)	Excellent oxygen and moisture barrier, formability	Coextruded with polyolefins for food packaging especially for moisture-sensitive and gas-sensitive foods
Ethylene-vinyl alcohol (EVOH)	Excellent oxygen and aroma barrier, stiffness	Coextruded with polyolefins and nylon for food packaging
Maleic-anhydride-modified polyolefins	Adhesive layer to bond dissimilar materials	Coextruded in a multilayer structure
Polyester (PET)	Clarity, heat resistance	Coextruded or laminated
Polycarbonate (PC)	Heat resistance, clarity	Laminated as a heat-resistant layer
Polyamide (PA)	Stiffness, heat resistance	Coextruded to provide stiffness and thermoformability

As the materials used in film extrusion become increasingly diverse, their rheology and processing windows also differ vastly. As a result, the die design requirements have become increasingly challenging; the dies need to have more flexibility than ever before to allow processors to design and process as many different types of structures as possible. In other words, the coextrusion die needs to have the flexibility for film producers to use different types of materials to meet their film property demands economically while at the same time providing the highest quality film that is free of such defects as interfacial instability, gels, and thickness nonuniformity.

3.1.2 Film Structures

To meet the flexibility requirement, the coextrusion dies are also moving towards having an increasing number of layers. Currently, nine-layer to eleven-layer extrusion systems are not uncommon in the industry. To demonstrate the advantages of systems with more layers, several examples will be shown below.

To package meat or fish, a common three-layer system used to consist of polyamide (PA) on the outside and an ionomer material on the inside with an adhesive layer to bond the two, that is, PA/adhesives/ionomer. An equivalent five-layer system would split the two polyamide layers: PA/adhesives/PA/adhesives/ionomer. The five-layer system would provide a film that has better flex crack resistance, better barrier properties, and better thermoformability.

For a package that requires a high oxygen barrier, a typical five-layer structure would be LLDPE with masterbatch/adhesives/EVOH/adhesives/ionomer. To reduce the cost of the package, a seven-layer structure can be LDPE/LLDPE with masterbatch/adhesives/EVOH/adhesives/EVA/ionomer, since EVA is a less expensive resin than an ionomer.

To improve the optics and surface printability, the LDPE on the outside layer can also be tailored. For a high barrier stand-up pouch or a thermoforming film, a seven-layer structure can consist of three layers of PA and one layer of EVOH, such as PA/adhesives/PA/EVOH/PA/adhesives/mLLDPE with mLLDPE being the sealant layer. In this structure, the mLLDPE is the most inexpensive material, but it is still more expensive than the conventional LLDPE or LDPE. As a result, to reduce the cost of this structure, a nine-layer film can be PA/adhesives/LLDPE or LDPE/adhesives/PA/EVOH/PA/adhesives/ mLLDPE. In addition, since LLDPE or LDPE separates the two PA layers, the film also tends to be flatter and have less tendency to curl. With the nine-layer coextrusion systems, the same die often needs to process film structures indicated in Table 3.2. Proper die designs are crucial in determining the flexibility of the system to reduce scrap and improve overall quality of the film.

Table 3.2 Typical Film Structures That Need to Be Processed on a Nine-Layer System [1]

Layer	Structure 1	Structure 2	Structure 3	Structure 4
1	PE or PP	PA	LLDPE+LDPE	PE
2	PE or PP	Adhesive	LLDPE+LDPE	PE
3	Adhesive	PE	LLDPE+LDPE	PE
4	PA	Adhesive	LLDPE+LDPE	Adhesive
5	EVOH	PA	LLDPE+LDPE	EVOH
6	PA	EVOH	LLDPE+LDPE	Adhesive
7	Adhesive	PA	LLDPE+LDPE	PE
8	PE or PP	Adhesive	LLDPE+LDPE	PE
9	PE or PP	PE	LLDPE+LDPE	PE

3.1.3 Equipment Producers

Table 3.3 lists the common film equipment producers as of this writing. With the technological advancements and globalization of the 21st century, these producers are capable of supplying equipment worldwide.

Table 3.3 Equipment Manufacturers

Manufacturer	Web Address
Windmöller and Hölscher	http://www.whcorp.com/
Reifenhäuser Kiefel Extrusion	http://www.reifenhauser-kiefel.com/
Macchi	http://www.macchi.it/
Hosokawa Alpine	http://www.halpine.com/
Kuhne Group	http://www.kuhne-group.com/index.php
Macro Engineering and Technology Inc.	http://www.macroeng.com/
Brampton Engineering Inc.	http://www.be-ca.com
Davis-Standard, LLC	http://www.davis-standard.com/
Gloucester Engineering	http://www.gloucesterengineering.com/
Cloeren Incorporated	http://www.cloeren.com/
Nordson EDI	http://www.extrusiondies.com/

3.2 Basic Coextrusion Film Die Types

3.2.1 Manufacturing Method

There are two methods of making film: cast film and blown film. Cast film is a process in which the polymer is extruded into a slit die, producing a thin flat film. The processes for manufacturing cast film and cast sheet are essentially identical and are only differentiated by the thickness. Materials with thickness of 0.003 in. (0.0762 mm) or less are considered film, and materials with thickness of 0.010 in. (0.245 mm) or greater are considered sheet [2]. A typical cast film process is illustrated in Fig. 3.2. A cast film process is typically produced by downward extrusion in which the polymer melt is extruded onto chilled chrome rolls. The film is cooled rapidly by the chilled rolls with the dimensions controlled predominantly by the extrusion rate, the die dimensions, and the takeoff speed. Cast film is only stretched in the machine direction (MD).

By comparison, the conventional blown film process involves melting the polymer in an extruder, and the polymer is then extruded through an annular die; the resulting tube or "bubble" is then inflated into its final desired size. The bubble is then cooled by using air from the outside with an air-ring and from the inside with internal bubble cooling (IBC). The film is stretched in both the machine direction (MD) and the transverse direction (TD), resulting in biaxial orientation of the film. A schematic of the blown film process is illustrated in Fig. 3.3.

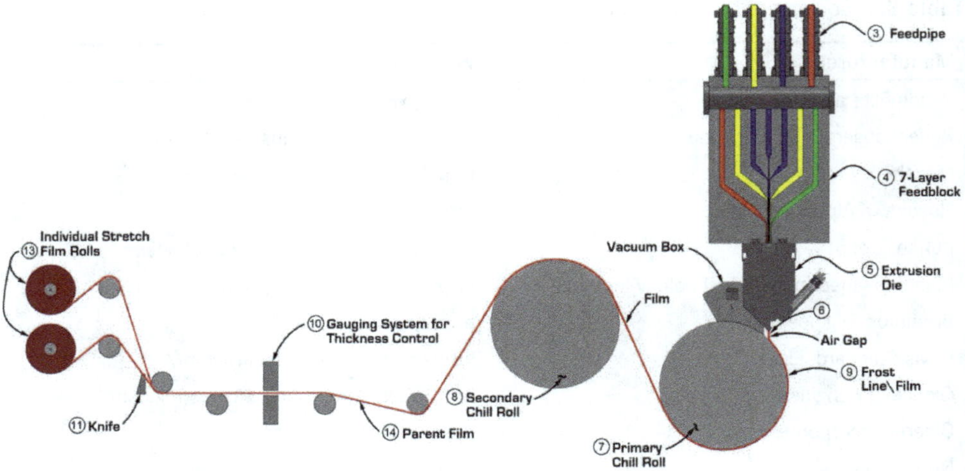

Figure 3.2 A typical line layout of a cast film process for stretch film (courtesy of Cloeren Incorporated)

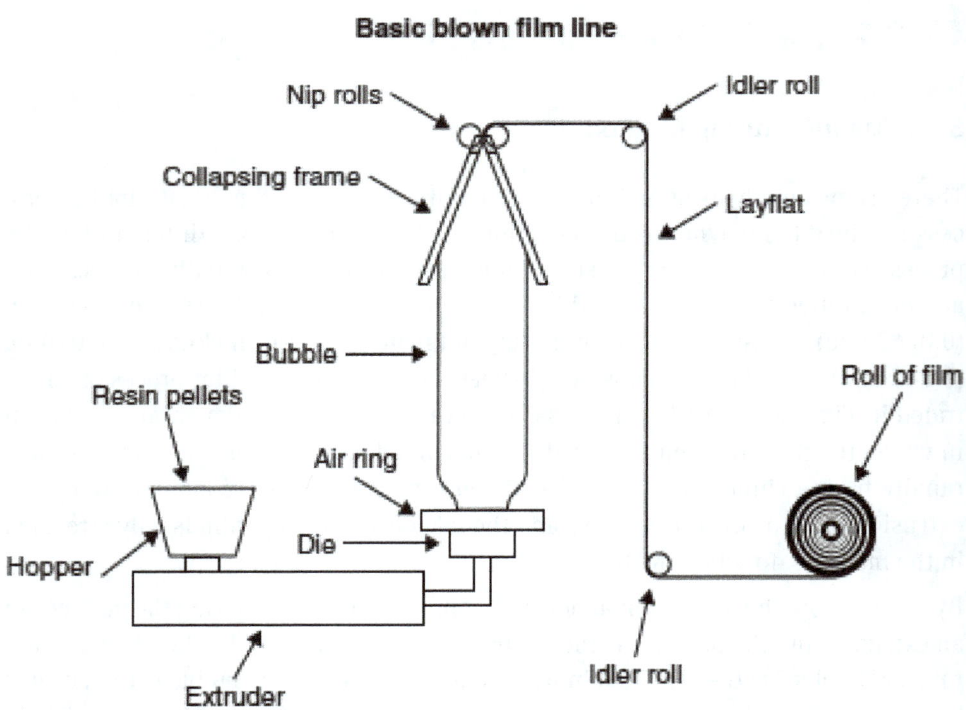

Figure 3.3 Major components of the blown film process [3]

There is also a relatively new blown film process that uses water instead of air to cool the film. This unique process marries the advantages of both a cast and a conventional blown film process; it retains the benefits of improved clarity, improved thermoformability, and reduced curl from a cast film process while maintaining the balanced orientation and the process flexibility of a blown film process. This system has also given blown film processors opportunities to use less expensive resins to achieve the same or superior properties. For example, in a typical thermoforming structure, processors would commonly use a copolymer polyamide or an amorphous polyamide to improve thermoformability and clarity; in a water-quench process, these more expensive copolymers can be reduced or eliminated completely [4–6]. Figure 3.4 gives a more detailed description of how the process works. As the film is extruded out of the blown film die, it is inflated to the desired bubble size as in a conventional blown film system. What is unique in this process is that the film is then quenched almost instantaneously as it passes through a water ring. Once the film is cooled, it is still collapsed and wound [4, 5].

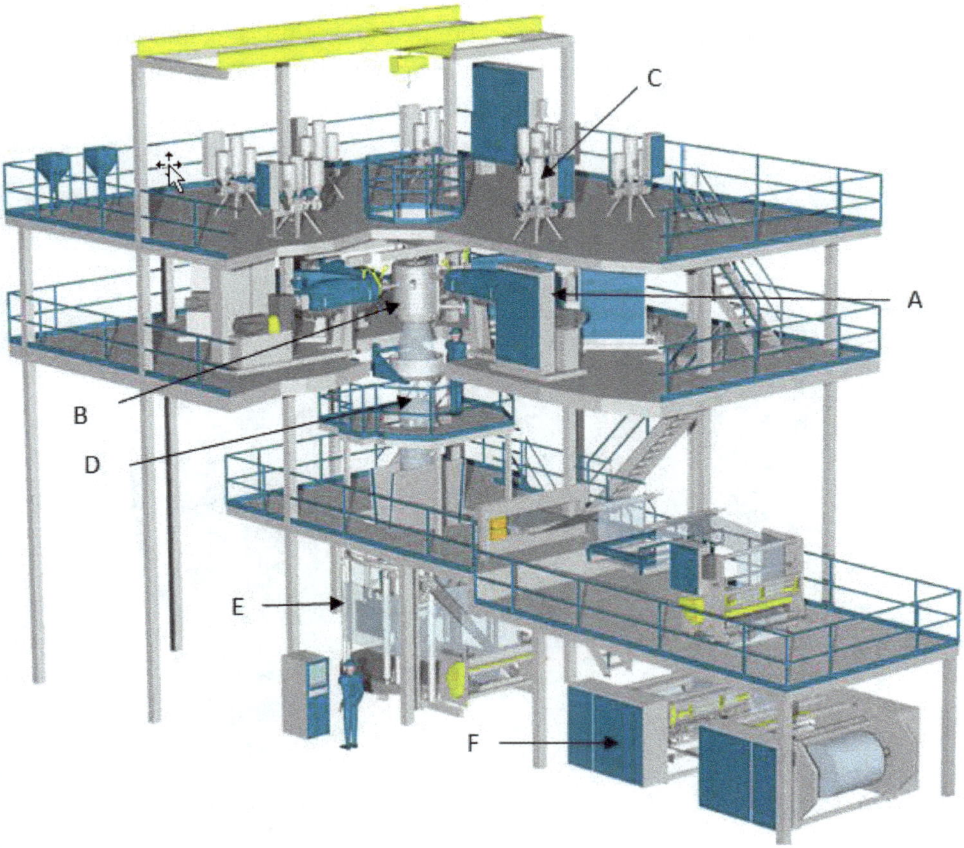

Figure 3.4 A schematic of a water-quench blown film line (courtesy of Brampton Engineering). The extruders and the blenders are on the top floors whereas the oscillating haul-off is on the ground floor; the film is extruded downwards: A) extruders; B) blown film die; C) blenders and hoppers; D) tubular film; E) haul-off; F) winders [3]

3.2.2 Cast Die Types

Cast film operation uses the typical planar dies. Uniform flow distribution is accomplished by using the proper manifold design, and a flexible lip is used to fine-tune the final thickness consistency. In multilayer coextrusion, there are two ways of combining the flow. A single manifold design can be used downstream of a coextrusion feedblock. Different materials are fed into the feedblock, where they are brought together into one melt stream. This system was developed by Shrenk and Chisholm [8, 9] and works well for materials of similar viscosities. This technique is illustrated in Fig. 3.5.

For materials with different viscosities, multimanifold designs should be considered [7]. In a multimanifold design, the materials are extruded into separate channels and are distributed. The materials are combined in the preland area and have a shorter coextrusion residence time, thereby reducing the propensity for interfacial instability to occur. This concept is illustrated in Fig. 3.6.

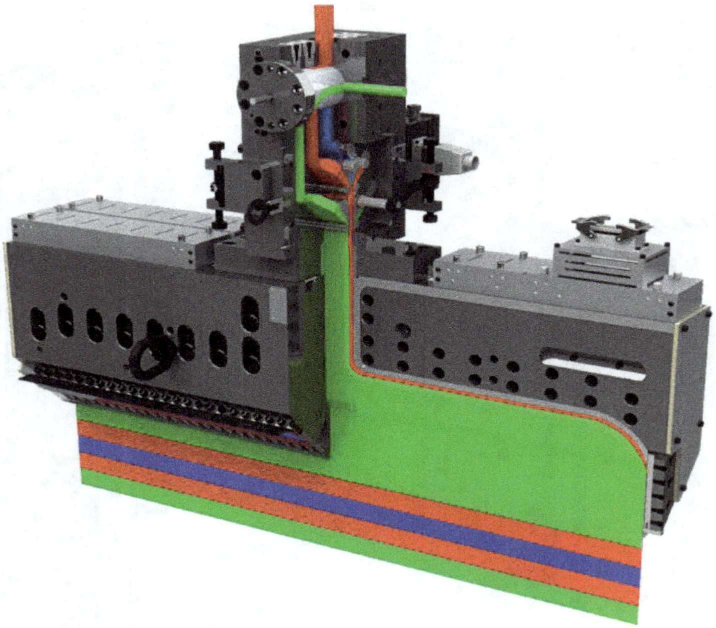

Figure 3.5 A single manifold die with feedblock (courtesy of Cloeren Incorporated)

Figure 3.6 A multimanifold die with feedblock (courtesy of Cloeren Incorporated)

Due to the advancement in computer-aided flow design software, the area of flat die design that has seen significant improvement over the past several years is the design of flow channels or manifold. There are two main types of planar dies, namely T-dies and coat-hanger dies. A T-die is essentially a large cylinder with sealed ends that is fed in the center with molten polymer that then flows out of a thin gap cut along the length of the cylinder. This thin gap is maintained through a constant-length channel downstream of the cylinder called the die land. This geometry is shown in Fig. 3.7. When examined from the top, this die looks like an inverted "T" and so the designation T-die.

The T-die geometry has existed for many years and has been used to make many planar polymeric structures [10]. The main drawback of this type of geometry is the difficulty in producing a planar structure with uniform thickness across the width of the product. This drawback was addressed with the introduction of the coat-hanger-style die [11].

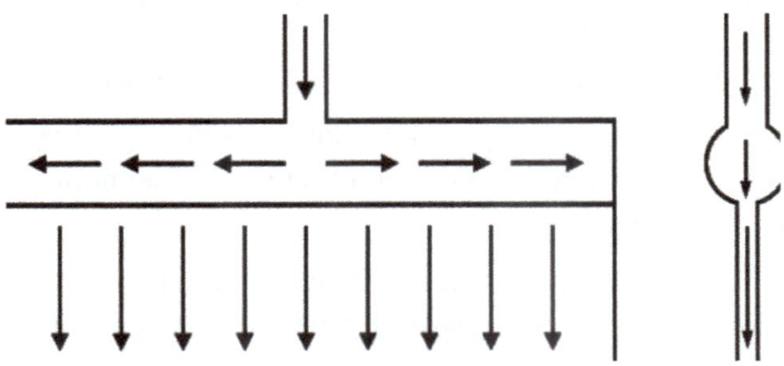

Figure 3.7 A schematic diagram showing a T-die from the top and side [11]

A coat-hanger die is similar to a T-die except the cylinder is tapered from a larger diameter in the center to a smaller diameter at the end, and the length of the die land is not constant from the center to the edges. This geometry was developed in order to help spread the flow of the polymer more uniformly across the die. In single layer, the manifold shape is typically a teardrop that reduces in size from the center to the end of the die [7]. This reduction in volume reduces residence time in the die, which is an important design consideration for most materials processed. After the preland, the polymer flows into the secondary manifold, an area of the flow channel that allows the material to move laterally again if required and is also used to control the total pressure drop of the die in conjunction with the final lip land. This geometry is shown in Fig. 3.8.

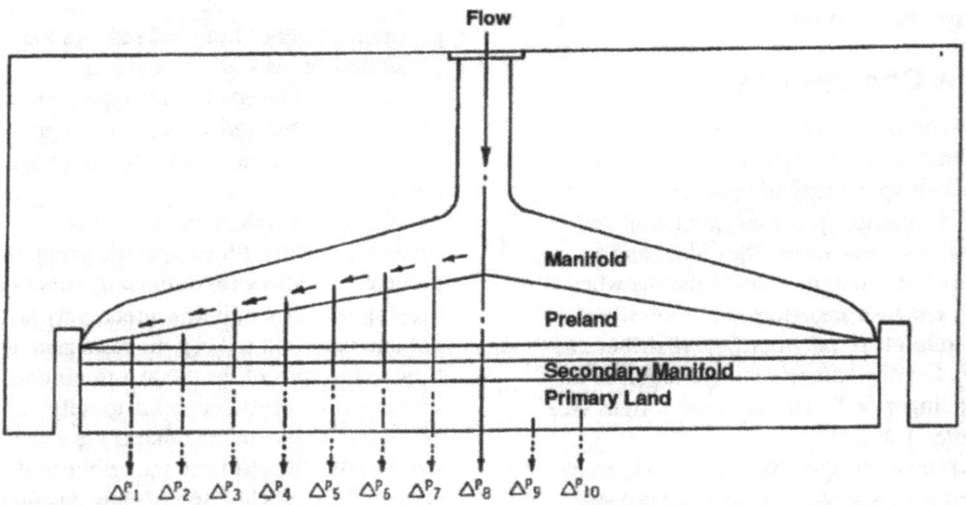

Figure 3.8 A schematic diagram showing a coat-hanger die geometry [12]

This die geometry was a significant improvement over the T-die and led to planar products with more uniform thickness from edge to edge [10]. This improvement in uniformity of thickness led to further product and process development, including down-gauging, which also improved the economics of the products in the industry.

There are a few disadvantages with coat-hanger manifolds. The channel of the manifold is farther from the exit of the die at the center than it is on the ends. This results in a die that will have a propensity to deflect more in the center, causing uneven melt distribution. "M" and "W" patterns are also common with coat-hanger manifolds. These can be a result of the entrance transition to the manifold or linear flow channels within this type of die [7].

The newest manifold designs are a hybrid design that addresses the drawbacks of the earlier mentioned designs. The teardrop shape in the conventional coat-hanger die is replaced with a large aspect ratio rectangular manifold that changes from the center to the end of the die. This manifold shape results in less interface distortion in coextrusion applications and reduced M and W patterns associated with narrow entrance teardrop-shaped manifolds [7]. The pressure drop in the manifold and the flow is balanced by adjusting the preland geometry. This manifold has a diminishing cross section to allow for reduced volume as leakage flow from the manifold to the preland occurs from the center to the end of the die.

3.2.3 Blown Film Die Types

Blown film operation uses annular dies. The initial single-layer dies used for annular products were spider dies. Spider dies are cylindrical dies that produce a tubular extrudate by passing the polymer through an annular flow channel. This annular flow channel is made by surrounding an inner mandrel with an outer die housing. The inner mandrel and the outer housing are held rigidly together by metal struts that bridge the annular gap [13]. These struts are sometimes referred to as "spiders," thus the name spider die. A typical spider die is shown in Fig. 3.9.

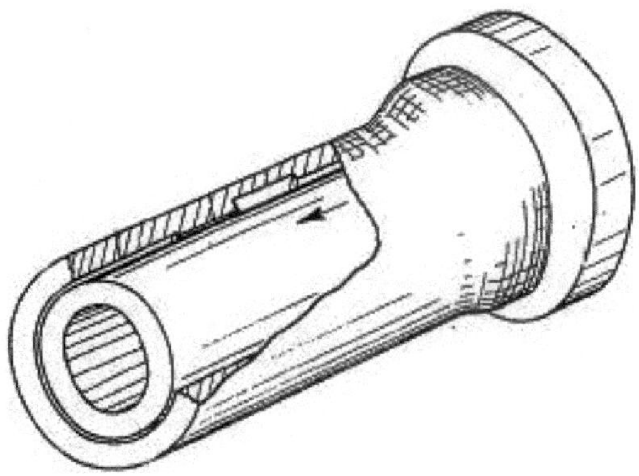

Figure 3.9 Spider die geometry

Some of the earliest uses for spider dies were not for polymer processing but for producing sausage skins [13]. This technology was subsequently successfully adapted for use with polymers [14]. The main concern with spider dies is the presence of "weld lines" in the product, produced when the polymer flows around the spiders and then recombines, leaving a defect where the material "welds" back together. These defects can lead to weak spots that can cause the product to fail. To address these concerns, spiral mandrel dies were introduced.

A schematic diagram of a single-layer spiral mandrel die is shown in Fig. 3.10. This style of die is a significant breakthrough and departure in thinking from the spider die described previously. In a spiral mandrel die, the inner mandrel is spirally cut with a groove that becomes shallower as you progress down the channel. This type of design allows for the removal of the supports used in the spider dies and essentially eliminates the weld line difficulties [11].

The early spiral mandrel dies were built for making pipes. This technology was subsequently used in making blown films as well [15]. With the increasing popularity of multilayer coextrusion, different styles of dies were introduced into the marketplace.

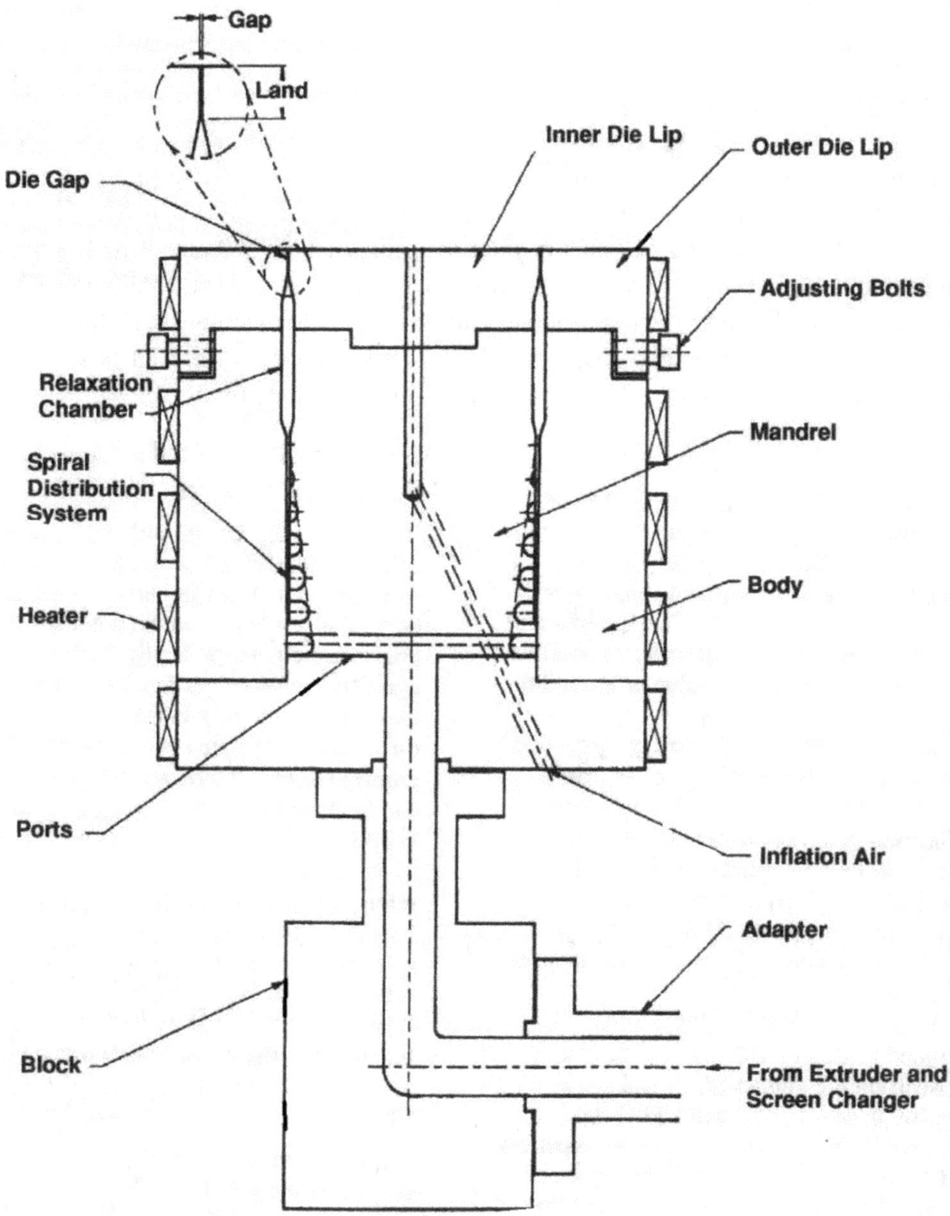

Figure 3.10 A single-layer spiral mandrel die [12]

A natural extension of the single-layer spiral mandrel die is a multilayer die with the number of mandrels equal to the number of layers. Figure 3.11 illustrates a cross section of a three-layer spiral mandrel die. Layer A indicates the outside layer of the film while layer C is the inside layer. Each layer is generally fed by a separate extruder for optimal versatility. As the need for productivity increased, demanding higher output per inch of circumference of the die, the bubble was not only cooled from the outside with an air-ring but also from the inside with the internal bubble cooling (IBC) system. This style of die also has enough space in the inner mandrel to allow for large enough air passages for efficient IBC, which, generally speaking, could increase outputs by as much as 20% or more when compared to a non-IBC system.

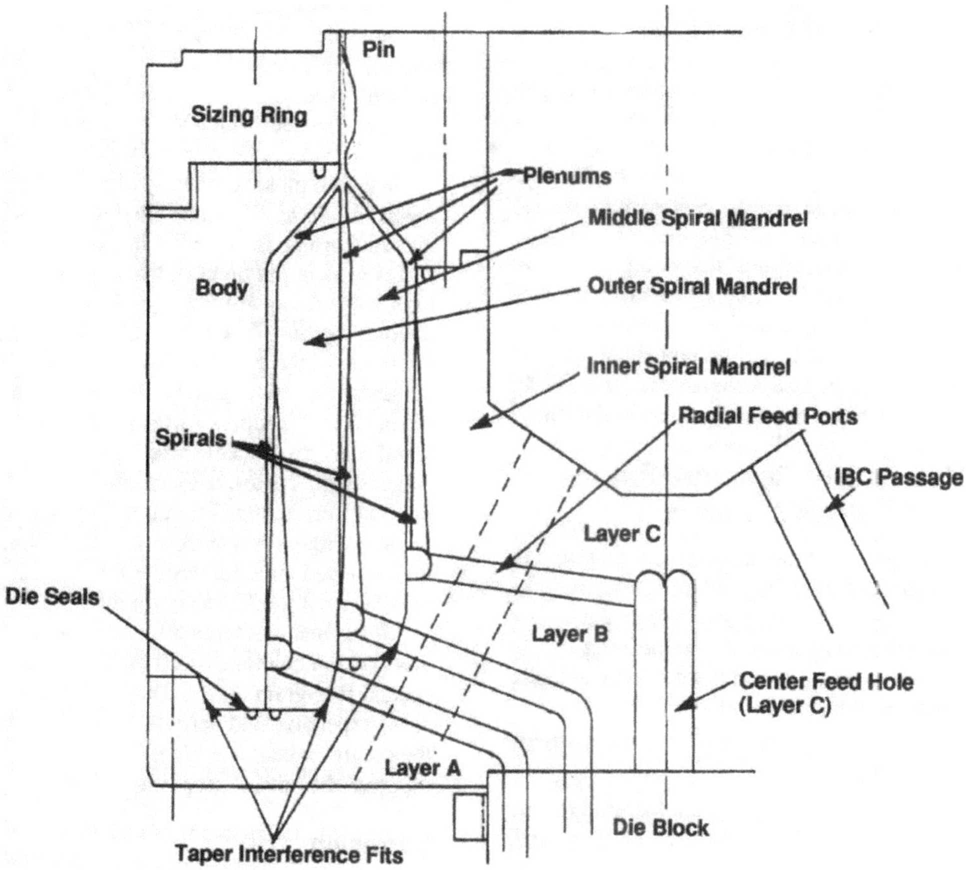

Figure 3.11 Three-layer spiral mandrel die [12]

Multilayer spiral mandrel dies were the most common ones used throughout the 1980s and 1990s. Figure 3.12 shows a nine-layer spiral mandrel die. As the number of layers increased, materials other than polyolefins were required to be processed in the same die. As a result, the processing temperatures of the materials in the same die became significantly different. One of the biggest disadvantages of the spiral mandrel die is that because the die is heated up by the heater bands wrapped around the outer body, the entire die needs to be operated at the highest processing temperature of the polymer in the structure. For example, a common occurrence in a nine-layer structure requires that the die process PA 6, LLDPE, adhesives, and EVA at the same time. Since the PA material is processed at 250°C, the entire die needs to be set at this temperature. If the temperature was too low, PA would solidify in the die. This is detrimental to a temperature-sensitive resin such as EVA, which becomes easily degradable at this elevated temperature. With the shortened run time, more frequent cleaning of the die would be required.

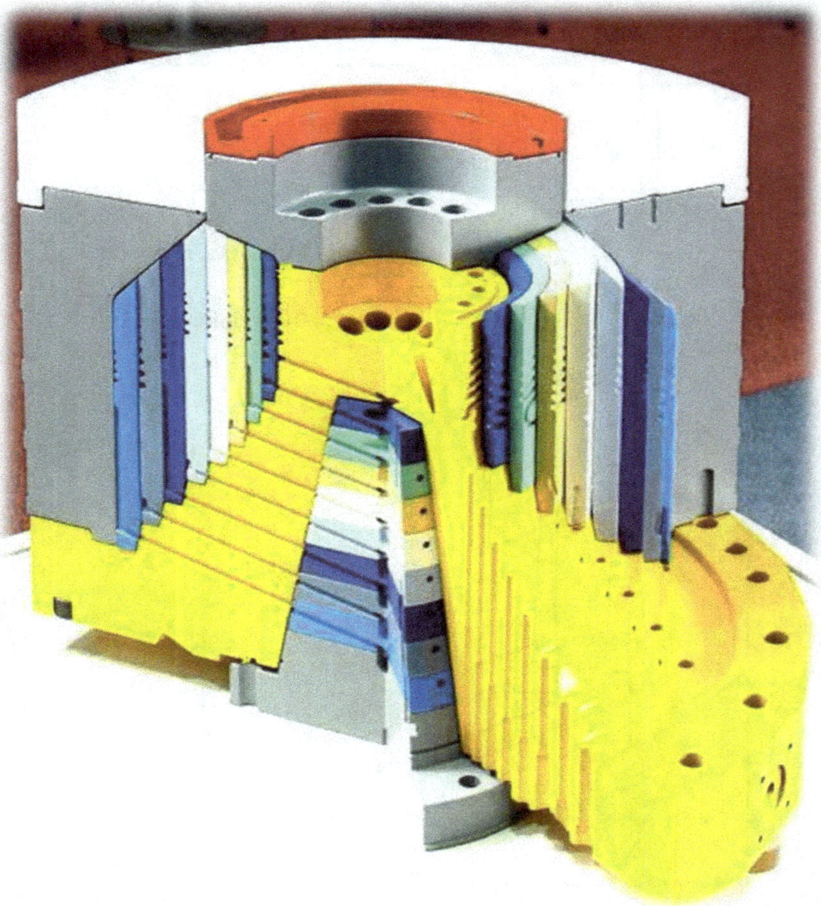

Figure 3.12 A nine-layer spiral mandrel die design (courtesy of Gloucester Engineering Inc.)

One method to try to alleviate this problem is thermal isolation of the materials in the die. One approach that has been used in annular dies is to thermally isolate the layers using air gaps as thermal barriers [16, 17]. This approach is shown in Fig. 3.13. Figure 3.13 shows how the design of this annular die geometry separates the individual layer plates with air gaps. Since the individual layers are not in intimate contact with each other, the transfer of heat from one layer to the next is minimized. This leads to the ability to process materials in adjacent layers at significantly different temperatures. This technique was originally designed for blow molding applications and was later adapted for blown film. Figure 3.14 shows a ten-layer blown film die with air gaps as the means for temperature isolation.

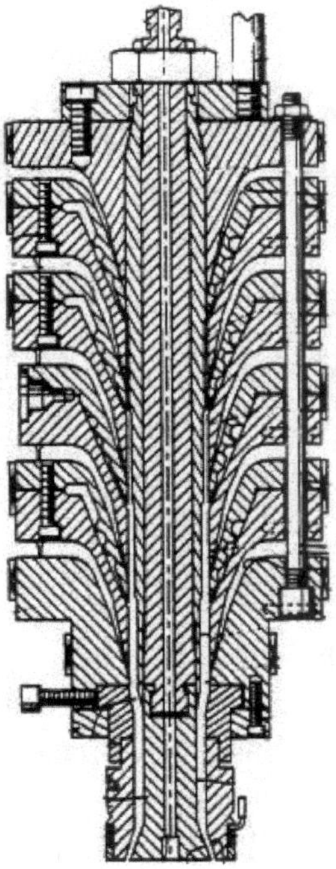

Figure 3.13 Thermal isolation of materials in an annular die geometry

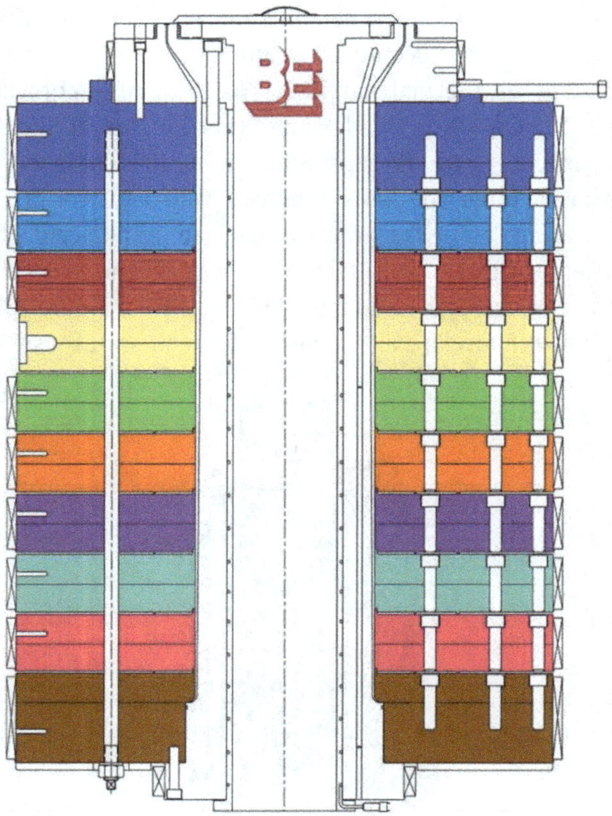

Figure 3.14 A cross section of a ten-layer stackable die (courtesy of Brampton Engineering Inc.)

One other technique that has been introduced to help thermally isolate layers in annular die geometries is the concept of encapsulation of a thermally sensitive material by a relatively thermally insensitive material [18, 19]. In this method, a thermally sensitive resin is extruded into a special adapter or feedblock first, where it is encapsulated by a more thermally stable material. The encapsulated structure is then extruded into the die. The thermally sensitive resin is, therefore, protected in the die, and the chance of degradation is minimized. This technique has proven very effective and is used in many commercial applications.

The stackable die design works very well for multilayer coextrusion. However, as the number of layers increased, the overall die height inherently increased also. This limits the total number of layers that this style of die can make. In order to reduce the die height, some producers have come up with a "combo" design in which some of the layers have tapered mandrels while others have straight mandrels. One example is illustrated in Fig. 3.15. This combo design yields shorter residence time in the coextrusion area at the expense of temperature isolation. Since polymer flow and polymer degradation have both a time and temperature dependency, the style of die that is most suitable depends on the application.

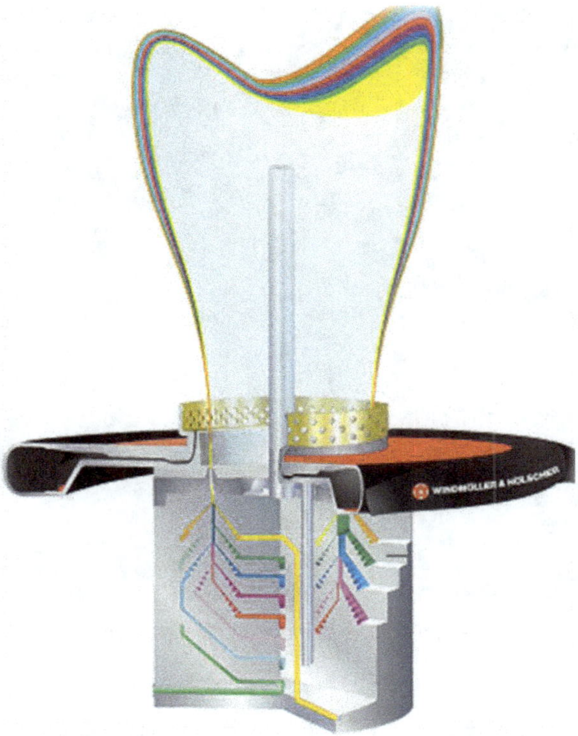

Figure 3.15 A picture of a nine-layer tapered die design (courtesy of Windmoeller and Hoelscher)

■ 3.3 Interfacial Instabilities

The coextrusion process can be used to produce multilayer sheet, blown film, cast film, tubing, wire coating, profiles, and others. It has been used since the early 1950s to improve product quality and process efficiency. However, under certain conditions, the flow of viscoelastic polymeric materials gives rise to unstable interfaces and undesirable layer distribution, which can significantly affect the product properties. In the open literature [20–51], two main types of interfacial instabilities have been detected: zigzag (high frequency, low amplitude), as in Fig. 3.16, and wave (high amplitude, low frequency), as in Fig. 3.17. Moreover, both types of interfacial instabilities can occur simultaneously on the film, as visible in Fig. 3.18.

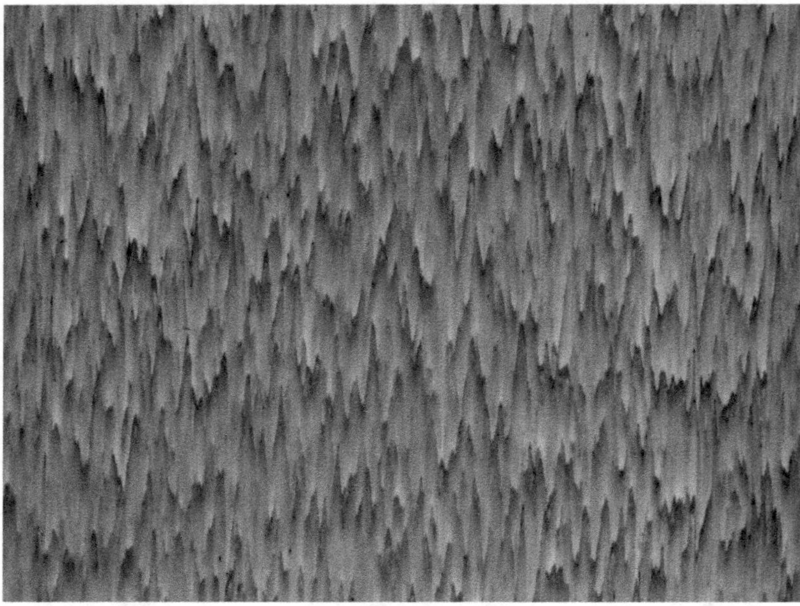

Figure 3.16 A photograph of a film sample exhibiting the zigzag-type instability [30–31]

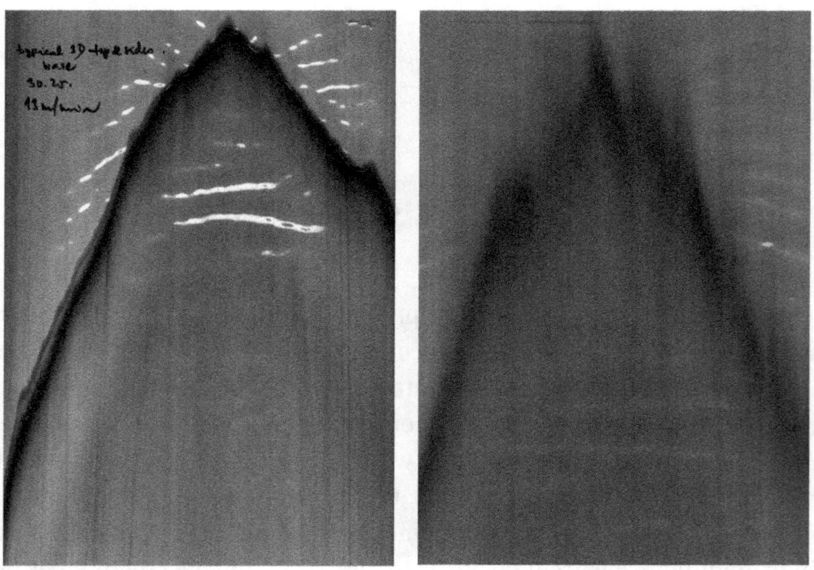

Figure 3.17 Photographs of two film samples exhibiting the wave-type instability [30–31]

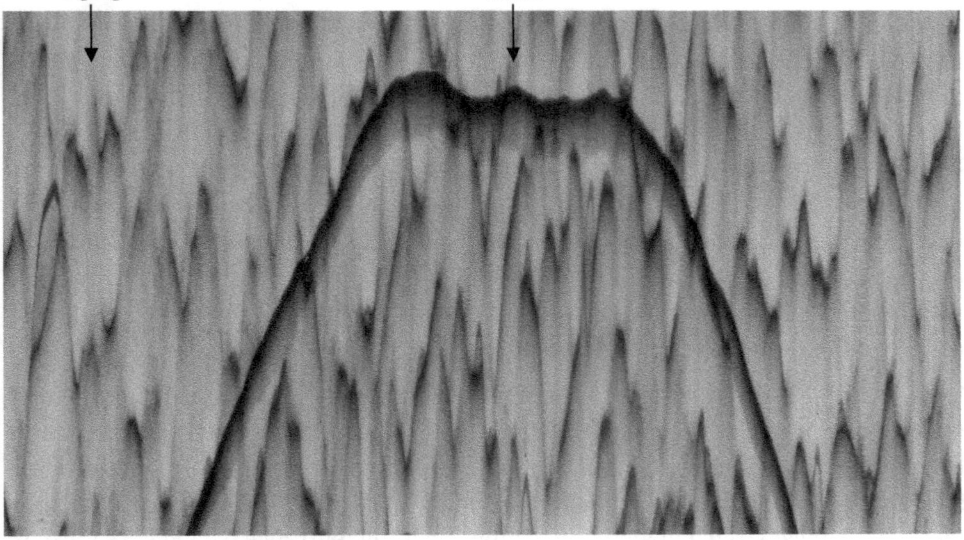

Figure 3.18 A photograph of a film sample exhibiting simultaneously wave and zigzag types of instabilities [30–31]

Zigzag instabilities have been shown to be connected with the end of the die, where the shear stresses usually achieve high values [21–25]. This type of instability can be captured via critical shear stress or recoverable shear on the interface [24–25]. A detailed experimental and theoretical study has been recently performed to determine the critical interfacial shear stresses for a variety of LDPEs with different MWD [34]. It has been found that the critical shear stress on the interface may vary significantly: materials with a broad MWD were found to be more sensitive to the zigzag instability compared to LDPEs with a narrow MWD.

On the other hand, wave instabilities seem to be linked with the merging area in which intensive layer stretching occurs [21–23, 27–33, 41–51]. To understand these instabilities, the TNSD (total normal stress difference) sign criterion, which quantifies the relative stretching of the coextrusion layers in the merging area, or extensional viscosity is usually used [22, 23, 27–29, 31–33, 40–46, 48].

3.3.1 Merging Area, Wave Type of Interfacial Instabilities, and the TNSD Sign Stability Criterion

In practice, the stability of the coextrusion interface is usually evaluated through the local interface stress state according to Schrenk et al. [24] and Mavridis and Shroff [25]. They concluded that minimizing the interfacial shear stress and matching the elastic properties of the adjacent layers at the interface are appropriate criteria for suppressing interfacial instabilities. However, Han and Shety [26] showed that although the critical shear stress theory is valid for a particular system, the critical shear stress value is affected by various factors, such as arrangement and thickness of the layers. Thus, it seems that the conditions for the occurrence of interfacial instabilities are more complicated than just the limitation on the interface shear stress. Recent experimental [20–23, 31, 41, 46], visualization [41–43, 49–51], and theoretical studies [27–29, 31, 41, 45, 46] suggest that the interfacial instability onset in the merging area is caused by intensive layer stretching. It has been observed experimentally [21–23, 31, 41–43] that if the minor flow decreases, the stable coextrusion flow becomes unstable. To understand this behavior, a FEM simulation of such a flow field has been performed [27], and some typical simulation results of this work are depicted in Fig. 3.19(a).

Here, lines 1, 2, and 3 represent the interface shapes for the corresponding minor mass flow rates $\dot{m}_1$, $\dot{m}_2$, and $\dot{m}_3$; $\dot{m}_1 > \dot{m}_2 > \dot{m}_3$. It is visible that the interface gets closer to the wall with the decreasing flow rate in the minor layer, which becomes intensively stretched. Such a flow situation is similar to capillary flow in which the polymer melt is most stretched along the centerline at the entrance (see Fig. 3.19(b)). In the coextrusion case, the maximum stretching does not need to be in the centerline because the minor flow is bounded by the moving interface and the die wall, in contrast to capillary flow, where the flow is bounded by two walls. In other words, the maximum stretching does not occur on the interface but somewhere in the middle of the minor layer, as shown in Fig. 3.20 [27].

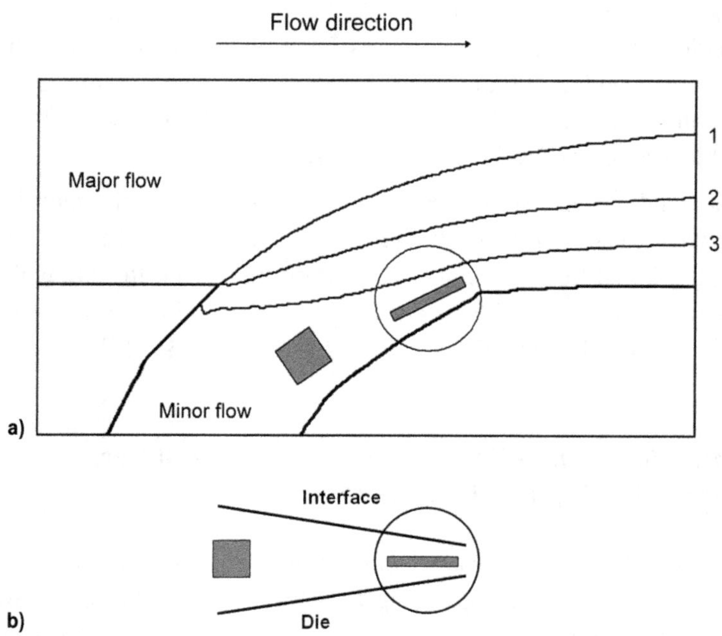

Figure 3.19 a) Typical shapes of predicted interfaces for decreasing mass flow rate in the merging point in the annular coextrusion die [27]; b) Detailed sketch of material element stretching in the merging area (capillary flow analogy) [27]

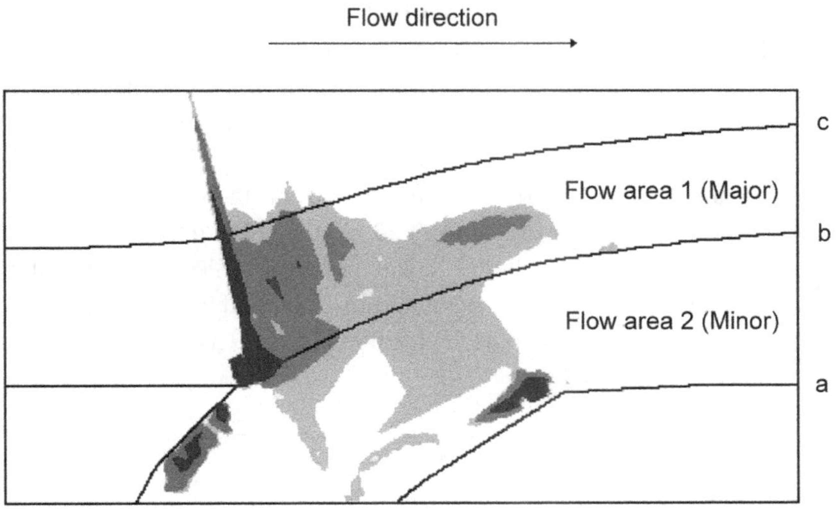

Figure 3.20 A typical shape of the predicted extensional rate field in the merge point during the flow in the annular coextrusion die [27]. Here, a is the die wall; b the interface, and c the streamline

In this figure, the extensional rate field is shown in terms of dark regions (the darker the area, the higher the values of extensional rate). It is nicely visible that intensive stretching does not occur only at the interface (at its beginning) but also within the thickness of the layers. Thus, it seems that the way to understand interfacial instabilities can be through quantification of relative stretching of the coextruded layers across the interface in the merging area. To enable such quantification, the total normal stress difference (TNSD) value has been defined [28]:

$$TNSD = \overline{N_{1,1}} - \overline{N_{1,2}} \tag{3.1}$$

where $\overline{N_{1,1}}$ and $\overline{N_{1,2}}$ are average normal stress differences computed as the mean value in flow area 1 (major layer) and 2 (minor layer), respectively, over all streamlines in these areas (see Fig. 3.20).

$$\overline{N_{1,1}} = \frac{1}{c-b}\int_b^c N_{1,1}(n)\,dn \tag{3.2}$$

$$\overline{N_{1,2}} = \frac{1}{b-a}\int_a^b N_{1,2}(n)\,dn \tag{3.3}$$

Flow area 1 is determined by the mass flow rate, which is the same as in flow area 2. Because mixed shear and extensional flow occurs in the merging area, TNSD is generated by both the extensional and the shear flow components and can be found through a numerical simulation.

The reason for introduction of the TNSD is that the amount of the first normal stress difference determines how particles in the flow area are stretched. Therefore, the TNSD is positive when the particles in the major layer are stretched more than the particles in the minor layer. The appearance of the negative sign, on the other hand, means that at this moment the minor layer is stretched more. When the TNSD value was calculated for two-layer coextrusion that goes from stable to wavy unstable flow conditions (the minor flow decreased from 22 to 10% of the total mass flow rate), it was revealed that for the stable state TNSD is positive, while for the unstable state TNSD changes sign or becomes negative within a particular relative position [28]; (0–1 represents the merging area) as depicted in Fig. 3.21. The physical meaning of TNSD is visualized in Fig. 3.22.

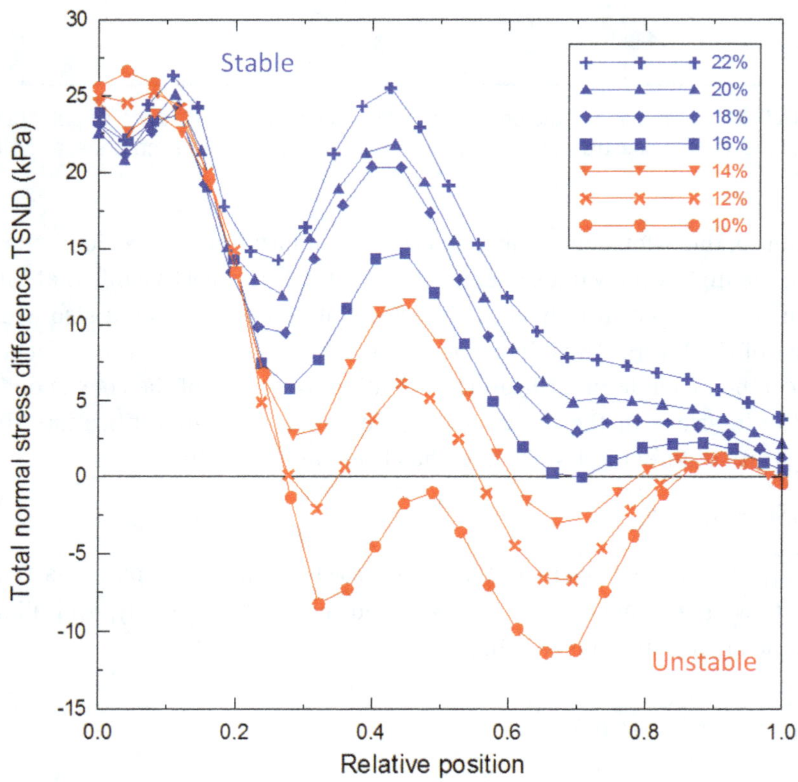

Figure 3.21 Comparison of the predicted TNSD vs. relative position along the interface for stable (minor layer 16 to 22%) and unstable (minor layer 10 to 14%) flow situations [28]

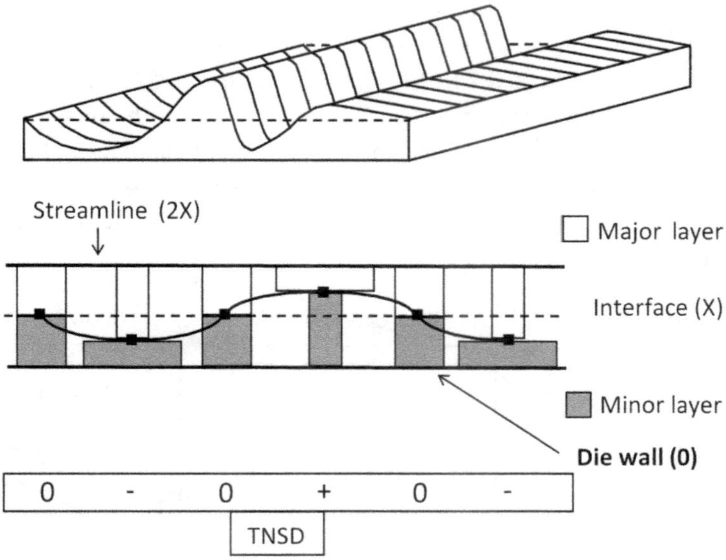

Figure 3.22 3-D and 2-D unstable interface shape under elastic aftereffect (TNSD changes sign) [28]

If TNSD changes sign from positive to negative, the coextrusion interface moves towards the die wall due to the minor layer stretching, and subsequently it moves from the die wall due to the major layer stretching. This elastic aftereffect evidently destabilizes the coextrusion flow. In an extreme case (when TNSD becomes strongly negative), the minor layer is stretched so much that it becomes very thin (the interface moves very close to the wall) and the layer breaks up. Of course, the question is why should a small region of negative TNSD along the interface, relatively close to the merging point, lead to instabilities and why does it not diminish. The explanation provided in reference 28 was the following: the wave is created due to the elastic aftereffect (Fig. 3.23(a)), and the velocity rearrangement in the merge area may cause intensive wave stretching, as depicted in Figs. 3.23(b) and 3.23(c). It is believed [28] that the irreversible nature of such drastic elongational deformations occurring just at the merge point of the layers may explain why wave instabilities do not die out. It should be mentioned here that Mavridis and Shroff [25] suggested that the stability of the interface can be evaluated through the difference in S_R (recoverable shear) for both coextruded layers ($\Delta S_R = S_{R,1} - S_{R,2}$), which can locally quantify shear elasticity differences (generated by pure shear flow) between adjacent layers according to the shear stress state of the interface. On the other hand, TNSD can globally (in bulk) quantify the flow history and layer stretching in the merging area, where mixed shear and extensional flow occur together. Therefore, the instability onset criteria based on TNSD and ΔS_R are in principle different [28].

The wavy unstable coextrusion flow in the merging area has been analyzed experimentally with the help of a tracer inside the coextrusion flow visualization cell having a 90° entrance angle (Fig. 3.24(a)) [41]. The sequence of the different positions of the tracer during the flow in the merging area is depicted in Figs. 3.24(b) to 3.24(i). As visible from Figs. 3.24(b) to 3.24(d), intensive stretching occurs in the merging area with a breaking melt stream (Fig. 3.24(e)) with consequent rotation movement, believably due to intensive shearing in the die land area (Figs. 3.24(f) to 3.24(i)). This intensive stretching of the minor layer with consequent rotation due to shear can explain the source of the wave instabilities as well as the TNSD concept for its detection.

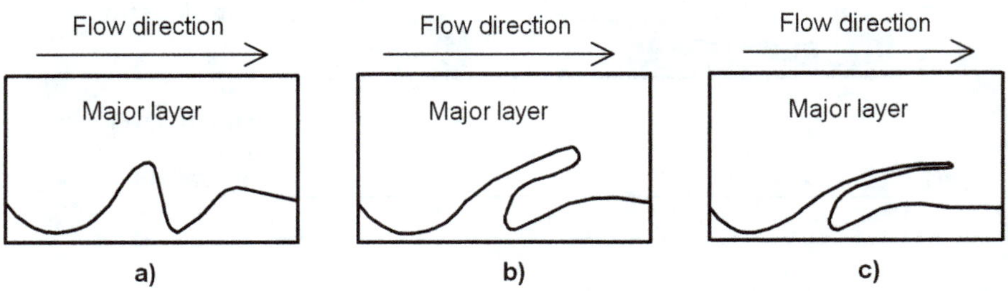

Figure 3.23 Mechanism of the wave instability propagation in the merging area proposed in reference 28

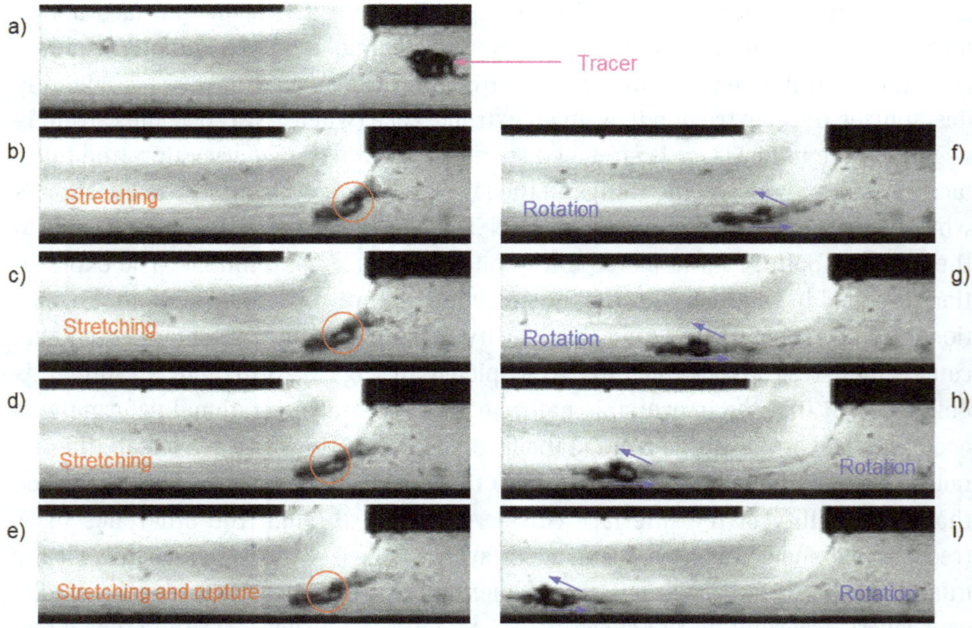

Figure 3.24 Flow of the tracer in the merging area of the 90° coextrusion visualization cell with the following stages [41]: (a) Trace shape before merging; (b–e) Intensive stretching of the tracer; (f–i) Rotation movement of the tracer

3.3.1.1 The Effect of Die Design on the Wave Type of Interfacial Instabilities

The effect of different film blowing coextrusion die design parameters depicted in Fig. 3.25 and Table 3.4 on the wave type of interfacial instabilities has been investigated by using the TNSD sign criterion [27].

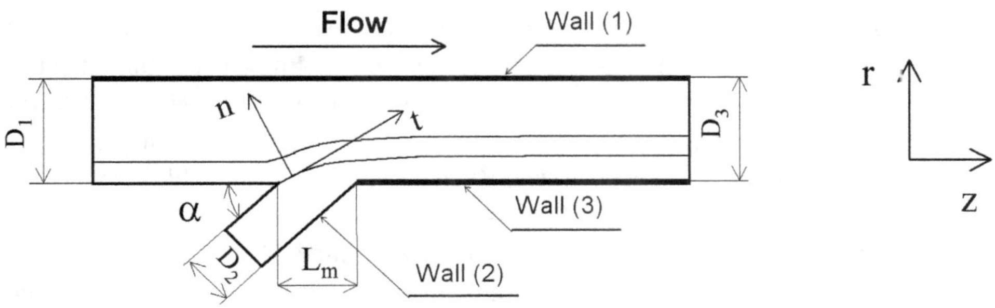

Figure 3.25 Sketch of the coextrusion flow in a wide annular die (geometry zero) [27]

Table 3.4 Different Types of Die Design (Fig. 3.25) [27]

Geometry	Inlet major flow	Inlet minor flow	Outlet	Merging angle
0	D_1	D_2	D_3	α
1	$0.5{*}D_1$	D_2	$0.5{*}D_3$	α
2	$1.5{*}D_1$	D_2	$1.5{*}D_3$	α
3	D_1	$0.5{*}D_2$	D_3	α
4	D_1	$1.5{*}D_2$	D_3	α
5	D_1	D_2	$0.75{*}D_3$	α
6	D_1	D_2	$1.25{*}D_3$	α
7	D_1	D_2	D_3	$2/3{*}\alpha$
8	D_1	D_2	D_3	$2{*}\alpha$

The changes $1.5{*}D_1$ and $0.5{*}D_1$ mean that the major channel is more open (wall (1) in Fig. 3.25 is moved up) and more closed (wall (1) is moved down), respectively. Similarly, the changes $1.5{*}D_2$ and $0.5{*}D_2$ mean that the minor channel is more open (wall (2) is moved right) and more closed (wall (2) is moved left), respectively. In the same way, changes $1.25{*}D_3$ and $0.75{*}D_3$ mean that the major channel is more open at the outlet area (wall (3) is moved down) and more closed (wall (3) is moved up), respectively. Finally, the merging angle a was increased ($2{*}\alpha$) and decreased ($2/3{*}\alpha$). Based on the TNSD sign criterion, it has been found that the stabilizing and destabilizing effects of the changed geometries increase with geometries 7, 6, 2, 3 and geometries 8, 5, 4, 1, respectively.

It is interesting to note that the geometries that permit higher preacceleration of the minor flow relative to the reference geometry zero have a greater stabilizing influence. Otherwise, the geometries destabilize the flow. Preacceleration means increasing the layer speed before the merging point relative to another layer. In the most stable geometry 3, the intensive preacceleration of the minor layer is done by decreasing the gap of the minor channel. Thus, at the same mass flow rate, the speed of the minor layer is increasing because of the well-known equation of continuity. The stabilizing effect of particular geometries can be explained by lower stretching of preaccelerated minor flow at the merging area (an elastic aftereffect does not occur because TNSD is positive in such a case): the most stable geometry 3. On the other hand, if the minor flow is deaccelerated, it becomes intensively stretched in the merging area, at least for a certain moment, and an elastic aftereffect (wave) occurs (TNSD changes the sign). It seems that in the extreme case, the stretching of the minor flow is so high that the layer will break (TNSD becomes more negative): the most unstable geometry 1. It should be pointed out that at the end of the minor channel not only the extensional rate but also the shear rate become two times higher than in the case of geometry 3. Therefore, the wave is not only stretched in this range but also intensively sheared, and thus deformation of the wave becomes more irreversible in this case. The above conclusions are in good agreement with the experimental work reported in the open literature [42–46, 52].

3.3.1.2 The Role of Extensional Viscosities

With the aim of understanding the role of extensional strain hardening in coextrusion flow from the interfacial instability point of view, simulations utilizing the modified Leonov model for the same materials flowing through the annular coextrusion die have been carried out [27]. The parameters of the modified Leonov model were chosen so that both LDPEs had identical shear viscosity but different extensional viscosity (Fig. 3.26). TNSD values were calculated for coextrusion of LDPE1/LDPE1 and LDPE2/LDPE2 structures under identical processing conditions (Fig. 3.27). Based on Figs. 3.26 and 3.27 and the TNSD sign criterion, it has been stated that extensional strain softening materials in both layers stabilize coextrusion flow (TNSD is positive), while strain hardening materials in both layers destabilize the flow (TNSD changes sign). This theoretical conclusion has already been confirmed experimentally in references 31 to 33 and 41.

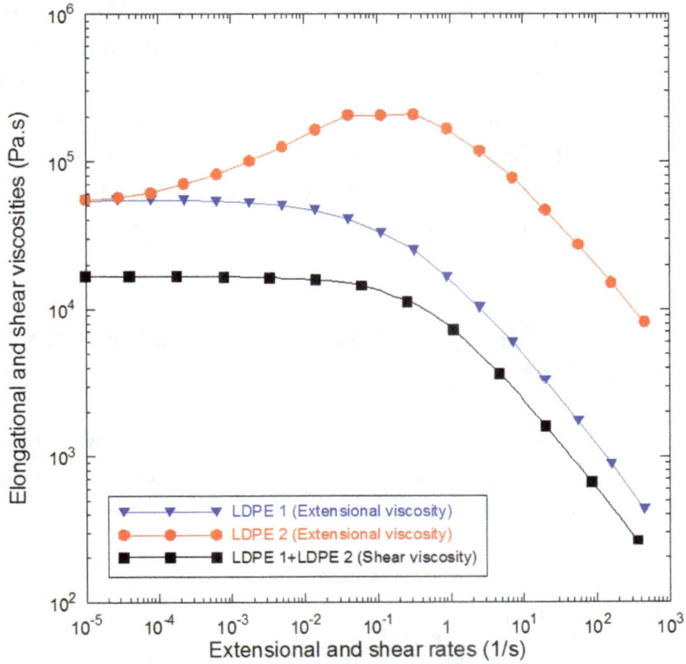

Figure 3.26 Modified Leonov model predictions of extensional viscosity for two different LDPE materials [27]

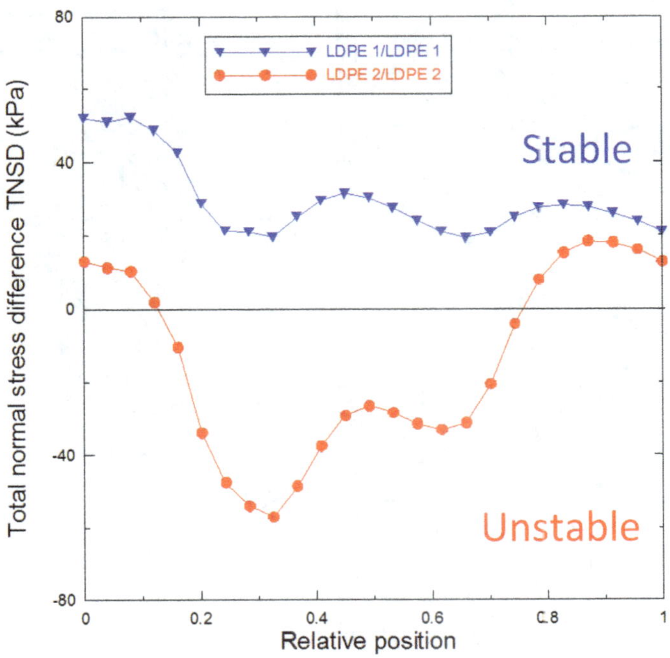

Figure 3.27 Comparison of the predicted TNSD vs. relative position along the interface for different combinations of the same LDPE materials in both layers (minor layer is 12%) [27]

For coextrusion of two different materials, it has been shown theoretically as well as experimentally that if the extensional strain hardening in the minor layer decreases or increases, the system becomes more stable or unstable, respectively [27, 32–33]. From the optimal resin point of view, coextruded materials with extensional thinning and no strain hardening should be preferred for coextrusion. Furthermore, the material with the lower extensional viscosity should be introduced into the minor flow and that with the higher extensional viscosity into the major layer.

3.3.2 Die Exit Region, Zigzag Interfacial Instabilities, and the Role of Polymer Processing Aids

Zigzag interfacial instabilities have been recently studied [35–39] for mLLDPE/HDPE/mLLDPE on a Collin laboratory three-layer coextrusion blown film line equipped with a flat spiral die system (Fig. 3.28).

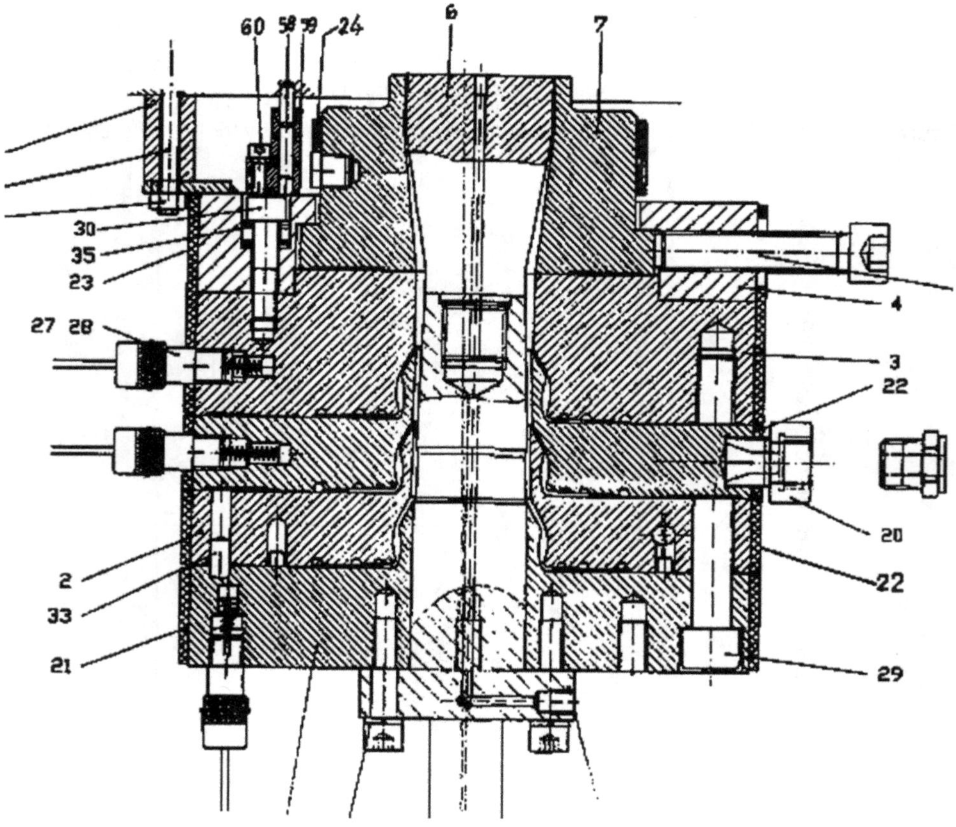

Figure 3.28 Sketch of the three-layer coextrusion die [37]

The thickness of the outer mLLDPE layers was gradually reduced by decreasing the mass flow rates in corresponding channels. Simultaneously, the mass flow rate in the HDPE channel was driven up to increase the thickness of the middle layer until interfacial instabilities were created. The minimum mass flow rate used in mLLDPE channels was 10% of the total output flow rate. In experiment number 1, the same resin (HDPE) was used in all three layers. It was found that in this case the film is hazy but without noticeable interfacial instabilities (see Fig. 3.29(a)). In experiment number 2, different resins were used in the coextrusion. In more detail, outer layers (A and C) were composed of mLLDPE whereas HDPE was used as the middle layer. If no interfacial instabilities occur, coextrusion results in a clear film. However, when the mass flow rate in mLLDPE channels was reduced to 16.5%, interfacial instabilities of the zigzag type were occurring in the film as shown in Fig. 3.29(b) (the chess piece was put behind the film to test the clarity). In experiment number 3, the conditions were kept the same but 0.05% wt% PPA was added in all three layers. Surprisingly, also in this case, interfacial instabilities were occurring in the film (Fig. 3.29(c)). In experiment number 4, the 0.05 wt% PPA was added in the middle layer of HDPE (Fig. 3.29(d)); surprisingly, the interface becomes more unstable. Suppression of the interfacial instabilities was achieved in the final experiment number 5, where the 0.05 wt% PPA was added only in the outer mLLDPE layers (Fig. 3.29(e)).

In the three-layer coextrusion, where PPA was used in the minor (outer) layers, that is experiment number 5, the stabilization effect can be explained through the occurrence of slip between mLLDPA and the PPA coated die wall, because in this case the PPA particles migrate toward the die wall. It reduces the stress state on the interface as well as stretching of the thin minor layer significantly, and according to the critical stress theory and the TNSD concept, this is suppressing interfacial instabilities. If the PPA is used in all of the three layers (experiment number 3) or just in the middle layer (experiment number 4), the PPA in the middle layer (HDPE) does not migrate towards the die wall because this layer is bounded by two moving interfaces and not by static die walls. In such a case the PPA particles migrate toward the interfaces, and they are flowing together with the HDPE. In this case, the utilized PPA type together with HDPE behaved as an extensional stress generator rather than stress dissipater because the extensional viscosity for HDPE + PPA was found to be higher than for HDPE [36, 39]. Thus, at the end of the converging section, a high stress state around the interface is generated. Because the stress state on the interface is increased in this case, the situation is more dangerous from an interfacial instability point of view. The discussion above is supported by FEM results provided in Fig. 3.30, where a high stress area was detected at the end of the converging section for the mLLDPE-PPA-HDPE interfaces (experiment number 3).

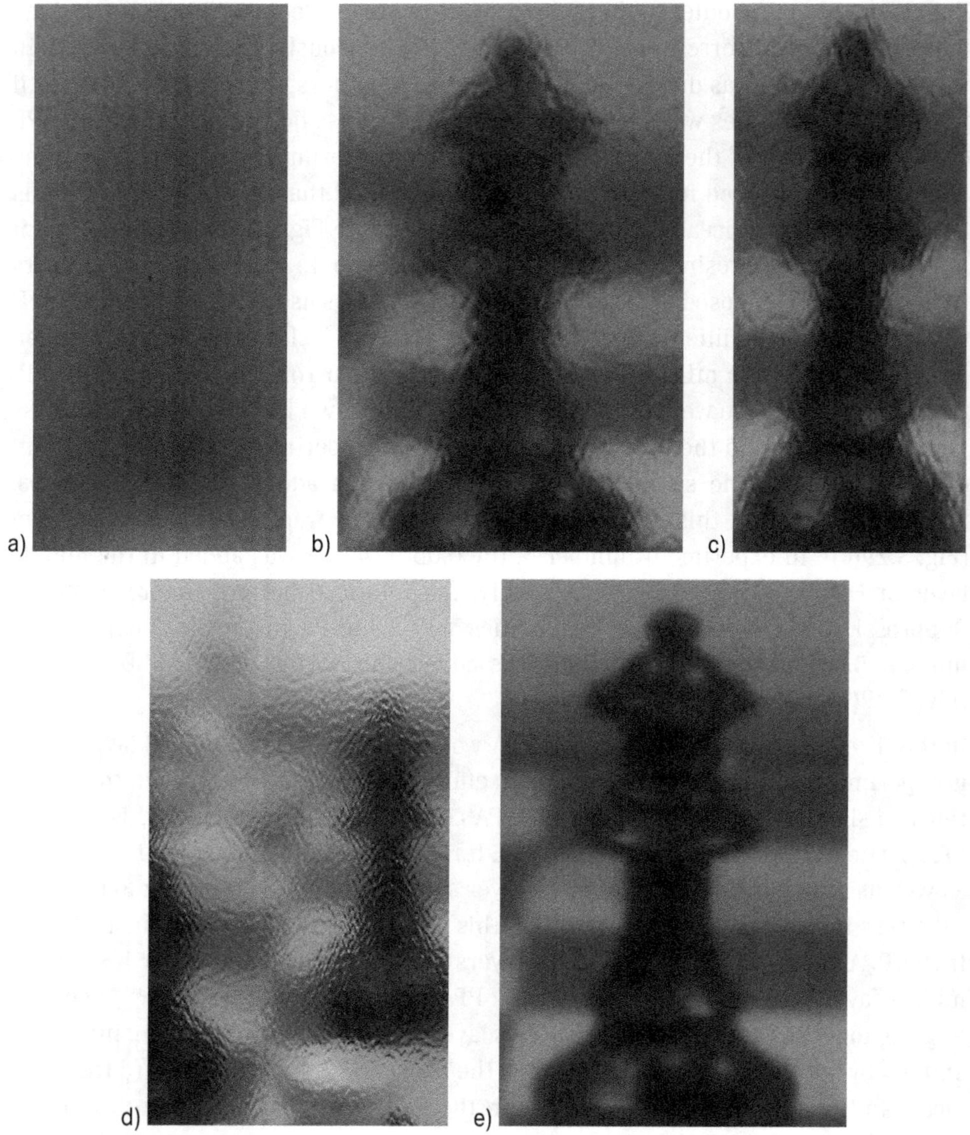

Figure 3.29 Clarity test of the coextruded A and B and C film under different conditions [35–39].
(a) Experiment number 1 (HDPE in A and B and C)
(b) Experiment number 2 (mLLDPE in A and C; HDPE in B)
(c) Experiment number 3 (mLLDPE + PPA in A and C; HDPE + PPA in B)
(d) Experiment number 4 (mLLDPE in A and C; HDPE + PPA in B)
(e) Experiment number 5 (mLLDPE + PPA in A and C; HDPE in B)

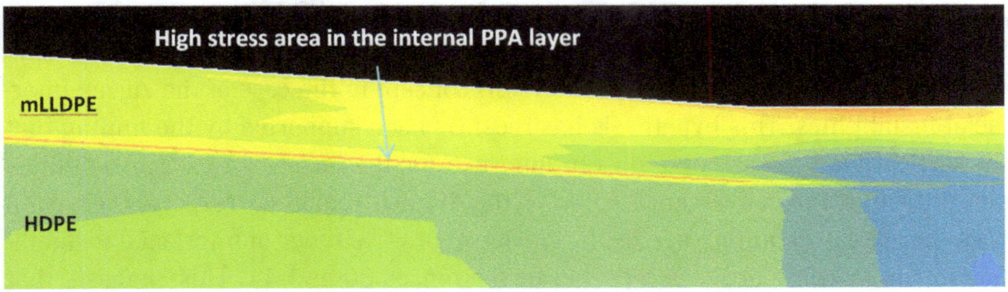

Figure 3.30 FEM modeling of the coextrusion experiment number 3, with PPA in all layers [36, 38]

The viscoelastic stress field calculation reported in reference 36 revealed that the maximum stress appears at the end of the converging section as result of both the shear and extensional flow components (see Fig. 3.31).

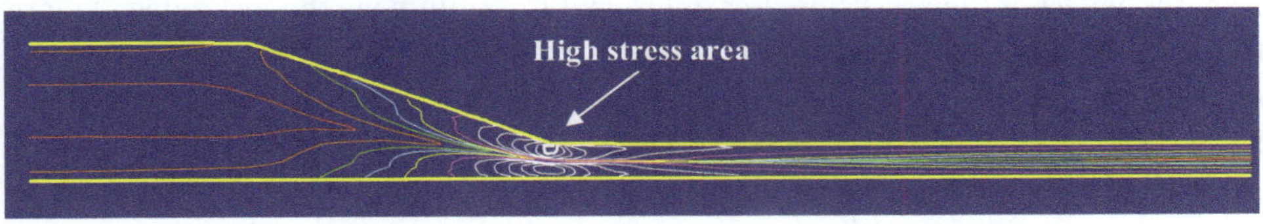

Figure 3.31 Example of the calculated total stress field at the end of the typical extrusion die [36, 38]

In this area, the interface crosses the constant stress line two times; that is, the stress is nonmonotonic along the interface in this region. This maximum stress then continuously disappears as the pure shear flow is being developed in the die land area (see Fig. 3.32).

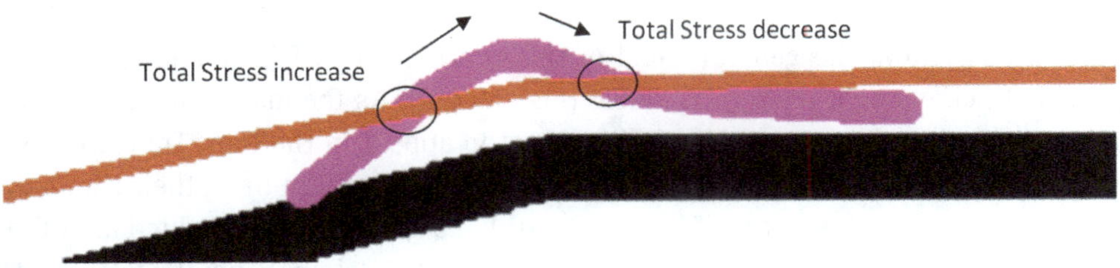

Figure 3.32 The detail of the end of the converging section during coextrusion. The brown line represents the interface; the purple line represents the constant value of the total stress [30, 44].

Clearly, the interface is stretched not only in the merging area as shown in Section 3.3.1 but also in the convergent section at the end of the die. Thus zigzag interfacial instabilities (known to be linked with the end of the die) can also be caused by intensive layer stretching according to the TNSD concept [27–28] with possible layer breakage, similar to the case of wave interfacial instabilities. A shorter residence time in the convergent section compared to the merging area can explain why zigzag instabilities have a much lower amplitude and higher frequency than wave-type instabilities; the zigzag instabilities have no time to fully develop themselves. This explanation may justify the use of the TNSD concept in the case of the zigzag interfacial instabilities. This hypothesis has recently been supported by the finding that the critical total stress value on the interface for the onset of zigzag instabilities determined at the convergent section at the die exit region corresponds very well with the critical condition needed for the onset of wave types of interfacial instabilities in the merging area, as reported in references 30 and 44. This suggests that both types of interfacial instabilities have the same source.

3.3.3 Development of Interfacial Instabilities in Cast Film and Film Blowing of LDPEs

Another aspect of the assessment of interfacial instabilities is the technology where they appear. The development of interfacial instabilities in a flat coextrusion die (visualized in Fig. 3.33) was investigated through continual reduction of the minor layer thickness at constant output mass flow rate [31].

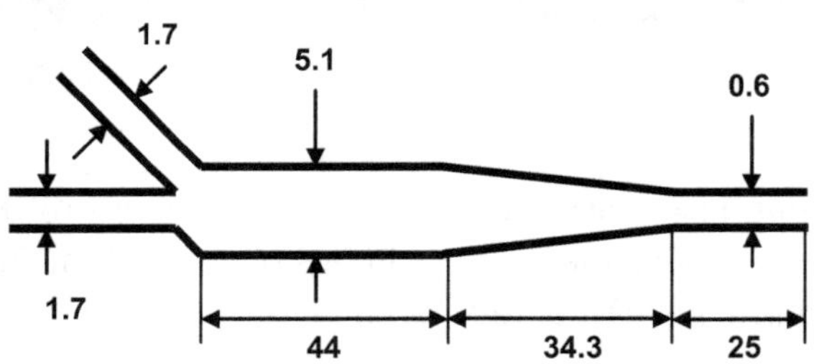

Figure 3.33 Flat multimanifold coextrusion die 300 mm wide [31]

Figure 3.34 shows the general trend of this process for LDPEs. When the minor layer is thick enough, the flow is stable (Fig. 3.34(a)). As the mass flow rate in the minor layer decreases, wave instabilities start to appear in the film. They are first weak and poorly developed, but when the minor layer is still thinner, their contours become sharper and better visible (Fig. 3.34(b)–3.34(f)). A further reduction of the layer causes the development of other waves inside the original one; the wave multiplies and penetrates the film thickness. Then zigzag instabilities appear gradually in the whole film area (Fig. 3.34(g)–3.34(i)). With a minimized minor layer the wave instabilities become significantly destroyed (Fig. 3.34(j)). Finally, for a very thin minor layer, the flow becomes "stable" again (last stage, Fig. 3.34(k)). The reason for this is probably that there is practically no material in the layer, and instead of coextrusion a single-layer extrusion starts.

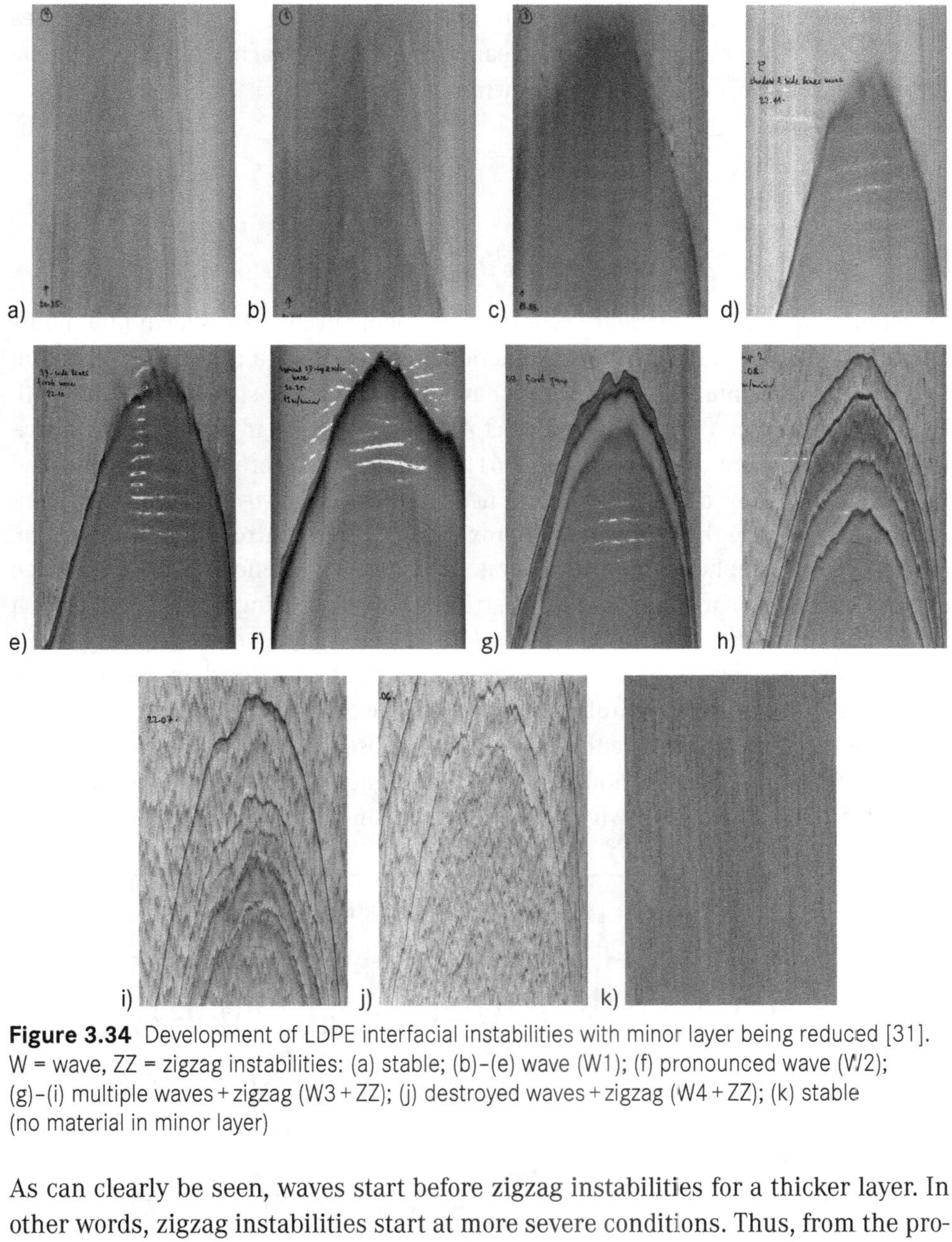

Figure 3.34 Development of LDPE interfacial instabilities with minor layer being reduced [31]. W = wave, ZZ = zigzag instabilities: (a) stable; (b)–(e) wave (W1); (f) pronounced wave (W2); (g)–(i) multiple waves + zigzag (W3 + ZZ); (j) destroyed waves + zigzag (W4 + ZZ); (k) stable (no material in minor layer)

As can clearly be seen, waves start before zigzag instabilities for a thicker layer. In other words, zigzag instabilities start at more severe conditions. Thus, from the production point of view, zigzag instabilities are not so important for cast film technology because the product is damaged as soon as the waves start. This can be explained by the fact that in the flat die (cast film), the temperature of the material is higher (230 to 270°C); that is, the shear viscosity is lower, which results in lower shear stresses. That is why the critical shear stress is hardly reached at the end of the die and the first type of interfacial instabilities to appear is waves. In a film blowing die, on the other hand, the temperatures are much lower (about 190°C), which causes higher shear stresses, so the critical shear stress is easily reached and zigzag instabilities are set before the waves, as shown in reference 21.

3.4 Flow Analysis for Flat Spiral Die

Spiral mandrel dies are standard types of dies widely used in the blown film industry for the last 40 years. Their main function is to distribute a polymer melt stream from an extruder into an annular flow channel. A number of simplified mathematical models have been proposed to predict their performance in terms of flow distribution and pressure consumption [53-61]. In one of the more accurate simplified models [61], the flow distribution is obtained by solving the two-dimensional momentum balances using the lubrication approximation and a control volume algorithm. Because of its simplicity, this model was employed in an effort to determine the optimal geometry of a spiral mandrel distribution system using a Taguchi approach [62]. In addition, control volume method (CVM)-based models have been improved by including entrance pressure drop effects to account for extensional deformation in leakage flows from the spiral channel into the overflow gap [55-56]. The model was able to correctly predict the experimental observation that a broadening of LDPE molecular weight distribution leads to a higher flow variation for an industrial-scale spiral mandrel blown film die, as visible in Figs. 3.35 and 3.36.

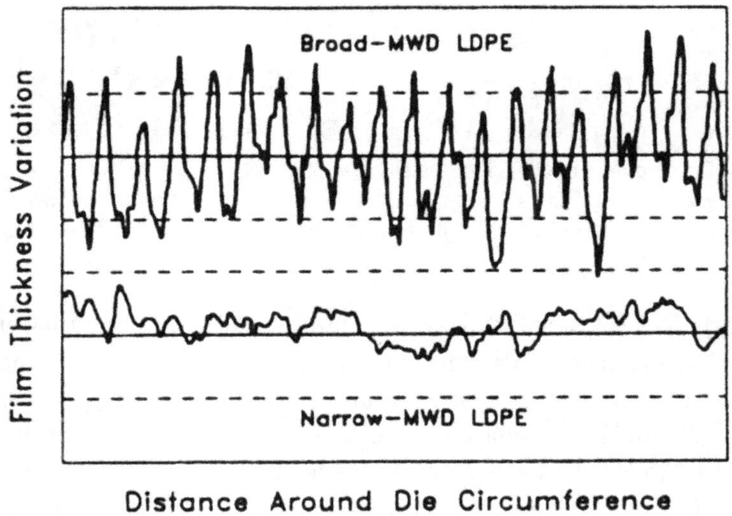

Figure 3.35 Experimentally determined film thickness variation for two LDPEs having different molecular weight distributions in the spiral mandrel die geometry (dashed lines represent ±10% variation around the respective mean thickness) [56]

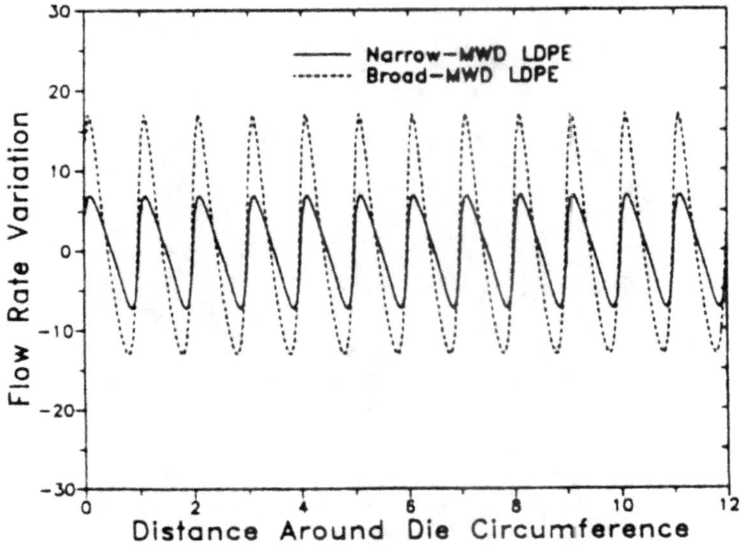

Figure 3.36 Predicted flow variation in the spiral mandrel die geometry for two LDPEs having different molecular weight distributions by the control volume method, which includes entrance pressure drop effects [56]

The effect of wall slip and melt viscosity on film thickness distribution in the circumferential direction in a film blowing spiral mandrel die was investigated in reference 63 considering nine different broad molecular weight distribution HDPEs. It was found that the film thickness distribution tends to decrease with decreasing HDPE wall slip, and its viscosity when comparing the samples having virtually the same shear thinning index.

Although the computations by the CVM are relatively simple, they need to be tested by a more realistic three-dimensional finite element method (FEM) because of the simplified assumptions in the CVM. Such comparisons for a typical spiral mandrel die have been carried out and they have demonstrated the validity of CVM algorithms [55-56, 64]. Although virtually all blown film dies employ a spiral distribution system, it is only recently that a new flat spiral distribution system used mainly in a "stacked" type of annular coextrusion die has been developed. In a flat spiral distribution system, the pressure requirements are reduced compared to a traditional spiral system, and the residence time of the material in the die is significantly reduced. This is especially significant for adhesives and barrier resins, which are more prone to thermal degradation. It should be pointed out that the idea of "stacking" the layers in coextrusion flows is not completely new. In fact, this "stacked" configuration has been previously used with blow molding dies [65-66] despite initial design drawbacks. The improved new flat spiral distribution system used in blown film dies is much more efficient at conveying the polymer melt while maintaining the distribution benefits of a traditional system. A typical very simple flat distribution system is shown in Fig. 3.37.

Figure 3.37 Simple flat spiral distribution system [21, 68]

In order to better understand the flow field within this type of die, both visualization experiments and isothermal 3-D FEM simulations were performed [67]. The results showed that for a Newtonian liquid numerical predictions agreed well with experimental data. For a shear thinning fluid it was found that exit flow variation increases as the power law index decreases due to faster leakage out of the channel [21]. The role of viscous dissipation and material viscoelasticity on flow distribution and pressure drop through the flat spiral die was analyzed through three-dimensional finite element simulations [68]. The isometric and top views of the FEM grid used to describe the flow in the flat spiral die can be seen in Figs. 3.38(a-b) respectively. Figure 3.38(c) presents a cross section of the grid showing the elements used. A close-up view of half the cross section of the flat spiral die is given in Fig. 3.38(d).

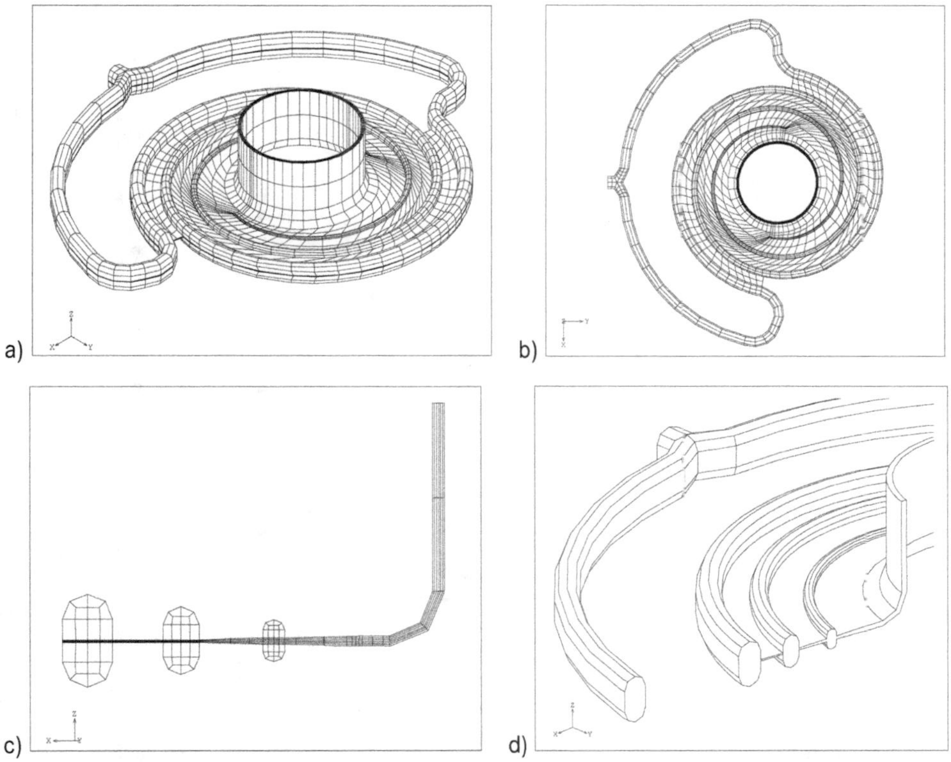

Figure 3.38 Detailed description of the simulated flat spiral die flow domain [68]: (a) isometric view of the FEM grid used for the flat spiral die; (b) top view of the FEM grid used for the flat spiral die; (c) cross-sectional view of the FEM grid; (d) close-up view of half of the cross section of the flat spiral die

It was found that the maximum temperature rise due to viscous dissipation surprisingly occurs within a localized region (0.65 to 0.75 of the reduced spiral length for this specific case), which may significantly increase the flow variation at the die exit. It is not surprising that the maximum temperature rise experienced by the material increases with the mass flow rate (see Fig. 3.39). The fact that the intensive viscous dissipation occurs particularly within the localized region of the reduced spiral length can be explained through particle tracking visualization depicted in Fig. 3.40 for the right-hand spiral channel. This reveals (due to improper die design) intensive leakage flow within a small area from the spiral channel into the small overflow gap at which high viscous dissipation takes place due to high shear rates in this region. In this context, it is necessary to note that a temperature rise of about 10 °C or higher can lead to degradation of the temperature-sensitive materials (such as EVOH, PVDC, and adhesive polymers) within the die. The intensive shearing and viscous heating of the polymer in a flat die may severely limit the maximum flow rate. Therefore, it is crucial to design a flat spiral die by taking into account nonisothermal temperature effects.

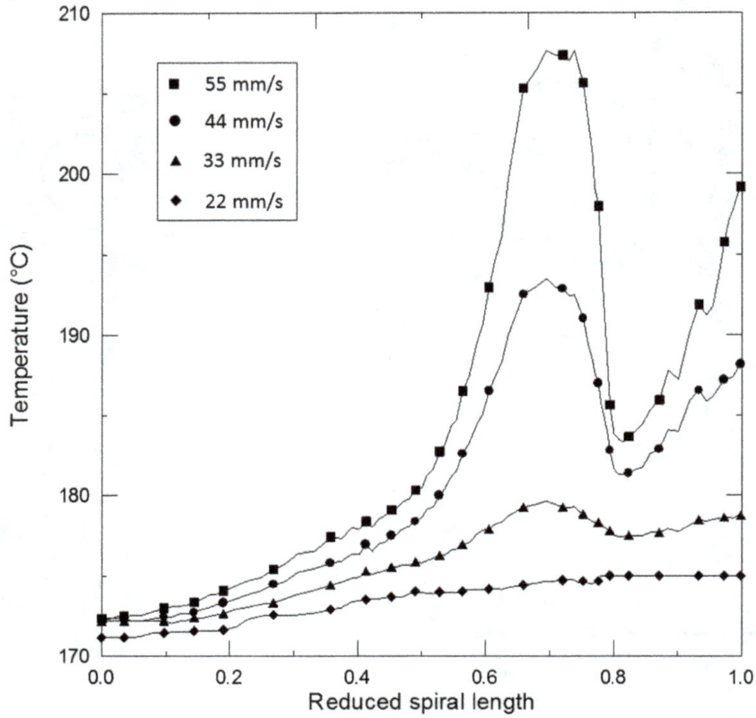

Figure 3.39 Temperature in the spiral channel for different inlet velocities at an inlet temperature of 170°C [68]

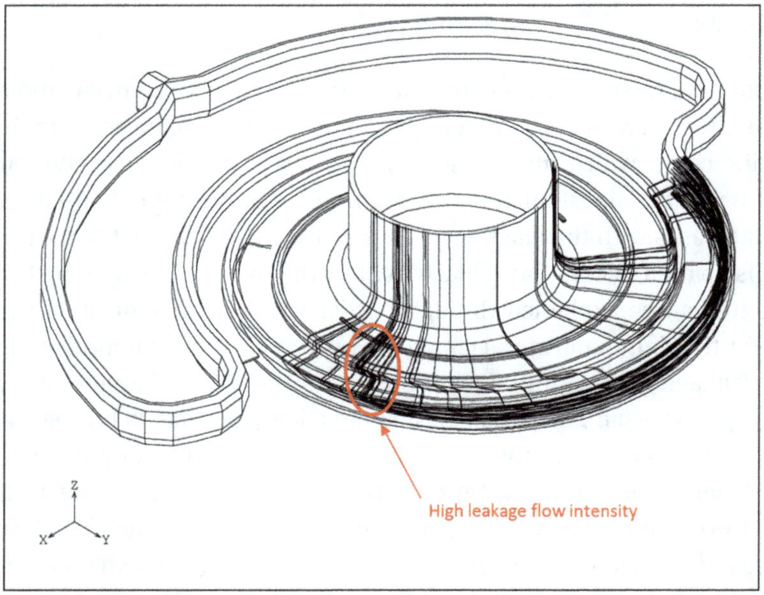

Figure 3.40 Isometric view of a particle plot for nonisothermal simulation (right-hand channel) at an inlet temperature of 170°C [68]

Acknowledgments

Martin Zatloukal wishes to acknowledge the Grant Agency of the Czech Republic (Grant No. P108/10/1325) and the Operational Program Research and Development for Innovations, cofunded by the European Regional Development Fund (ERDF) and national budget of Czech Republic, within the framework of project Centre of Polymer Systems (reg. number CZ.1.05/2.1.00/03.0111) for the financial support.

References

1. Xiao, K., "Current and Future Trends in Multilayer Film Packaging," Canadian Chemical Engineering Conference, Keynote Presentation, Ottawa, Ontario, Canada (2008)
2. Hernandez, R., Selke, S., Culter, J., *Plastics Packaging* (2000) Hanser, Cincinnati, OH
3. Xiao, K., Gammell, S., "Blown Film," Chapter 7 in *Handbook of Troubleshooting Plastics Processes* (2012) John Wiley & Sons, New York, NY
4. Xiao, K., Armstrong, R., Lee, I., "Comparison of Water-Quench Versus Air-Quench Blown Film Processes. Part I: Flat Film Properties," SPE ANTEC Proceedings, Orlando, FL (2012)
5. Xiao, K., Lee, I., Armstrong, R., "Comparison of Water-Quench Versus Air-Quench Blown Film Processes. Part II: Thermoformability," SPE ANTEC Proceedings, Orlando, FL (2012)
6. Armstrong, R., Lee, I., Xiao, K., "Performance Comparison of Water-Quench Versus Air Quench Blown Films," TAPPI Conference Proceedings, Seattle, WA (2012)
7. Oliver, G., Chapter 7.3, "Film and Sheet Dies," *Extrusion of Polymers: Theory and Practice* (2011) Hanser, Cincinnati, OH
8. Chisholm, D., Schrenk, W.J., U.S. Patent 3,557,265 (January 19, 1971)
9. Schrenk, W.J., Chisholm, D.S., Cleereman, K.J., Alfrey Jr., T., U.S. Patent 3,759,647 (September 18, 1973)
10. Carley, J.F., *J. Appl. Phys.* (1954) 25, no. 9, p. 1118
11. Dooley, J., Oliver, G., Xiao, K., "Breakthrough Inventions and Innovations in Die Design for Polymer Processing," SPE ANTEC Proceedings, Chicago, IL (2009)
12. *Film Extrusion Manual: Process, Materials, Properties*, TAPPI Press, (1992)
13. Becker, O.W., U.S. Patent 2,345,086 (March 28, 1944)
14. Goodrum, R.W., Cook, C.A., Swenson, R.L., Raley, G.E., U.S. Patent 3,488,669 (January 6, 1970)
15. Upmeier, H., U.S. Patent 3,689,192 (September 5, 1972)
16. Teutsch, E.O., U.S. Patent 5,069,612 (December 3, 1991)
17. Teutsch, E.O., Coyle, D.J., U.S. Patent 5,393,216 (February 28, 1995)
18. Dooley, J., Jenkins, S.R., Naumovitz, J.A., U.S. Patent 6,685,872 (February 3, 2004)
19. Xiao, K., Jenkins, S.R., Dooley, J., TAPPI PLACE Proceedings (2008)

20. Ramanathan, R., Shanker, R., Rehg, T., Jons, S., Headley, D.L., Schrenk, W.J., "'Wave' Pattern Instability in Multilayer Coextrusion: An Experimental Investigation," *Annu. Tech. Conf. - Soc. Plast. Eng. (ANTEC)* (1996) 1, pp. 224-228

21. Perdikoulias, J., "Analysis and Design of Annular Dies for Mono- and Multilayer 22. Polymer Flows," Ph.D. thesis, University of Waterloo, Waterloo, Ontario, Canada (1997)

22. Perdikoulias, J., Tzoganakis, C., "Interfacial Instabilities during Coextrusion of LDPEs," *Annu. Tech. Conf. - Soc. Plast. Eng. (ANTEC)* (1997) 1, pp. 351-355

23. Tzoganakis, C., Perdikoulias, J., "Interfacial Instabilities in Coextrusion Flows of Low-Density Polyethylenes: Experimental Studies," *Polym. Eng. Sci.* (2000) 40, no. 5, pp. 1056-1064

24. Schrenk, W.J., Bradley, N.L., Alfrey Jr., T., Maack, H., "Interfacial Flow Instability in Multilayer Coextrusion," *Polym. Eng. Sci.* (1978) 18, no. 8, pp. 620-623

25. Mavridis, H., Shroff, R.N., Multilayer Extrusion: Experiments and Computer Simulation," *Polym. Eng. Sci.* (1994) 34, no. 7, pp. 559-569

26. Han, C.D., Shetty, R., "Studies on Multilayer Film Coextrusion II. Interfacial Instability in Flat Film Coextrusion," *Polym. Eng. Sci.* (1978) 18, pp. 180-186

27. Zatloukal, M., Saha, P., Vlcek, J., Tzoganakis, C. "Viscoelastic Stress Calculation in Multi-layer Coextrusion Dies: Die Design and Extensional Viscosity Effects on the Onset of 'Wave' Interfacial Instabilities," *Polym. Eng. Sci.* (2002) 42, no. 7, pp. 1520-1533

28. Zatloukal, M., Tzoganakis, C., Vlcek, J., Saha, P., "Numerical Simulation of Polymer Coextrusion Flows: A Criterion for Detection of 'Wave' Interfacial Instability Onset," *Int. Polym. Process.* (2001) 16, no. 2, pp. 198-207

29. Zatloukal, M., Perdikoulias, J., Tzoganakis, C., Saha, P., "The Influence of Die Design and Extensional Rheology on the Onset of 'Wave' Interfacial Instabilities in Coextrusion Flows," *TAPPI: Polymers, Laminations and Coatings Conference* (2001) pp. 529-535

30. Zatloukal, M., Kopytko, W., Vlcek, J., Saha, P., "Investigation of Zig-zag Type of Interfacial Instabilities in Coextrusion," *Annu. Tech. Conf. - Soc. Plast. Eng. (ANTEC)* (2005) 1, pp. 96-100

31. Zatloukal, M., Kopytko, W., Lengalova, A., Vlcek, J., "Theoretical and Experimental Analysis of Interfacial Instabilities in Coextrusion Flows," *J. Appl. Polym. Sci.* (2005) 98, no. 1, pp. 153-162

32. Zatloukal, M., Kopytko, W., Vlcek, J., "The Effect of Different Batches of the Same Polymer on the Flow in Flat Coextrusion Dies," *Annu. Tech. Conf. - Soc. Plast. Eng. (ANTEC)* (2003) 3, pp. 3509-3513

33. Zatloukal, M., Martyn, M.T., Coates, P.D., Vlcek, J., "Modelling of Viscoelastic Coextrusion Flows in Multi-manifold Flat Dies," *Plast., Rubber Compos.* (2004) 33, no. 7, pp. 305-309

34. Kopytko, W., Zatloukal, M., Vlcek, J., "Coextrusion of LDPEs in the Cast Film Process (Zig-zag Interfacial Instabilities)," *Annu. Tech. Conf. - Soc. Plast. Eng. (ANTEC)* (2004) 1, pp. 372-375

35. Zatloukal, M., De Witte, J., Lavallee, C., Saha, P., "Investigation of PPA Interactions with Polymer Melts in Single Layer Extrusion and Coextrusion Flows," *Plast., Rubber Compos.* (2007) 36, no. 6, pp. 248–253

36. Zatloukal, M., De Witte, J., "Influence of Process Aids on Zig-zag Type of Interfacial Instabilities in Multilayer Flows: Theoretical and Experimental Investigation," *Plast., Rubber Compos.* (2006) 35, no. 4, pp. 149–154

37. Zatloukal, M., De Witte, J., Lavallee, C., "Investigation of PPA Interactions with Polymer Melts in Coextrusion Flows," *Annu. Tech. Conf. - Soc. Plast. Eng. (ANTEC)* (2006) 2, pp. 846–850

38. Zatloukal, M., De Witte, J., "Effect of Process Aids on Interfacial Instabilities in Coextrusion," *Annu. Tech. Conf. - Soc. Plast. Eng. (ANTEC)* (2005) 1, pp. 101–105

39. Zatloukal, M., De Witte, J., "The Effect of Process Aids on Interfacial Instabilities in Coextrusion Flows: Theoretical and Experimental Investigation," *Annu. Tech. Conf. - Soc. Plast. Eng. (ANTEC)* (2004) 1, pp. 381–385

40. Zatloukal, M., Martyn, M.T., Vlcek, J., Coates, P.D., "Modeling of Viscoelastic Multilayer Flows for Different Polymer Melts," *Annu. Tech. Conf. - Soc. Plast. Eng. (ANTEC)* (2006) 2, pp. 851–855

41. Zatloukal, M., Martyn, M.T., Coates, P.D., Saha, P., "Wave Interfacial Instabilities at Coextrusion in Flat Dies: Simulation and Experimental Results," *Annu. Tech. Conf. - Soc. Plast. Eng. (ANTEC)* (2003) 1, pp. 295–299

42. Martyn, M.T., Spares, R., Coates, P.D., Zatloukal, M., "Imaging and Analysis of Wave Type Interfacial Instability in the Coextrusion of Low-Density Polyethylene Melts," *J. Non-Newtonian Fluid Mech.* (2009) 156, no. 3, pp. 150–164

43. Martyn, M.T., Spares, R., Coates, P.D., Zatloukal, M., "Visualisation and Analysis of Interfacial Instability in Coextrusion of LDPE Melt," *Plast., Rubber Compos.* (2004) 33, no. 1, pp. 27–35

44. Zatloukal, M., Kopytko, W., Saha, P., Martyn, M., Coates, P.D., "Theoretical and Experimental Investigation of Interfacial Instability Phenomena Occurring During Viscoelastic Coextrusion," *Plast., Rubber Compos.* (2005) 34, no. 9, pp. 403–409

45. Zatloukal, M., Martyn, M.T., Coates, P.D., Vlcek, J., "Influence of Die Design on Interfacial Instabilities in Coextrusion," *Annu. Tech. Conf. - Soc. Plast. Eng. (ANTEC)* (2005) 1, pp. 86–90

46. Zatloukal, M., Xue, A., Amon, M., "Theoretical and Experimental Investigation of Interfacial Instabilities in Coextrusion Feed-Block Dies," *Annu. Tech. Conf. - Soc. Plast. Eng. (ANTEC)* (2010) 1, pp. 721–727

47. Zatloukal, M., Martyn, M. T., Coates, P.D., Vlcek, J., Modeling of Viscoelastic Coextrusion Flows in Multi-manifold Flat Dies," *Annu. Tech. Conf. - Soc. Plast. Eng. (ANTEC)* (2004) 1, pp. 101–105

48. Zatloukal, M., Martyn, M.T., Coates, P.D., Vlcek, J., "Modelling of Viscoelastic Coextrusion Flows in Multi-manifold Flat Dies," *Plast., Rubber Compos.* (2004) 33, no. 7, pp. 305–309

49. Martyn, M.T., Gough, T., Spares, R., Coates, P.D., "Visualization of Melt Interface in a Co-extrusion Geometry," *Annu. Tech. Conf. - Soc. Plast. Eng. (ANTEC)* (2001) 2
50. Martyn, M.T., Gough, T., Spares, R., Coates, P.D., "Visualization of Melt Interface in a Co-extrusion Geometry," in *Polymer Process Engineering*, 01, IOM Communation, Coates, P.D. (Ed.) (2001) University Press, Cambridge, UK
51. Martyn, M.T., Gough, T., Spares, R., Coates, P.D., Zatloukal, M., "Visualization and Analysis of LDPE Melt Flows in a Coextrusion Geometry," *Annu. Tech. Conf. - Soc. Plast. Eng. (ANTEC)* (2002) 1, pp. 937-941
52. Yamaguchi, H., Yasumoto, T., Yamamoto, H., "Classification of Flow Modes of Viscoelastic Fluids at a Junction of Two Stratified Laminar Flow Layers," *J. Rheol.* (1999) 43, pp. 1373-1390
53. Benkhoucha, K., Sebastian, D.H., "Design and Analysis of Spiral Blown Film Dies," *Annu. Tech. Conf. - Soc. Plast. Eng. (ANTEC)* (1989) pp. 1774-1778
54. Fahy, E.J., Gilmour, P.W., "Finite Element Analysis of Generalized Newtonian Fluids through Spiral Mandrel Dies," *Int. J. Numer. Methods Eng.* (1986) 23, no. 1, pp. 1-11
55. Mavridis, H., *TAPPI PLC Conf. Proc.* (1997) pp. 657-666
56. Mavridis, H., "Model of Polymer Melt Flow in Spiral Mandrel Dies," *Annu. Tech. Conf. - Soc. Plast. Eng. (ANTEC)* (1998) 1, pp. 41-44
57. Procter, B., "Flow Analysis in Extrusion Dies," *SPE J.* (1972) 28, no. 2, pp. 34-41
58. Rauwendaal, C., Flow Distribution in Spiral Mandrel Dies," *Polym. Eng. Sci.* (1987) 27, no. 3, pp. 186-191
59. Saillard, P., Agassant, J.F., "Polymer Flow in a Spiral Mandrel Die for Tubular Films," *Polym. Process Eng.* (1984) 2, no. 1, pp. 37-52
60. Vlcek, J., Kral, V., Kouba, K., "Calculation of the Form of a Spiral Mandrel," *Plast. Rubber Process. Appl.* (1984) 4, no. 4, pp. 309-315
61. Vlcek, J., Vlachopoulos, J., Perdikoulias, J., "Determination of Output Uniformity from Spiral Mandrel Dies," *Int. Polym. Process.* (1988) 3-4, no. 2, pp. 174-181
62. Huang, C.C. "A Systematic Approach for the Design of a Spiral Mandrel Die," *Polym. Eng. Sci.* (1998) 38, no. 4, pp. 573-582
63. Higuchi, H., Sato, M., Koyama, K., "Effect of Wall Slip on Blown Film Thickness Distribution," *Polym. Eng. Sci.* (2003) 43, no. 11, pp. 1788-1797
64. Coyle, D.J., Perdikoulias, J., "Flow Simulation and Visualization in Spiral Mandrel Dies," *Annu. Tech. Conf. - Soc. Plast. Eng. (ANTEC)* (1991) pp. 2445-2447
65. Colombo, R., U.S. Patent 2,820,249 (1958)
66. Schrenk, W.J., U.S. Patent 3,308,508 (1967)
67. Perdikoulias, J., Tzoganakis, C., "Flow Visualization in a Stacked-Type Spiral Die," *Annu. Tech. Conf. - Soc. Plast. Eng. (ANTEC)* (1997) 1, pp. 303-307
68. Zatloukal, M., Tzoganakis, C., Perdikoulias, J., Saha, P., "Numerical Simulations of Polymer Flow in Flat Spiral Dies," *Polym. Eng. Sci.* (2001) 41, no. 10, pp. 1683-1694

4 Die Flow Analysis and Mathematical Modeling of Film Blowing

J. Vlachopoulos and V. Sidiropoulos

4.1 Introduction

In blown film extrusion, the polymer is melted in an extruder and the hot melt is pumped through a die to form a thin-walled tube, which is simultaneously axially drawn and radially expanded. In most installations the extruders are horizontal and the blown film bubble is formed vertically upward as shown in Fig. 4.1.

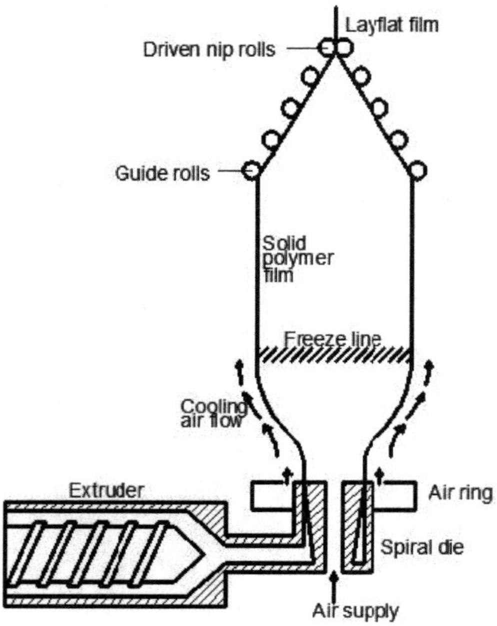

Figure 4.1 Schematic of the film blowing process

The blow-up ratio (BUR) is defined as the ratio of the final bubble diameter to the die diameter and is a good indicator of the amount of expansion in the radial direction. In typical installations BUR ranges from 1.2 to 4. The extension in the machine direction is characterized by a different dimensionless number, the take-up ratio (TUR), which is defined as the ratio of the final film velocity over the velocity at the die lips. In the literature the TUR is often called drawdown ratio (DDR). Film blowing of high molecular weight high-density polyethylene (HMW-HDPE), typically having melt index MI < 0.1, is usually carried out in "high stalk" bubbles, where a long and narrow neck extends from 5 to 9 die diameters above the die followed by a quick bubble expansion (BUR ranging from 3.5 to 5) as shown in Fig. 4.2.

The molten polymer is pumped through die lip gaps usually ranging from 0.76 to 3 mm and is typically drawn down to a film thickness ranging from 0.01 to 0.5 mm. Coextrusion of 3 to 7 layers (lately up to 12) is often used for production of film for food packaging.

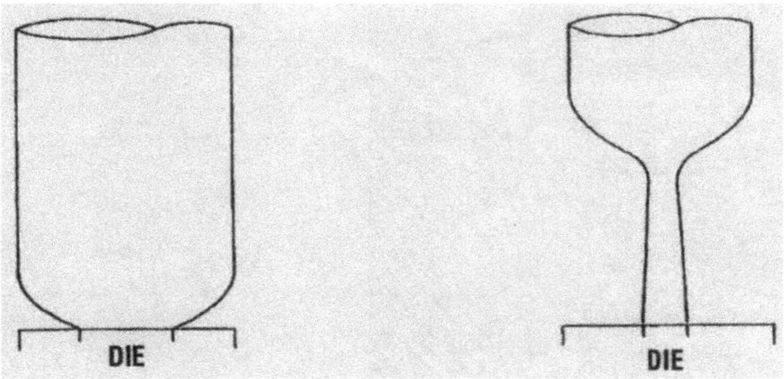

Figure 4.2 A standard bubble shape (typical for LDPE) is shown on the left and a high-neck wine glass bubble (typical for HMW-HDPE) is shown on the right

The hot melt is cooled externally (and sometimes internally) by annular streams of high-speed air from air rings. The maximum achievable output rate is often limited by the air-cooling capability of the line. The die-specific output rate for lines equipped with dual-lip air rings varies from 0.45 to 1.1 kg/hr per mm of die diameter. Internal bubble cooling (IBC) is finding continuously increasing applications, increasing throughput rates to 0.7 to 1.7 kg/hr per mm of die diameter [1].

Due to poor distribution in the die and uneven cooling after the melt exits the die lips there is always some gauge variation. Systems have been developed to reduce this variation through flow modulation of the melt or the cooling air, but it is virtually impossible to eliminate it. Gauge randomization (that is distribution of thick and thin bands) can be accomplished by employing various techniques involving rotation or oscillation of the nip collapser and bubble. Actually, some gauge variation may be caused by the collapsing process, because the distances from the points where the bubble first touches the collapser to the nip line are not all the same around the bubble circumference.

4.2 Die Flow Analysis

The thin tubular film is formed as the melt flows through the die lips. Die diameters range from a few centimeters for laboratory lines to more than one meter for industrial installations producing more than one ton of film per hour. The annular flow is formed in the gap between the inner mandrel and the outer die body. Several types of dies have been proposed and built [2, 3] involving mandrels supported by spider legs, screen packs, or a breaker plate. Also, side-fed dies are sometimes used. The problem with these designs is that they result in the formation of weld lines [4] in the machine direction, along which the polymer is poorly bonded. By far the most common die geometry for blown film production is the spiral one, which is shown schematically in Fig. 4.3 from Perdikoulias [5].

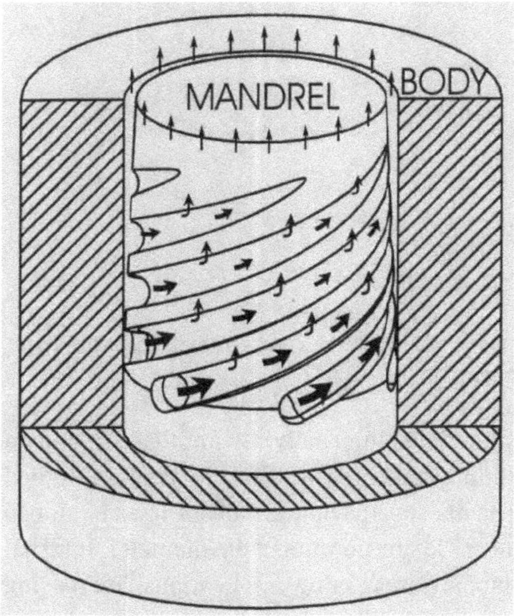

Figure 4.3 Schematic of spiral die [5]

The polymer melt is fed by a number of melt tubes ending with a "port" at the start of each spiral. The melt flows both along the spirals and in the gap between the mandrel and the die body. The flow rate becomes progressively uniform around the circumference towards the die exit. After the end of the spirals the melt may pass through a low shear relaxation chamber [4] for the purpose of reducing the memory of its complex strain history. Spiral geometry is also used in the so-called stacked coextrusion dies, where side-fed disks are stacked one on top of another [3, 4].

Above the die lips the outgoing polymer stream must have uniform thickness. Thickness variation up to 5% can be perhaps tolerated. Uneven thickness tends to be exaggerated by the subsequent inflation and stretching operations, resulting in unacceptable film production. Polymer melts tend to flow preferentially directly above the ports, resulting in periodic thick and thin sections [3–6].

For simulation of polymer melt flow through the channels of a spiral die, the equations of conservation of mass, momentum, and energy under creeping flow conditions (Reynolds number $Re \ll 1$) must be solved. Usually the polymer melt is assumed to be incompressible, having a viscosity that is a function of temperature but not of pressure [3].

The equation of conservation of mass is

$$\nabla \cdot \overline{V} = 0 \tag{4.1}$$

and the equation of conservation of momentum is

$$-\nabla P + \nabla \cdot \overline{\overline{\tau}} = 0 \tag{4.2}$$

and the equation of conservation of energy is

$$\rho c_p \overline{V} \cdot \nabla T = k \nabla^2 T + \overline{\overline{\tau}} : \nabla \overline{V} \tag{4.3}$$

where $\overline{V}$ is the velocity vector, P pressure, $\overline{\overline{\tau}}$ stress tensor, ρ density, c_p specific heat capacity, T temperature, k thermal conductivity, and the term $\overline{\overline{\tau}} : \nabla \overline{V}$ represents the frictional heating (viscous dissipation). The stress tensor is usually expressed in terms of the Generalized Newtonian Fluid model (GNF) in the form

$$\overline{\overline{\tau}} = \eta\left(II_D\right) 2\overline{D} \tag{4.4}$$

where II_D is the second (scalar) invariant of the strain rate tensor (given in Table 4.1).

Table 4.1 The Second Invariant of the Strain Rate Tensor in Rectangular, Cylindrical, and Spherical Coordinates

Rectangular:	$II_D = 2\left[\left(\frac{\partial \upsilon_x}{\partial x}\right)^2 + \left(\frac{\partial \upsilon_y}{\partial y}\right)^2 + \left(\frac{\partial \upsilon_z}{\partial z}\right)^2\right] + \left[\frac{\partial \upsilon_y}{\partial x} + \frac{\partial \upsilon_x}{\partial y}\right]^2 + \left[\frac{\partial \upsilon_z}{\partial y} + \frac{\partial \upsilon_y}{\partial z}\right]^2 + \left[\frac{\partial \upsilon_x}{\partial z} + \frac{\partial \upsilon_z}{\partial x}\right]^2$
Cylindrical:	$II_D = 2\left[\left(\frac{\partial \upsilon_r}{\partial r}\right)^2 + \left(\frac{1}{r}\frac{\partial \upsilon_\theta}{\partial \theta} + \frac{\upsilon_r}{r}\right)^2 + \left(\frac{\partial \upsilon_z}{\partial z}\right)^2\right] + \left[r\frac{\partial}{\partial r}\left(\frac{\upsilon_\theta}{r}\right) + \frac{1}{r}\frac{\partial \upsilon_r}{\partial \theta}\right]^2 + \left[\frac{1}{r}\frac{\partial \upsilon_z}{\partial \theta} + \frac{\partial \upsilon_\theta}{\partial z}\right]^2 + \left[\frac{\partial \upsilon_r}{\partial z} + \frac{\partial \upsilon_z}{\partial r}\right]^2$
Spherical:	$II_D = 2\left[\left(\frac{\partial \upsilon_r}{\partial r}\right)^2 + \left(\frac{1}{r}\frac{\partial \upsilon_\theta}{\partial \theta} + \frac{\upsilon_r}{r}\right)^2 + \left(\frac{1}{r\sin\theta}\frac{\partial \upsilon_\varphi}{\partial \varphi} + \frac{\upsilon_r}{r} + \frac{\upsilon_\theta \cot\theta}{r}\right)^2\right] + \left[r\frac{\partial}{\partial r}\left(\frac{\upsilon_\theta}{r}\right) + \frac{1}{r}\frac{\partial \upsilon_r}{\partial \theta}\right]^2 + \left[\frac{\sin\theta}{r}\frac{\partial}{\partial \theta}\left(\frac{\upsilon_\varphi}{\sin\theta}\right) + \frac{1}{r\sin\theta}\frac{\partial \upsilon_\theta}{\partial \varphi}\right]^2 + \left[\frac{1}{r\sin\theta}\frac{\partial \upsilon_r}{\partial \varphi} + r\frac{\partial}{\partial r}\left(\frac{\upsilon_\varphi}{r}\right)\right]^2$

$$\overline{\overline{D}} = \frac{1}{2}\left(\nabla\overline{V} + \nabla\overline{V}^T\right) = \frac{1}{2}\left(\frac{\partial V_i}{\partial x_j} + \frac{\partial V_j}{\partial x_i}\right) \tag{4.5}$$

The viscosity is usually expressed in terms of the power-law, Carreau-Yasuda, and Cross models.

The power-law equation is

$$\eta = K\dot{\gamma}^{n-1} \tag{4.6}$$

where K is the consistency index (that is, the value of the viscosity at shear rate $\dot{\gamma}$ = 1 s^{-1}), which is temperature dependent in the same way as zero shear viscosity. The power-law exponent is usually not a function of temperature.

The Carreau-Yasuda equation is

$$\eta = \eta_o\left[1 + (\lambda\dot{\gamma})^{\alpha}\right]^{\frac{n-1}{a}} \tag{4.7}$$

The Cross equation is

$$\eta = \frac{\eta_o}{1 + (\lambda\dot{\gamma})^{1-n}} \tag{4.8}$$

where η_o is the zero shear viscosity, which is a function of temperature, usually expressed either by an Arrhenius expression, mostly used in polymer physics and rheology

$$\eta = \eta_{ref}\left[\frac{E}{R}\left(\frac{1}{T} - \frac{1}{T_{ref}}\right)\right] \tag{4.9}$$

or a simple exponential, mostly used in equipment design

$$\eta = \eta_{ref}\exp[-b(T - T_{ref})] \tag{4.10}$$

where η_{ref} is a viscosity measured at a reference temperature, E is the activation energy, and R is the gas constant. The temperature sensitivity coefficient b is usually between 0.01 and 0.1 °C^{-1} for most commercial polymers. For HDPE (linear polymer) the value of b is roughly 0.01, while for LDPE (branched) it may reach 0.03. The parameter λ is a time constant and represents some sort of material relaxation. In the Carreau-Yasuda model, λ determines the shear rate at which a transition occurs from the zero-shear-rate plateau to the shear-thinning portion of the viscosity curve. In the Cross model, when $\lambda = 1/\dot{\gamma}$ then the viscosity becomes $\eta = \eta_o/2$. In some simulations λ is considered to obey the same temperature dependence as the zero shear viscosity and with the same parameters, either Arrhenius or single exponential in the form

$$\lambda = \lambda_{ref}\exp[-b(T - T_{ref})] \tag{4.11}$$

The fully 3-D solution is very demanding computationally. Several simplifications are possible using the Hele-Shaw flow approximation [3] or a control volume method (flow analysis network) for flow in the axial and radial directions [4]. Numerous publications using simplified flow analysis methods have been appearing since the 1970s, and references can be found elsewhere [3–7]. Fully 3-D simulations of spiral dies have also been published [5, 8–10]. The authors are not aware of any published work on simulation of spiral dies using viscoelastic constitutive equations. However, the work of Sun and Gupta [9] involves a special consideration of elongational viscosity.

■ 4.3 Computer-Aided Die Design

If a suitable equation solver is available, the die designer must choose a spiral die geometry to perform a flow simulation to obtain pressure drop, output mass flow rate (which corresponds to film thickness) variation, wall shear rate and wall shear stress, temperature distribution, and other quantities that will be used to decide whether a proposed design is satisfactory. The first die geometry choice is likely to be based on previous experience or knowledge of rules of thumb: spiral dies usually have one spiral per 25 mm of die diameter, and each spiral usually travels over six to eight ports. After a simulation for the first geometry is completed, the designer must examine the results to see how far the predictions are from the design constraints. Then the geometry must be updated and another simulation must be performed and so forth, until all of the design constraints are satisfied. The usual constraints in spiral die design are the following [3]:

- The predicted output thickness variation around the circumference after the end of the spirals should be no more than 5%.
- The pressure drop is below a maximum value. Usually, single screw extruders can generate up to 50 MPa, but because of pressure drop in melt pipes and other melt distributors, the actual pressure drop in the die would probably have to be less than the maximum in the extrusion line.
- The wall shear rate should not be less than a minimum value, which is frequently quoted as 8 s^{-1}, but for some temperature-sensitive materials it should be higher. Low wall shear rate means that the fluid moves slowly over the wall and it may have time to degrade or the corresponding shear stress is low and the material will get stuck on the wall.

Other constraints might also be imposed, which may involve a maximum allowable wall shear stress at the die lips (to avoid shark skin instability), a maximum residence time (to prevent degradation), and a maximum allowable temperature. The temperature of the melt increases beyond the set values on the die wall due to viscous dissipation (frictional heating), and the designer must examine the simulation results for potential hot spots, which should always be avoided.

A good software package can predict all of the above variables, enabling the designer to carry forward an iterative trial-and-error procedure on the computer screen until all of the specifications of satisfactory performance are met. In this methodology the search is for good or satisfactory solutions instead of optimal ones. As Simon [11] put it (in his book on the philosophy of design), "no one will settle for good or better if he can have best. But that is not the way the problem usually poses itself in actual design situations. In the real world we usually do not have a choice between satisfactory and optimal solutions, for we only rarely have a method for finding the optimum."

In spiral die design the number of variables is so large that it is impossible to find a globally optimum design. Of course, it is possible to determine an optimum for a limited set of variables and constraints. In fact, a doctoral thesis has recently been published describing how an optimization algorithm can be combined with numerical simulations for automated design of spiral dies [12]. The present industrial practice, however, for spiral die design, is based on iterative simulation procedures that eventually lead to satisfactory design. The results shown in Fig. 4.4 were obtained from such an iterative simulation procedure using the SPIRALCAD ADVANCE software [3, 13].

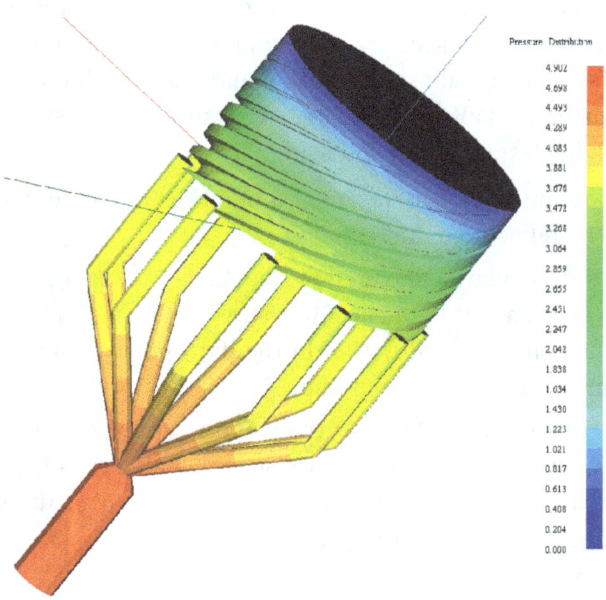

Figure 4.4 Pressure distribution predicted by SPIRALCAD ADVANCE [3, 13] for a 304.8 mm diameter die having 10 ports at 150 kg/hr. Maximum pressure at the feeding tube is 4.902 MPa

4.4 Modeling of Bubble Forming

As the molten polymer emerges from the annular die lips, the thin tube undergoes nonuniform biaxial stretching. The deformation effectively ceases at the frost line, because the stresses are not large enough to deform the solid film. For typical film production rate and thickness, it is reasonable to neglect inertia forces, surface tension, air drag, and gravity on the highly viscous melt. Perhaps the neglect of gravity is the weakest assumption when high production rates are involved.

Figure 4.5 shows a sketch of an axisymmetric thin tubular film (of thickness h) with an embedded Cartesian system having unit vectors in the direction of flow m (meridional) in the tangential t and normal n directions.

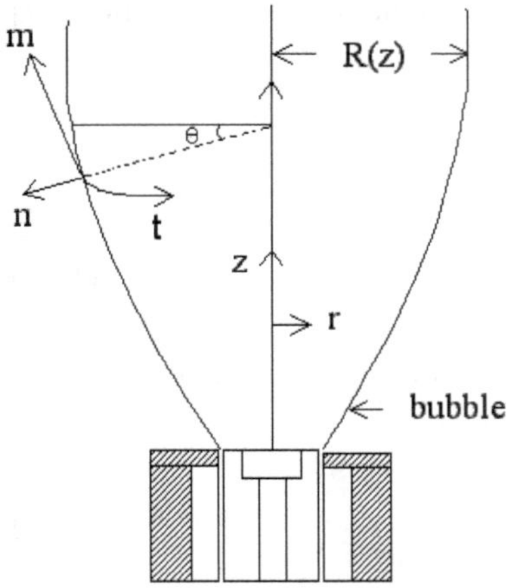

Figure 4.5 Blown film bubble and intrinsic coordinate system (n, m, t)

Mass conservation gives

$$Q = 2\pi R h V_m = \text{constant} \tag{4.12}$$

where Q is the volumetric flow rate, R is the bubble radius, h the bubble thickness, and V_m the velocity in the meridional direction. Taking the derivative with respect to the meridional direction m (the distance along the film surface)

$$\frac{dV_m}{dm} = -\frac{V_m}{h}\frac{dh}{dm} - \frac{V_m}{R}\frac{dR}{dm} \tag{4.13}$$

The rate-of-strain tensor is defined as

$$\dot{\gamma}_{ij} = \frac{Q\cos\theta}{2\pi Rh}\begin{bmatrix} -\frac{1}{R}\frac{dR}{dz}-\frac{1}{h}\frac{dh}{dz} & 0 & 0 \\ 0 & \frac{1}{h}\frac{dh}{dz} & 0 \\ 0 & 0 & \frac{1}{R}\frac{dR}{dz} \end{bmatrix} \tag{4.14}$$

For the coordinates shown in Fig. 4.5 we have $\gamma_{11} = \gamma_m$, $\gamma_{22} = \gamma_n$, and $\gamma_3 = \gamma_t$.

The "thin shell" approximation has been extensively used to describe the balance of forces on the film bubble. This approach was apparently suggested by Alfrey [14] for the first time. The subsequent derivations are based on the work of Pearson and Petrie [15, 16], which are also described in a number of books including Tanner [17] and Kanai and Campbell [18].

The equilibrium of forces in the normal direction for the thin membrane assumption is

$$\frac{\Delta P}{h} = \frac{\sigma_m}{\rho_m} + \frac{\sigma_t}{\rho_t} \tag{4.15}$$

where ρ_m and ρ_t are the curvatures in two directions m and t, and σ_m and σ_t are the corresponding stresses, and ΔP is the inflation pressure.

$$\rho_m = -\left(1+\left(\frac{dR}{dz}\right)^2\right)^{3/2}\left(\frac{d^2R}{dz^2}\right)^{-1} \tag{4.16}$$

$$\rho_t = R\left(1+\left(\frac{dR}{dz}\right)^2\right)^{1/2} = \frac{R}{\cos\theta} \tag{4.17}$$

The force balance on any plane z gives

$$-\pi R^2 \Delta P + 2\pi Rh\sigma_m \cos\theta = F_T = \text{constant} \tag{4.18}$$

where F_T is the total force exerted on the thin membrane.

If F_T and ΔP are known, the stress σ_m can be directly determined from Equation 4.18 followed by stress σ_t from Equation 4.15 (that is, from force equilibrium equations) irrespective of any constitutive equation.

The Newtonian constitutive equation for the stresses gives

$$\sigma_m = 2\eta\left[\gamma_m - \gamma_n\right] \tag{4.19}$$

$$\sigma_t = 2\eta\left[\gamma_t - \gamma_n\right] \tag{4.20}$$

The Newtonian model can be combined with the thin shell approximation of the bubble by substituting Equations 4.19 and 4.20 into Equations 4.15 and 4.18, respectively.

By introducing dimensionless parameters $r = R/R_0$, $w = h/h_0$, Equation 4.18 reduces to

$$\frac{w'}{w} = -\frac{r'}{2r} - \frac{Br^2 + F_c}{4}\left(1 + r'^2\right) \tag{4.21}$$

where

$$F_c = \frac{F_T R_o}{\eta Q} \quad \text{and} \quad B = \frac{\pi R_o^3 \Delta P}{\eta Q}$$

By substituting Equation 4.21 into 4.15 a differential equation containing only the dimensionless radius (r) as the unknown is obtained:

$$2r^2\left(Br^2 + F_c\right)r'' = 6r' + r\left(F_c - 3Br^2\right)\left(1 + r'^2\right) \tag{4.22}$$

Integration of Equation 4.22 provides a prediction of the bubble shape. At the die, the dimensionless radius and thickness have unit value and can be used as boundary conditions.

The standard thin shell approximation can be simplified if the bubble is approximated by a "quasi-cylindrical" geometry [19]. The representation of the bubble by a series of "tubular elements" can be seen in Fig. 4.6.

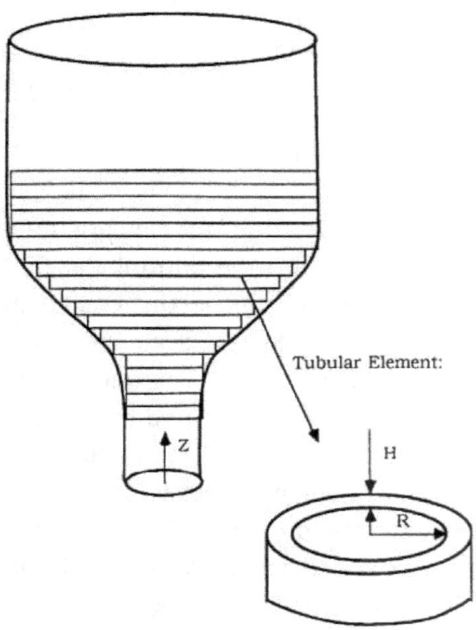

Figure 4.6 Diagram showing the "quasi-cylindrical" approximation of a blown film bubble [19]

The proposed approximation formulates a model with the same general structure as the thin shell model but with the bubble curvatures fixed to $\rho_m = \infty$ and $\rho_t = R$. Then Equations 4.15 and 4.18 are simplified to

$$\Delta P = \frac{h}{R}\sigma_t \tag{4.23}$$

and

$$F = 2\pi R h \sigma_m \tag{4.24}$$

The simplifications shown above significantly reduce the nonlinearity of the ensuing system of differential equations and guarantee a stable solution. The quasi-cylindrical approximation has been used in combination with flow-induced crystallization, viscoelasticity, and bubble cooling by Doufas and McHugh [20].

4.5 Effect of Viscoelasticity

Modeling of bubble formation for a viscoelastic fluid starts again from the equilibrium of forces (Equation 4.15) with the stresses expressed in terms of the velocity components as determined from a suitable constitutive equation. Luo and Tanner [21] and Beaulne and Mitsoulis [22] have obtained identical results for an Upper-Convected Maxwell (UCM) model. The elasticity level is determined by the dimensionless Weissenberg Number, defined as the product of a characteristic (relaxation) time and the strain rate:

$$W_i = \lambda \dot{\gamma} \tag{4.25}$$

For $W_i = 0$ (Newtonian) the bubble diameter predicted by the UCM model [21, 22] attains its largest value. As the Weissenberg Number increases the predicted bubble diameter becomes smaller, and in the limiting case of very large Weissenberg numbers it becomes equal to the diameter of the die from which the thin tubular film emerges. Beaulne and Mitsoulis [22] also discussed their results obtained using an integral K-BKZ-type constitutive equation. Cain and Denn [23] have dealt with some problems of instability in modeling viscoelastic bubble forming. The question of stability of solution was also examined by Andre et al. [24]. Muslet and Kamal [25] employed a Phan-Tien and Tanner constitutive equation [17] and a neo-Hookean model with crystallization effects. Henrichsen and McHugh [26] used a Giesekus constitutive equation for the amorphous phase and a rigid rod model for the semi-crystalline phase. An improved 2-D version was also published by Xu and McHugh [27], focusing on crystallinity development. Sarafrazi and Sharif [28] used a multi-mode extended Pom-Pom model and crystallization kinetics.

4.6 Modeling of Bubble Cooling

The rate at which the tubular film is being cooled (mainly by the radial air jet(s) issuing from the air ring) has a great influence on the process because it determines the distance above the die where solidification is complete (freeze line height, FLH). Radiative heating may be responsible for as much as 20% of the heat transfer and must be taken into account. A heat balance [29] gives

$$\rho C_p Q \cos\alpha \frac{dT}{dz} = -2\pi R\left(h_t\left(T - T_a\right) + \kappa\varepsilon\left(T^4 - T_a^4\right)\right) \tag{4.26}$$

with $T/T_0 = 1$ at the die lips (boundary condition), where ρ is the density, C_p is the heat capacity, Q is the polymer flow rate, h_t is the convective heat transfer coefficient, κ is the Stefan-Boltzmann constant, and ε is the emissivity.

In the above equation, special consideration must be given to the local heat transfer coefficient h_t and the viscosity dependence on temperature in the bubble formation model of Equation 4.22. Bubble shape and other variables can be reasonably well predicted [30] by using the purely viscous approximation (Equations 4.22 and 4.26) and employing a modification of the Arrhenius expression for temperature dependence of viscosity (Figs. 4.7 and 4.8). These equations were the basis of the B-FILM-CAD [31] software package. It was shown that the local heat transfer coefficient and the variation of the molten film viscosity with temperature are crucial parameters that determine the bubble shape. Consequently, the speed of the air jet(s) emerging from the air ring(s) and the structure of the turbulent airflow play a very important role in the determination of the final bubble shape, the strains, the stresses (which are eventually frozen-in), and the final mechanical and optical properties.

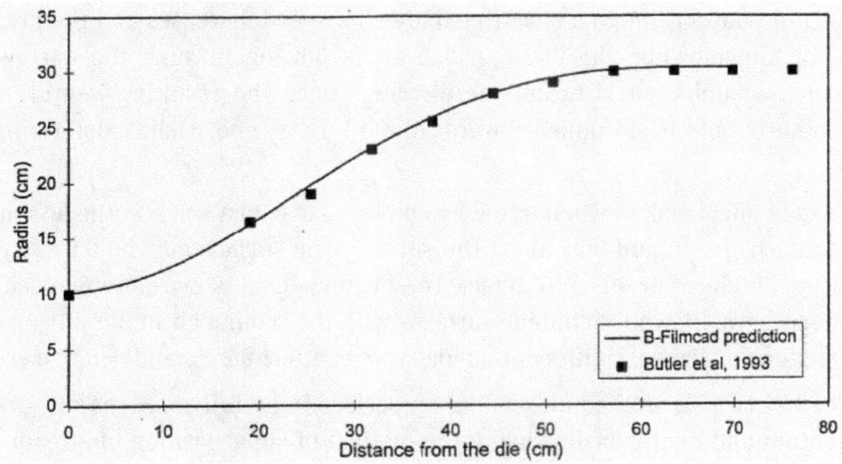

Figure 4.7 Comparison between experimental data for an LLDPE bubble and B-FILMCAD prediction, for bubble radius [30]

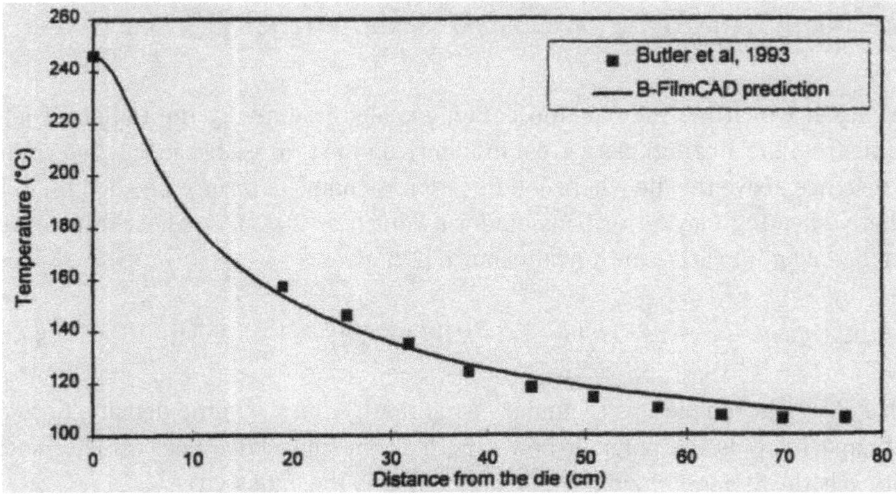

Figure 4.8 Comparison between experimental data for an LLDPE bubble and B-FILMCAD prediction, for the temperature variation with distance from the die lip exit [30]

Most conventional air rings are dual orifice, though single orifice implementations are still used for high-stalk, HDPE bubbles. In a dual orifice setup, some of the airflow is injected through the lower orifice and flows for a while between the cone and the bubble surface. Most of the air volume is injected through the higher orifice onto the bubble surface. Some air rings also include a chimney (and possibly more complex devices like an IRIS) to redirect the cooling air. An IRIS is a mechanical diaphragm composed of metal plates that fold or expand one on top of another for controlling the aperture for airflow around the bubble.

Two important aerodynamic phenomena may be associated with the cooling airflow, namely the *Venturi* and *Coanda* effects [32]. The well-known Venturi effect is caused when a fluid flows through a constricted area: its speed increases and the pressure drops. In film blowing, the lower orifice air is flowing through the narrow gap between the bubble surface and the air ring cone. The resulting Venturi effect causes the bubble to be pulled towards the air ring cone, visibly deforming the bubble.

The Coanda effect occurs when a free jet emerges close to a surface: the jet tends to bend, “attach” itself, and flow along the surface. The surface may be flat or curved and located inclined or offset to the jet. The Coanda effect is more pronounced near curved surfaces. Blown film bubble surfaces with the cooling air impinging on them at an angle offer the possibility of Coanda-type jet attachments and detachments.

Campbell et al. [33] studied air cooling by performing a full aerodynamic analysis (momentum and heat transfer) using the method of superposition of stream functions as well as macrobalances of mass and energy. Later, Sidiropoulos and Vlachopoulos [34] presented a study of the aerodynamics of dual orifice air rings using a commercially available finite-volume software package to solve the Reynolds-averaged Navier-Stokes equations.

$$\frac{\partial(\rho U_i)}{\partial t}+\frac{\partial(\rho U_i U_j)}{\partial x_j}=\frac{\partial}{\partial x_j}\left[\mu\left(\frac{\partial U_i}{\partial x_j}+\frac{\partial U_j}{\partial x_i}\right)-\left(\frac{2}{3}\mu\frac{\partial U_k}{\partial x_k}\right)\right]-\frac{\partial p}{\partial x_i}+\rho g_i+\frac{\partial}{\partial x_j}\left(-\rho\overline{u_i'u_j'}\right) \quad (4.27)$$

In these equations the effect of turbulence is included through the Reynolds stresses $(-\rho\overline{u_i'u_j'})$, which involve products of the velocity fluctuations. A variation of the so-called k-ε turbulent model was used to achieve closure of the system of differential equations. In this approach the contribution of Reynolds stresses to the momentum balance is introduced through the concept of the "effective" viscosity. Although there are more accurate models to simulate the airflow, the results are often only marginally different; not enough to justify the increased complexity of the simulation.

Figure 4.9 shows a simulation example depicting the calculated airflow streamlines around a typical blown film bubble [32, 34].

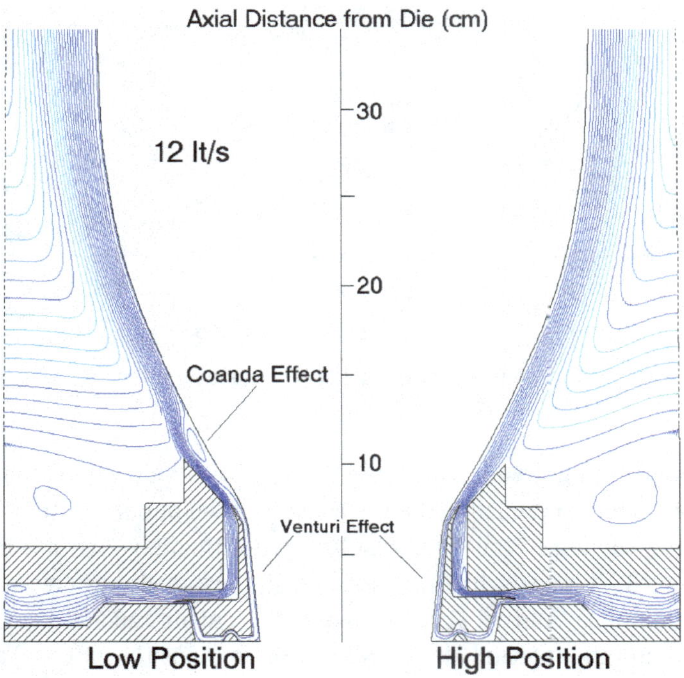

Figure 4.9 Cooling air streamlines around an LLDPE blown film bubble for different setups of the adjustable ring (moderate airflow) [34]; left side: low position, right side: high position

The two simulated airflows (presented on the left and right side of the same figure) correspond to slightly different setups for the adjustable part of the air ring. Although the geometrical differences are minute, the simulation predicts significant differences in the local airflow pattern, which consequently result in differences in local heat fluxes, as shown in Fig. 4.10.

The numerical simulations explained the importance of underpressure generated by the Venturi effect, which is large enough to force the bubble to take the shape of the air ring. It appears that the Venturi effect overwhelms any tendency for different bubble shapes dictated by viscoelasticity. Generally, the Venturi effect is used to stabilize the process. More information on the various aspects of bubble cooling by air jets can be found in Spirgatis [35], Peters and Ruppel [36], Zhang et al. [37], and Gao and Ewing [38, 39]. Housiadas [40] developed a model involving the interaction between bubble formation and the air jet.

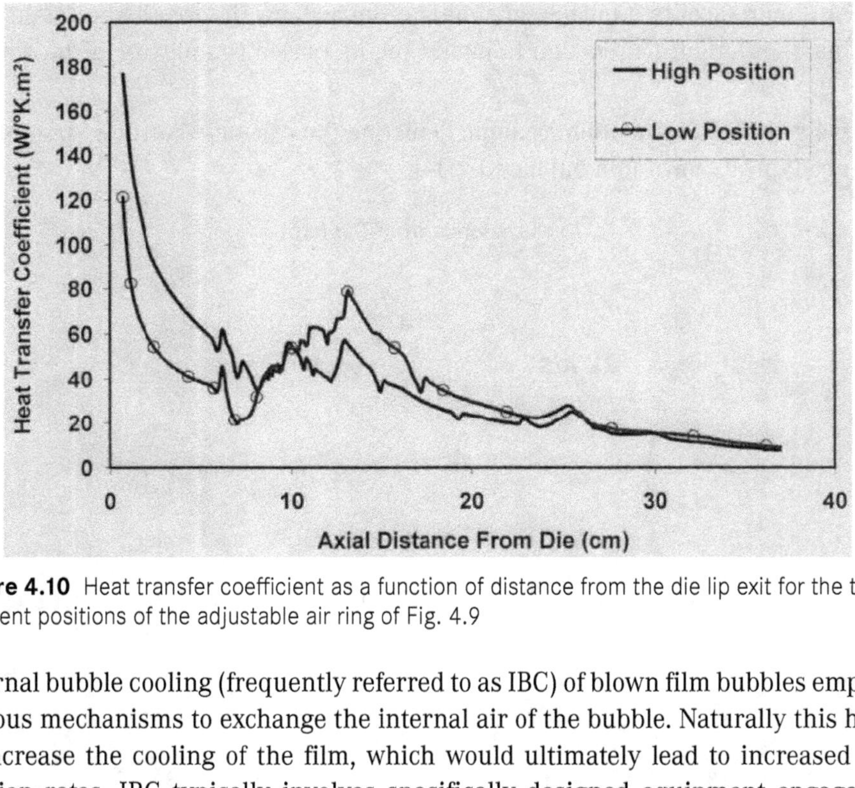

Figure 4.10 Heat transfer coefficient as a function of distance from the die lip exit for the two different positions of the adjustable air ring of Fig. 4.9

Internal bubble cooling (frequently referred to as IBC) of blown film bubbles employs various mechanisms to exchange the internal air of the bubble. Naturally this helps to increase the cooling of the film, which would ultimately lead to increased production rates. IBC typically involves specifically designed equipment engaged in exchanging the warm internal air with colder external air and also constantly circulating and mixing the internal air [41, 42]. In some cases the external air is chilled before it is injected inside the bubble to maximize the cooling benefit. With IBC, the expected production rate improvement becomes increasingly important as die size and film layflat width increase. Production rate increases range from 20% for small bubbles (up to 8 in. in die diameter) to as much as 80% for very large bubbles.

Although IBC systems decisively increase the production rate of numerous blown film bubbles, the expectations and the excitement that was generated with the inception of the technology has only been partially met. The reason appears to be the increased complexity of the technology and the introduction of many additional variables that must be controlled.

4.7 Bubble Collapsing

Once solidified, the tubular film must be collapsed to form a layflat tube. This particular task is handled by the collapsing system. Some films need sophisticated collapsing systems (for example "tacky" films need low-friction or even noncontact collapsers) while others do fine with simple, roller-equipped collapsing frames.

Most collapsing systems induce some form of distortion to the film. Simple triangulation reveals that the distances the film has to travel in the collapsing system are not even across the tube circumference, and the film has to stretch in some areas before assuming the final layflat shape. Figure 4.11 shows the relative distances (*L*) that the film has to travel in the collapsing frame.

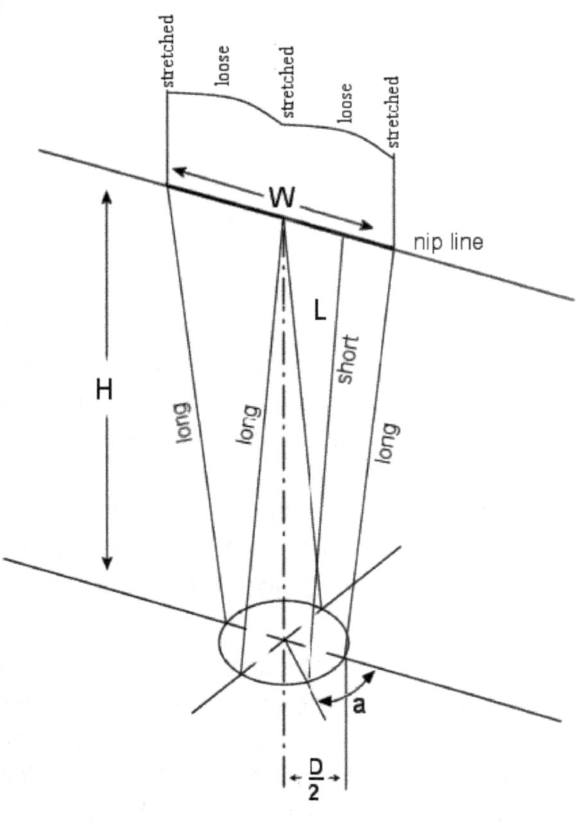

Figure 4.11 Film stretching during collapsing process

The following equation can be used to calculate the relative distance [43] as a function of the position across the tube circumference:

$$L = \sqrt{H^2 + \left[\left(1 + \tfrac{\pi}{2} - a\right)\pi D + 4 - D \cos\left(a\right) + 2\right]^2 + \left[D \sin\left(a\right) + 2\right]^2} \tag{4.28}$$

The uneven stretching creates "stretched" and "loose" regions in the final layflat roll (see Fig. 4.11). During the collapse, soft, pliable polymers (like LDPE) extend and the film becomes slightly thinner in the "stretched" regions. In higher modulus materials (like HDPE and PET) the strength of the film prevents stretching, resulting in broad wrinkles in the "loose" regions.

A reasonable approach to the problem would be to extend the height (*H*) of the collapsing frame, minimizing the % variation of *L*. This approach is difficult to implement for bubbles with large diameter, which would require a very long collapsing frame. Additionally, this solution increases the contact area between the bubble and the collapsing frame. The additional drag force can distort the flatness of the film, making it baggy in the center and difficult to postprocess. Some solutions have been proposed to minimize this problem, like tall noncontact air-slide collapsers or geometrically modified versions of the conventional collapsing frames. To date, none has achieved industry-wide acceptance, and wrinkles in blown film still remain a major problem [44].

4.8 Critique on Flow Analysis and Bubble Modeling

Mathematical modeling and computer simulation of creeping non-Newtonian (shear-thinning) and nonisothermal flow are currently used for routine design purposes. The impact on die design technology has been enormous. Thickness variations have been reduced and productivity has been increased significantly. The film properties and overall quality have been improved as a result of better design of flow channels for reduction of output variations from the die lips and elimination of nearly stagnant flow areas, which could promote material degradation. The procedure nearly always involves iterative simulations until a satisfactory design is reached on the computer screen.

One challenge in the near future will be to move from the iterative simulations on the computer screen to numerical optimization of die design. Such methodologies have barely started for research purposes, and the challenges are significant.The iterative simulations will still have to be carried out within the overall optimization process as a sequence is constructed from initial guesses towards the optimum.

Using fully 3-D flow analysis in each step would be very demanding on computer resources.

Another challenge is the use of viscoelastic constitutive equations for flow simulations of industrial relevance. While there have been numerous publications on viscoelastic simulations, they are usually limited to flows through channels of simple geometrical shape, and the results are frequently not in agreement with experimental data. Using viscoelastic equations for nonisothermal flows through the channels of a spiral die would be a major challenge, and another challenge would be to prove that the results are physically meaningful and quantitatively useful. In such dies, there are regions of large strain and stress gradients that result in some difficult numerical problems. It would be unreasonable to decide the die channel dimensions on the basis of computer simulations, which are likely to be plagued by numerical issues of stability and convergence. However, it might be possible in the near future to make informed engineering judgements, on the basis of qualitative trends, based on viscoelastic models for the description of rheological behavior of polymer melts as they flow through spiral and other types of die channels for the production of tubular film from the die lips.

The mathematical modeling of blown film bubbles started with the work of Pearson and Petrie [15, 16] on the basis of the thin membrane approximation. The models predict bubble shapes, velocity, thickness, temperature, strain rates, and stresses. It turns out that this modeling is sensitive to the viscosity variation with temperature [30]. Kanai [45] showed that modeling of the bubble for polymers that have small activation energy and small power index predicts the wine glass shape typical of HMW-HDPE. The purely viscous formulation described in previous sections is capable of predicting both bubble shapes shown in Fig. 4.2. Beaulne and Mitsoulis [22] show that viscoelasticity tends to decrease the radius of the standard (LDPE-like) bubbles. However, the determination of shape on the basis of the rheological properties might be meaningful for well-controlled laboratory experiments but greatly at variance with industrial practice. As Kurtz [46] put it: "The shape of the bubble in real film blowing processes can be seen, often adjusted ... controlling the shape is very important in controlling the stability and the properties of the film." The bubble shape is primarily determined by the fabrication conditions and the cooling jet aerodynamics and not by material rheology. The long molecular length of HMW-HDPE (MI usually less than 0.1) prevents its drawdown to very thin film when blown in the standard bubble. The high-neck (wine glass) bubble is obtained by reducing the cooling air to allow the freeze-line to rise and lengthen its neck. The long neck permits the molecules to relax, and the subsequent high blow-up ratio (BUR) results in optimization of properties in both MD and TD. From industrial practice it is known that optimum bubble configuration usually involves BUR between 3 and 4.5 and neck height of 6 to 9 die diameters [47], but there is no way to prove this with mathematical models developed thus far.

While this chapter is on mathematical modeling, the reader must not lose sight of the recent technological developments for maintaining stable bubbles, frequently involving 11 to 12 coextruded layers and very high output rates, which, for a die of 500 mm diameter, can exceed 750 kg/hr, especially with internal bubble cooling. There have been improvements in air ring design, and in some installations three lip air rings are used. Gauge control technologies have been developed involving differential cooling or heating around the bubble via segmented die lips or segmented air rings. With these innovations film blowing is approaching cast film quality.

Numerous publications (some of them cited in this chapter) have contributed to improving the understanding of some important mechanisms that control the bubble formation process. However, to the best knowledge of the authors of this chapter, mathematical models and computer simulations of bubble forming are not used for blown film bubble design, film property optimization, or industrial troubleshooting purposes. Bubble equipment design is mainly based on previous experience and trial-and-error. Process optimization by carrying out small modifications known as "tweaking" is the usual practice in the blown film industry. To change the present unsatisfactory state of a dichotomy between mathematical modeling of bubble forming and technological innovations it would be necessary to develop new methodologies. The aerodynamics of cooling require simulations and experiments involving large production lines and study of the interaction between bubble and airflow. The continuing modeling focus on the bubble shape is probably counterproductive, because the bubble shape is determined by the mechanical manipulations and the aerodynamics of cooling. The suggestion of Kurtz [46] of "using the shape of the bubble as a parameter" could be the starting point for developing new modeling approaches for industrial blown film equipment design and film quality improvement.

4.9 Concluding Remarks

This chapter started off with a description of the mathematical modeling for flow analysis of spiral mandrel dies, which represent the dominant type used to form a tubular film. This is followed by a brief description of computer-aided design, based on iterative simulations until a satisfactory design is reached. The advent of computer simulation has had a significant impact on die design technology and improvement of productivity and quality of blown film production lines. It is argued that the first challenge for the forseeable future is to combine the solution of the mass, momentum, and energy conservation equations with numerical optimization techniques for automated computer die design. The mathematical modeling of bubble forming is described on the basis of the thin membrane approximation for the determination of bubble shape, velocity, and film thickness. This is followed by a review of the methodologies used for the study of cooling by turbulent impinging air jets. A brief description of bubble collapsing has also been included. It is argued that the present approach, which starts off with bubble shape prediction, has had very limited impact on industrial blown film technology, and new approaches are called for.

References

1. Butler, T. I., in *The SPE Guide on Extrusion Technology and Troubleshooting*, Vlachopoulos, J., Wagner J. (Eds.) (2001) Society of Plastics Engineers, CT, p. 11.1
2. Michaeli, W., *Extrusion Dies* (1992) Hanser, Munich
3. Vlachopoulos, J., Castillo, R., Polychronopoulos, N., Tanifuji, S., in *Design of Extrusion Forming Tools*, Carneiro, O. S., Nobrega, J. M. (Eds.) (2012) Smithers Rapra, Shawbury, UK, pp. 141–168
4. Perdikoulias, J., Vlachopoulos, J., Vlcek, J., in *Film Processing*, Kanai, T., Campbell, G. A. (Eds.) (1999) Hanser, Munich, pp. 39–57
5. Perdikoulias J., *Analysis and Design of Annular Dies for Mono- and Multilayer Polymer Flows* (1997) University of Waterloo, Waterloo, ON, Canada, Doctoral Thesis
6. Perdikoulias, J., Vlcek, J., Vlachopoulos, J., *Adv. Polym. Technol.* (1987) 7, pp. 333–341
7. O'Brien, K. T., *Applications of CAE in Extrusion and Other Continuous Processes* (1992) Hanser, Munich
8. Coyle, D. J., Perdikoulias, J., *Annu. Tech. Conf. - Soc. Plast. Eng. (ANTEC)* (1991) p. 2445
9. Sun, Y., Gupta, M., *Adv. Polym. Technol.* (2006) 25, pp. 90–97
10. Zatloukal, M., Tzoganakis, C., Perdikoulias, J., Saha, P., *Polym. Eng. Sci.* (2001) 41, pp. 1683–1694
11. Simon, H. A., *The Science of the Artificial* (1969) MIT Press, Cambridge, MA
12. Saul, K., *Automatisierte Auslegung von Extrusionswerkzeugen* (2011) University of Duisburg-Essen, Germany, Doctoral Thesis
13. SPIRALCAD ADVANCE software from H.A.S.L. Tokyo, Japan, and Polydynamics Inc., Dundas, ON, Canada
14. Alfrey, T., Jr., *SPE Trans.* (1965) April, pp. 68–74
15. Pearson, J. R. A., Petrie, C. J. S., *J. Fluid Mech.* (1970) 40, pp. 1–9
16. Pearson, J. R. A., Petrie, C. J. S., *J. Fluid Mech.* (1970) 42, pp. 609–625
17. Tanner, R. I., *Engineering Rheology* (1985) Clarendon Press, Oxford
18. Kanai, T., Campbell, G. A., *Film Processing* (1999) Hanser, Munich
19. Liu, C. C., Bogue, D. C., Spruiell, J. E., *Int. Polym. Process.* (1995) 10, pp. 230–236
20. Doufas, A. K., McHugh, A. J., *J. Rheol.* (2001) 45, pp. 1085–1104
21. Luo, X., Tanner, R. I., *Polym. Eng. Sci.* (1985) 25, pp. 620–629
22. Beaulne, M., Mitsoulis, E., *J. Appl. Polym. Sci.* (2007) 105, pp. 2098–2112
23. Cain, J. J., Denn, M. M., *Polym. Eng. Sci.* (1988) 28, pp. 1527–1541

24. Andre, J.M., Agassant, J.F., Demay, Y., Haudin, J.M., Monasse, B., *Int. J. Form. Processes* (1998) 1, pp. 187-210
25. Muslet, I.A., Kamal, M.R., *J. Rheol.* (2004) 48, pp. 525-550
26. Henrichsen, L.K., McHugh, A.J., *Int. Polym. Process.* (2007) 22, pp. 179-189
27. Xu, F., McHugh, A.J., *Chem. Eng. Sci.* (2009) 64, pp. 4787-4795
28. Sarafrazi, S., Sharif, F., *Int. Polym. Process.* (2008) 23, pp. 30-37
29. Han, C.D., Park, J.Y., *J. Appl. Polym. Sci.* (1975) 19, pp. 3277-3290
30. Sidiropoulos, V., Tian, J. J., Vlachopoulos, J., *J. Plast. Film Sheeting* (1996) 12, pp. 107-129
31. B-FILMCAD software from Polydynamics Inc., Dundas, ON, Canada
32. Sidiropoulos, V., Vlachopoulos, J., *Int. Polym. Process.* (2000) 15, pp. 40-45
33. Campbell, G.A., Obot, N.T., Cao, B., *Polym. Eng. Sci.* (1992) 32, pp. 751-759
34. Sidiropoulos, V., Vlachopoulos, J., *Polym. Eng. Sci.* (2000) 40, pp.1611-1618
35. Spirgatis, J., *Untersuchung des Einflusses des instationären konvektiven Wärmeübergangs bei der Follienherstellung auf die Produktqualität,* (2004) University of Duisburg-Essen, Germany, Doctoral Thesis
36. Peters, F., Ruppel, C., *Int. Polym. Process.* (2006) 21, pp. 319-327
37. Zhang, Z., Lafleur, P.G., Bertrand, F., *Int. Polym. Process.* (2006) 21, pp. 527-535
38. Gao, N., Li., S., Ewing, D., *Int. Polym. Process.* (2006) 20, pp. 68-77
39. Li, S., Gao, N., Ewing, D., *Int. Polym. Process.* (2006) 20, pp. 432-440
40. Housiadas, K.D., *Polym. Eng. Sci.* (2011) 51, pp. 2115-2117
41. Sidiropoulos, V., Vlachopoulos, J., *Int. Polym. Process.* (2001) 16, pp. 48-53
42. Sidiropoulos, V., *The Effects of Air Cooling on the Film Blowing Process* (2000) McMaster University, Hamilton, ON, Canada, Doctoral Thesis
43. Knittel, R.R., in *Film Extrusion Manual,* Butler, T.I., Veazy, E.W. (Eds.) (1992) TAPPI Press, Atlanta, GA, pp. 311-340
44. Knittel, R.R., *Plast. Technol.* (2007) June, accessed on www.ptonline.com
45. Kanai, T., *Int. Polym. Process.* (1987) 1, p. 137
46. Kurtz, S.J., *Int. Polym. Process.* (1995) 10, pp. 148-154
47. Bourgeois, J.R., in *Film Extrusion Manual,* Butler, T.I., Veazy, E.W. (Eds.) (1992) TAPPI Press, Atlanta, GA, pp. 491-500

5 T-die Film Casting

Toshitaka Kanai

5.1 Introduction

Recently, extrusion products such as films and sheets have come to be used in various fields including industrial products, food packaging, daily life products, and so on, and their application has been expanding even further with the introduction of new types of polymer materials. The T-die casting method, the most widely used for flat film forming, gives high productivity and good uniformity of film thickness. The T-die casting method is widely used in producing sheets and as a step in manufacturing sheets for thermoforming or biaxially oriented film.

Furthermore, new demands for highly functional films for use with multilayer technology, optical films for liquid crystal displays, and the like facilitate such expansions. Along with these expansions, there is a need to improve the quality of films. In particular, thickness uniformity is very important, which depends on the elongational condition.

In the film casting process, the molten polymer extruded through a slit die is solidified and taken up by a chill roll. The take-up velocity is faster than the extrusion velocity at the die exit, and the molten polymer is elongated between the die exit and the chill roll. The film properties are heavily dependent on this elongational condition.

In this process, the instability known as draw resonance sometimes occurs at a high drawdown ratio. The necking phenomenon occurs and the formation of thick edges called edge beads is observed. These edge beads, which need to be trimmed from the film, inhibit the thickness uniformity and the productivity of films.

Thus, due to the industrial importance of this process, many rheological models for the film casting process have been studied. Yeow [1] introduced an isothermal Newtonian model, and Aird and Yeow [2] introduced an isothermal generalized Newtonian (power law) model. Anturkar and Co [3] and Iyengar and Co [4] studied a viscoelastic model. Cotto et al. [5], Barq et al. [6, 7], Kanai [8], and Smith and Stolle [9] studied a nonisothermal model for the film casting process. d'Halewyn et al. [10] and Sakaki et al. [11] studied an isothermal Newtonian and membrane (two-dimensional) model that was mainly concerned with the simulation of the necking and the edge beads phenomena. Recently, Debbaut et al. [12], Silagy et al. [13, 14], and Sakaki et al. [15] studied an isothermal viscoelastic and membrane (two-dimensional) model.

As for experimental studies for the film casting process, Kase [16] introduced the first experimental research. Kanai and Funaki [17, 18] investigated the velocity profile, and Dobroth and Erwin [19] studied the necking phenomenon. Recently, Ito et al. [20] and Agassant et al. [21] experimentally studied the membrane behavior for the film casting process.

5.2 Film Casting Process

Figure 5.1 shows the equipment used in the film casting process. In this equipment, a molten polymer was continuously fed by an extruder, and plasticized molten polymer was extruded from a slit die. The extruded molten polymer (molten film) was stuck to the casting roll using a slit nozzle, called an air knife or air chamber, and cooled to solidify on the chill roll, and then it was wound by a winder.

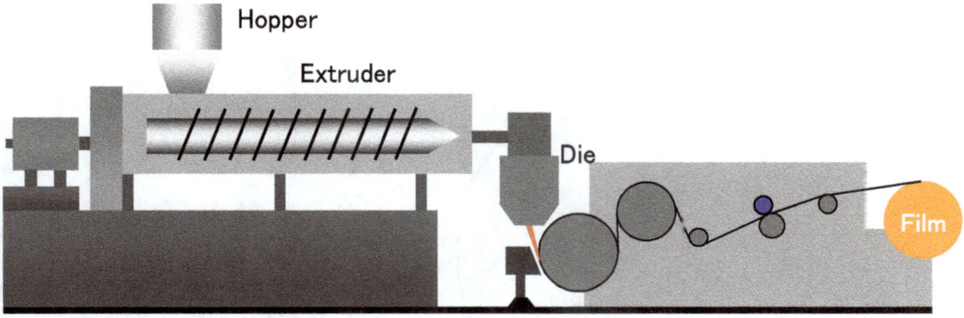

Figure 5.1 Schematic view of T-die casting film extrusion

Film thickness is controlled by output rate and the take-up speed. In general, the T-die casting method can produce film with good transparency and uniform thickness with a high production rate compared to the tubular film process, but the machine cost of T-die casting is more expensive than that of tubular film extrusion. Film physical properties are not very much influenced by the process conditions except for the impact strength, heat sealing temperature, optical properties, and friction coefficient. Recently, high-speed production has been required to supply the increasing demand and to reduce costs, so the film orientation in the machine direction (MD) is dominant and reduces the impact strength. For this reason, the polymer design that reduces the long relaxation time portion and the orientation of film is important. In the case of optical usage, the film is required to have uniform thickness, surface flatness, no scars, small necking, and low optical anisotropy.

5.3 Theoretical Analysis of Film Deformation under a Steady State

As the production rate has been increasing and the film thickness has been getting thinner, there are many problems such as instability, film breakage during continuous production, and the necking phenomena causing narrower film width. In order to explain these phenomena, a theoretical analysis of the film casting process using a simple model is introduced [16, 20].

A polyolefin melt such as polypropylene (PP) and polyethylene (PE) is usually cast vertically from the die onto a chill roll. The mathematical model for such vertical casting as shown in Fig. 5.2 is derived by Kanai [8].

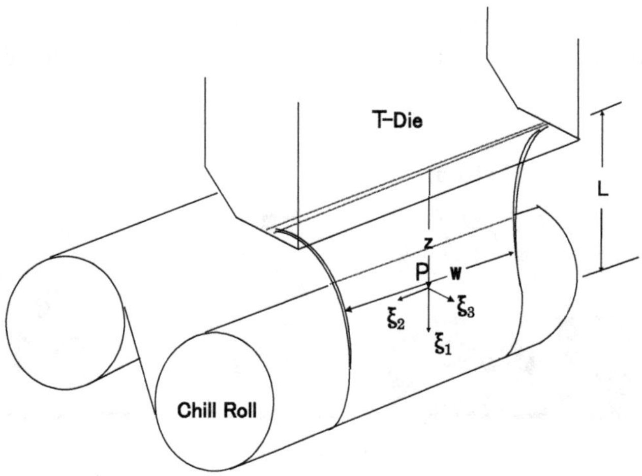

Figure 5.2 Vertical casting process of film

The schematic view of T-die casting film and sheet is shown in Fig. 5.2. ξ_1, ξ_2, and ξ_3 are Cartesian coordinates at an arbitrary point P; the subscripts 1, 2, and 3 denote the machine (flow), transverse (width), and thickness directions, respectively; w, h, and v are width, thickness, and velocity of film at P; z is the distance from a die to point P.

The force balance equation and energy balance equation during the T-die casting process are described as follows. These equations assume that the molten polymer, which follows the power law fluid, deforms between the die and the chill roll located downwards.

$$\frac{4Q\eta(\Pi)}{h}\frac{dh}{dz} = F + \int_{Z}^{L} wh\rho dz \tag{5.1}$$

$$\rho QC_p \frac{dT}{dz} = -2wU\left(T - T_{air}\right) - 2\varepsilon\lambda w\left(T^4 - T_{room}{}^4\right) \tag{5.2}$$

$$\eta = A\exp\left(E / RT\right)\Pi_d{}^{(n-1)/2} \tag{5.3}$$

where F is a take-up force at the roll, h is the film thickness, w is the film width, T is the film temperature, Q is the output rate, ρ is the film density, C_p is the specific heat, and R is the gas constant. η is the molten polymer viscosity; the strain rate is defined as $\Pi_d = d_{11}{}^2 + d_{22}{}^2 + d_{33}{}^2$; A is the viscosity constant; E is the activation energy; U is the heat transfer coefficient; n is the power law constant of elongational viscosity; ε is the emissivity; λ is the Stefan-Boltzmann constant, and T_{room} is the room temperature.

Heat transfer coefficient U is obtained from the dimensionless Equation 5.4 [8]:

$$\frac{UL}{K_{air}} = 0.664\left(\frac{v_{ave}L\rho_{air}}{\eta_{air}}\right)^{1/2}\left(\frac{C_{air}\eta_{air}}{K_{air}}\right)^{1/3} \tag{5.4}$$

where K_{air} is the thermal conductivity, ρ_{air} is the air density, η_{air} is the air viscosity, L is the air gap, v_{ave} is the film mean velocity, and C_{air} is the specific heat of air.

By solving the above equations, deformation velocity and rate, temperature, and stress distributions can be obtained as a function of activation energy E and power law index n.

5.4 Deformation Behavior of T-die Casting

The predicted results for the deformation behavior of polypropylene melt when casting it vertically from the die to the chill roll under a steady state are shown in Fig. 5.3 and Fig. 5.4. The process conditions are output rate Q = 1 kg/h, take-up velocity v_L = 3 m/min, die width w = 155 mm, lip clearance h_0 = 1.26 mm, air gap L = 80 mm, resin temperature at the die exit T_0 = 227°C, resin viscosity at die exit η_0 = 3000 poise, activation energy E = 10.7 kcal/mole, and power law index n = 1. As seen from Fig. 5.3, the film velocity v increases exponentially with increasing distance from the die exit z = 0, and as a result, the film thickness h decreases rapidly. Then the film temperature T decreases gradually because of the convection of air and radiation from molten polymer.

Figure 5.4 shows the strain rate d_{11} and stress σ_{11} in the machine direction. From those figures, it can be seen that the strain rate d_{11} increases with distance from the die exit and its increment decreases gradually near the landing point, while the stress σ_{11} increases exponentially because of the increase in the strain rate and viscosity.

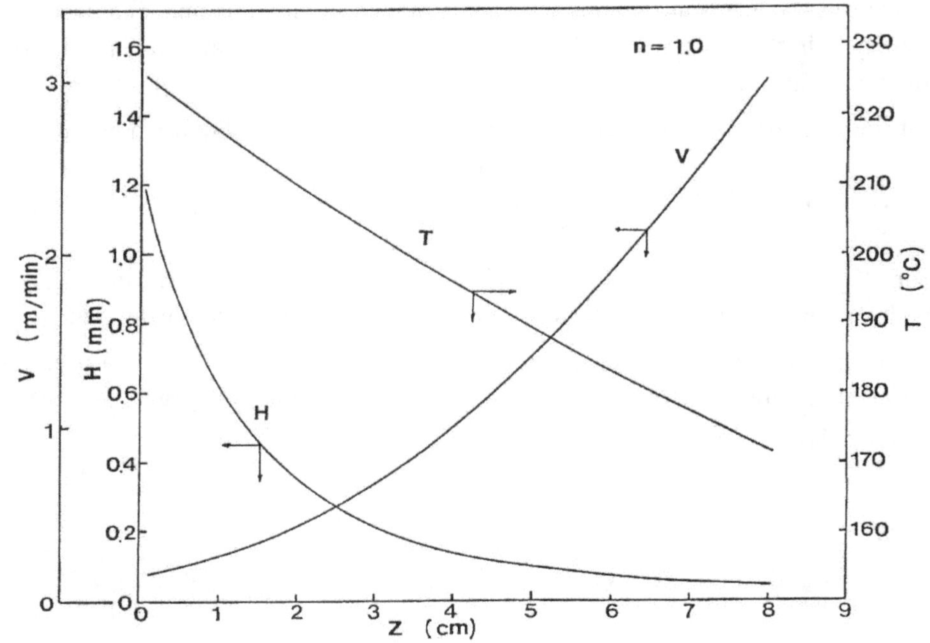

Figure 5.3 Predicted velocity, film thickness, and temperature along the length of flat film for PP with Q = 1 kg/hr, W = 155 mm, H_0 = 1.26 mm, T_0 = 227°C, air gap L = 80 mm

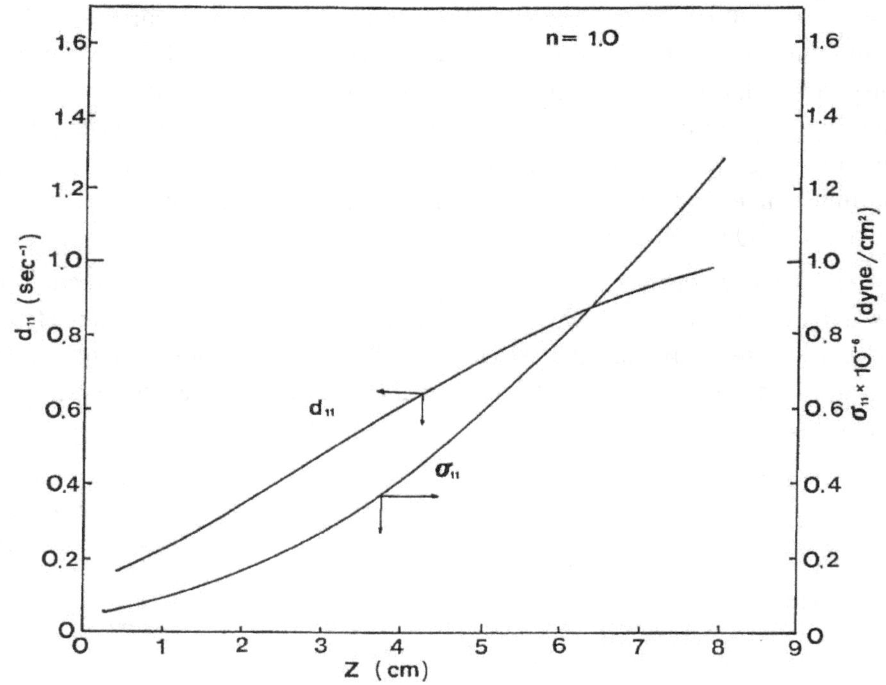

Figure 5.4 Predicted deformation rate d_{11} and take-up stress σ_{11} along the length of flat film for PP with Q = 1 kg/hr, W = 155 mm, H_0 = 1.26 mm, T_0 = 227°C, air gap L = 80 mm

Figures 5.5 and 5.6 show the change in the film velocity v, strain rate d_{11}, and stress σ_{11} as a function of take-up speed v_L. Each curve of the film velocity and strain rate similarly increases with increasing take-up speed. The curve of stress is similar to that of film velocity, but the change of stress with increase in take-up speed is greater than that of film velocity.

These data, such as stress, strain rate distribution, and temperature distribution, are closely related to film physical properties, film breakage, and instability.

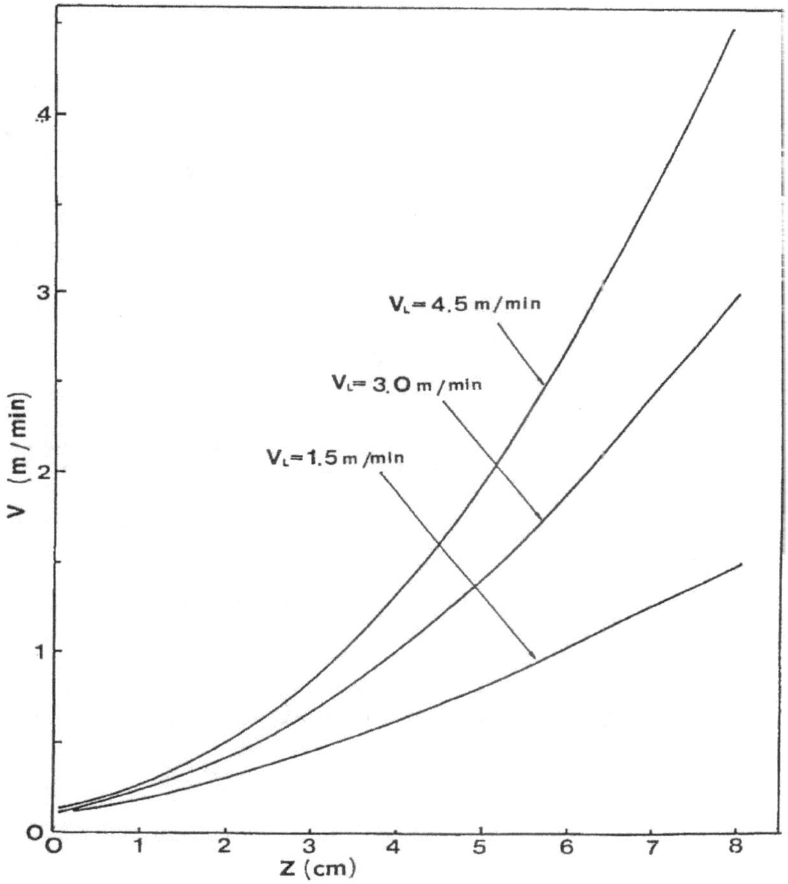

Figure 5.5 Predicted velocity profiles of flat film for different take-up velocities for PP (η_0 = 30,000 poise, n = 1.0) with Q = 1 kg/hr, W = 155 mm, H_0 = 1.26 mm, T_0 = 227°C, air gap L = 80 mm

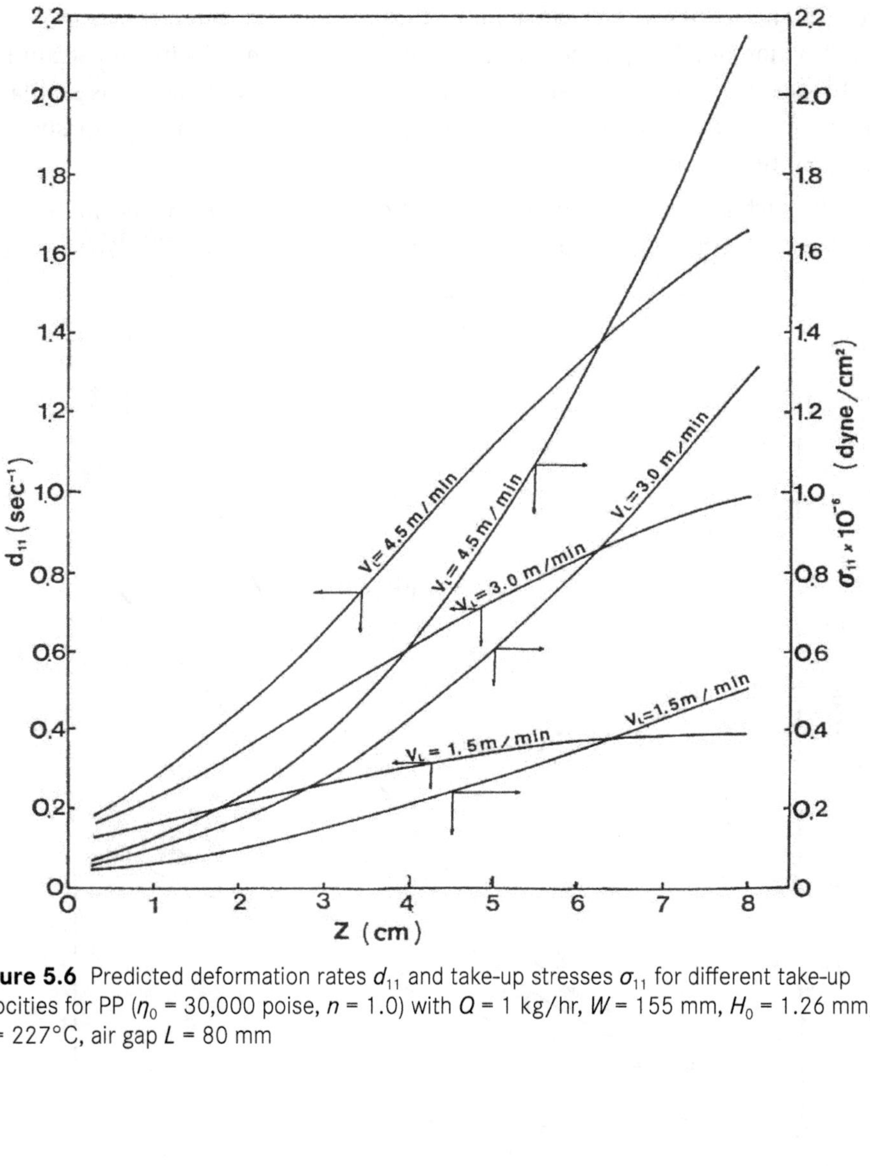

Figure 5.6 Predicted deformation rates d_{11} and take-up stresses σ_{11} for different take-up velocities for PP (η_0 = 30,000 poise, n = 1.0) with Q = 1 kg/hr, W = 155 mm, H_0 = 1.26 mm, T_0 = 227°C, air gap L = 80 mm

■ 5.5 Draw Resonance

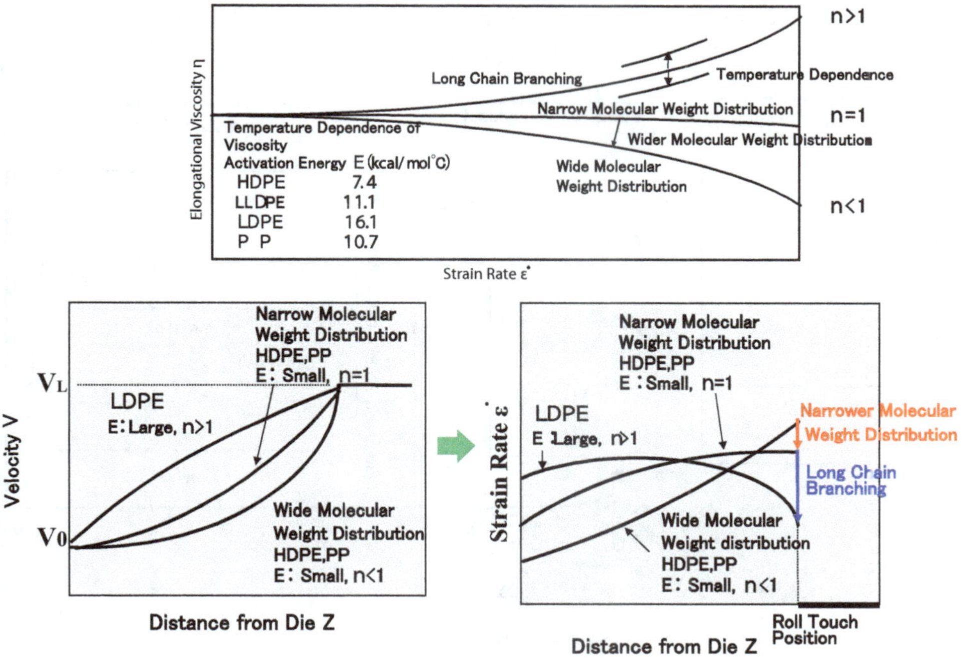

Figure 5.7 Relationship between rheological properties and deformation behavior

The predicted results of deformation velocity and strain rate obtained by using the above theoretical equations are shown in Fig. 5.7 for various polyolefins such as PP, LDPE, LLDPE, and HDPE. Under high-speed conditions, instability, known as draw resonance, occurs [17, 18, 23, 24]. This oscillation instability [23, 24] is very much influenced by the factors of the drawdown ratio and strain rate parameter defined by Equation 5.5, which refers to the dimensionless parameter of strain rate at the roll touch position [18]. The high strain rate change occurs at the roll touch position, because the strain rate of molten polymer is very high before the roll touch position and then is almost zero after the roll touch. The discontinuity of high strain rate causes instability of the film (Fig. 5.8) [17, 18].

$$\varepsilon_L = \frac{L}{v_L}\left(\frac{dv}{dz}\right)_{z=L} \tag{5.5}$$

Following this concept, a polymer that is easy to deform near the die, exhibiting strain rate hardening and high activation energy of elongational viscosity (power law index *n* is large), has a wider process window and can be produced at a higher speed without instability (Fig. 5.9). This could be a long chain branching polymer such as LDPE, or, if no long branching exists, a polymer with narrow molecular weight distribution or low molecular weight. In contrast, a polymer with wide molecular weight distribution shows the strain rate softening and easily becomes unstable. High cooling speed by an air knife leads to better stability due to deformation occurring near the die rather than near the roll, because elongational viscosity increases rapidly by cooling from high speed air [18, 25].

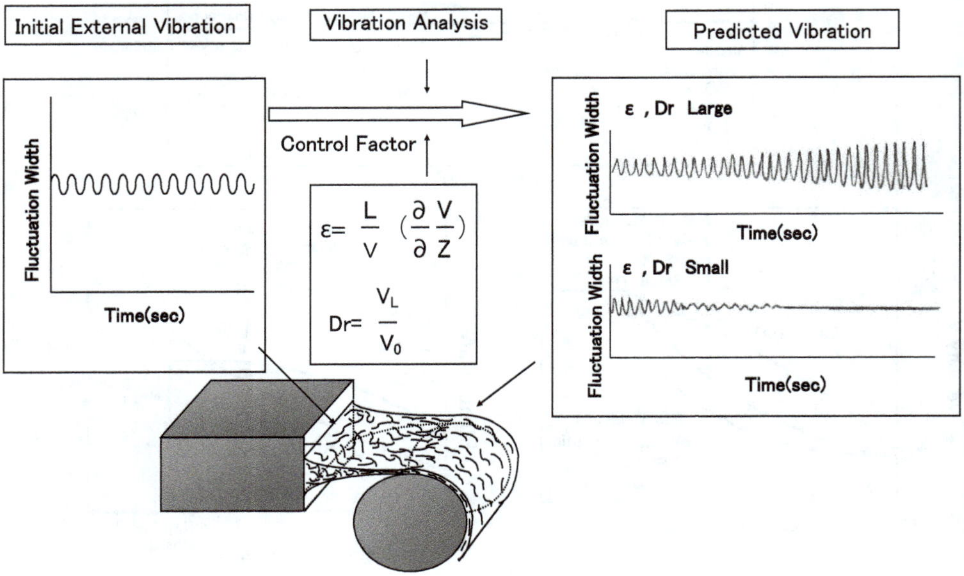

Figure 5.8 Draw resonance

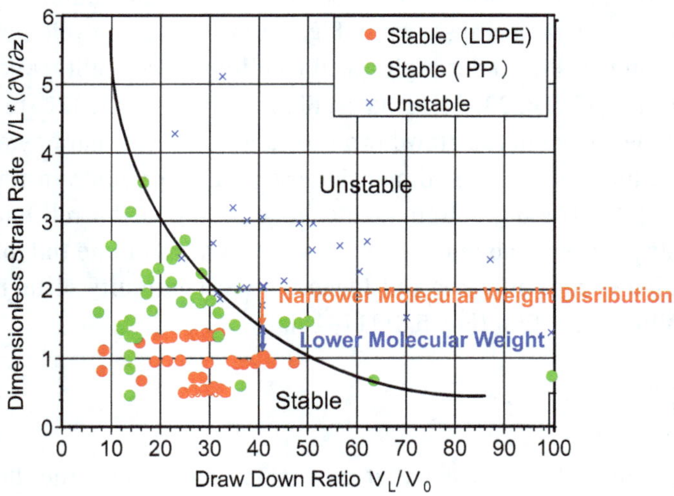

Figure 5.9 Relationship between strain rate, draw down ratio, and stability

■ 5.6 Film Breakage

Cohesive failure of film during T-die film casting of polypropylene occurs above the critical stress shown in Fig. 5.10; molecular chains can no longer reorganize to relieve the local stress in the time frame imposed by the deformation. The elastic molecular chain entanglement may cause film breakage.

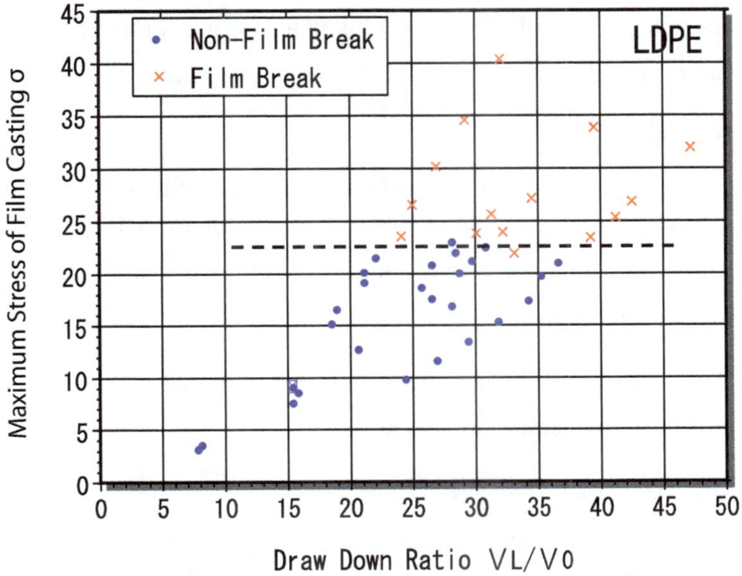

Figure 5.10 Relationship between maximum stress and film breakage

For this reason, polymer having long chain branching or a high molecular weight portion breaks under high stretching stress conditions during the T-die casting process. High air speed cooling leads to good stability and high stretching stress. So it can be seen that if high stretching stress occurs in the film, it will break if the stress is over the critical stress [18, 25].

Recently, high production speed has been increasing and this condition causes a higher stretching stress in MD, so the film possesses anisotropic properties between the MD and TD (transverse direction). In particular, the residual stress that is frozen in during film processing is relaxed with difficulty for a film with a high molecular weight portion, so the highly oriented film in MD is produced and has low impact strength. For this reason, the polymer having a narrow molecular weight distribution is preferable. It is important to design the polymer properly for each application.

■ 5.7 Necking Phenomenon

The phenomenon in which film width becomes narrower than the die width is called "necking," as shown in Fig. 5.11. As the necking phenomenon makes film width narrower and the film thickness is thus worse near both the edges, this necking phenomenon is undesirable. This necking is very much influenced by the viscoelastic property of the polymer. If the viscoelastic model such as the Giesekus model [26, 27] is used, a good prediction of necking behavior is obtained [22, 23].

The constitutive equation of the Giesekus model is expressed as

$$\left[\mathbf{I}+\left(\alpha\lambda/\eta\right)\mathbf{T}\right]\cdot\mathbf{T}+\lambda\mathbf{T}=2\eta\mathbf{D} \tag{5.6}$$

where η is viscosity, α is a parameter of elongational viscosity, λ is the relaxation time, $\mathbf{T}$ is the extra stress tensor, and $\mathbf{D}$ is the deformation rate tensor.

The relaxation time with shear rate dependence is expressed as

$$\lambda(\dot{\gamma})=N_1(\dot{\gamma})/2\eta(\dot{\gamma})/\dot{\gamma} \tag{5.7}$$

where $N_1(\dot{\gamma})$ is the primary normal stress and $\eta(\dot{\gamma})$ is viscosity.

Then the relaxation time equation is expressed as

$$\lambda(\dot{\gamma})=\lambda_{\infty}+\left(\lambda_0-\lambda_{\infty}\right)(1+\varsigma_2{}^2\dot{\gamma}^2)^{(m-1)/2} \tag{5.8}$$

where λ_0 is the zero shear rate relaxation time, λ_{∞} is the infinite shear rate relaxation time, ζ_2 is the constant number, and m is the power law index.

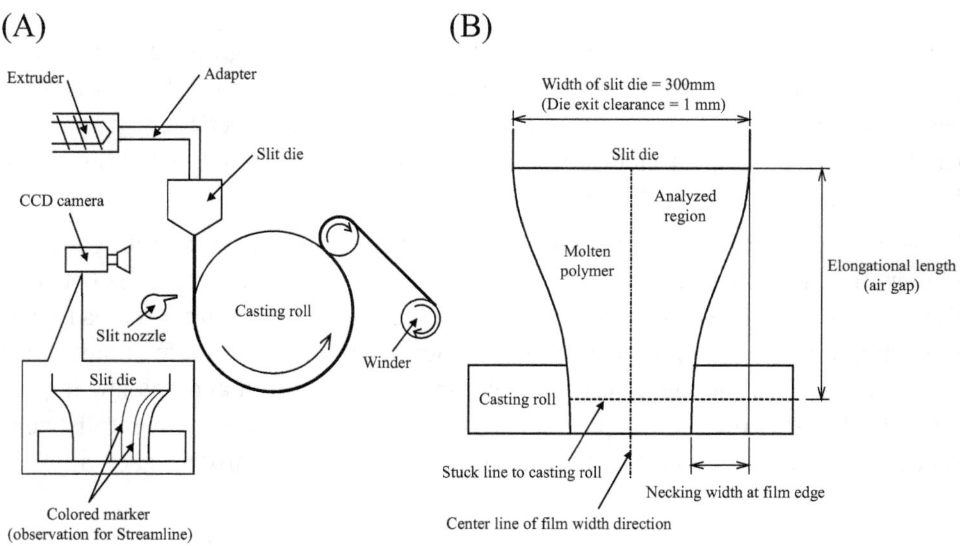

Figure 5.11 Equipment for film casting process: (A) measuring system, (B) analyzed region

The theoretical prediction suggests that a polymer with strong strain hardening in elongational viscosity (parameter α is small) shows weak necking, that is, wider film width (Fig. 5.12) [28]. Figure 5.13 shows that half the area of the film casting process is between the die and the chill roll, and the upper portion is the die and the lower portion is the chill roll. The figures on the left are experimental results of PP and LLDPE, and the ones on the right are the theoretical results [28].

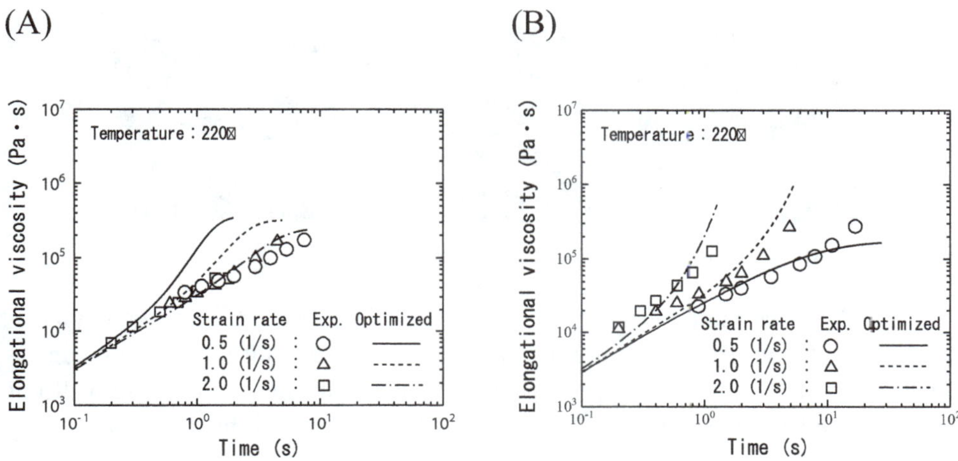

Figure 5.12 Relationship between time and elongational viscosity for (A) polypropylene, (B) low density polyethylene

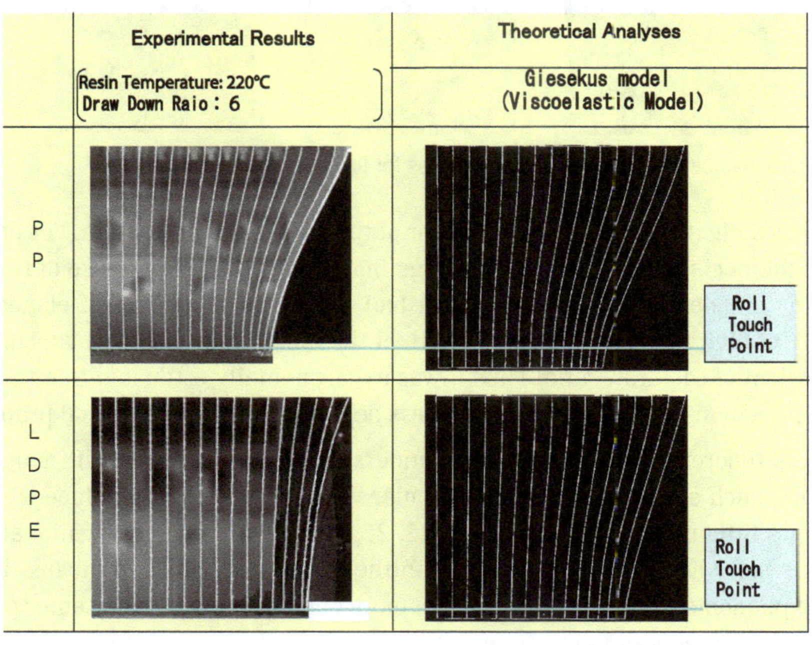

Figure 5.13 Comparison of experimental results with theoretical ones using the Giesekus model for necking phenomena

Figure 5.14 shows the typical behavior of the casting process for polypropylene. In Fig. 5.14, the casting speed (draw ratio) was changed under these fixed conditions: die width 300 mm, die exit (die lip) clearance 1 mm, elongational length (air gap) 150 mm, and extruded flow rate 15 kg/h. The necking width at the film edge increased with the increase of the draw ratio.

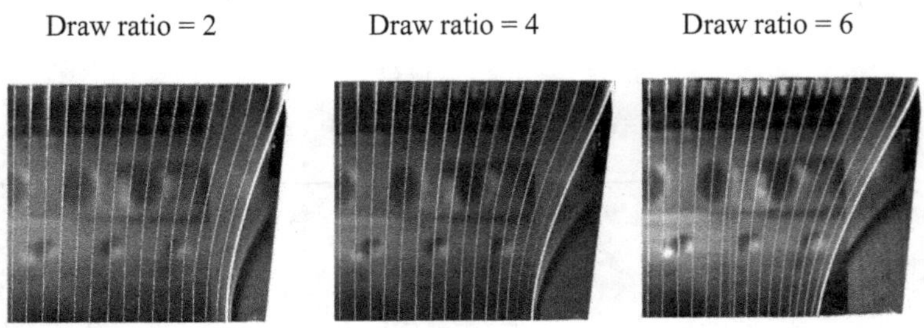

Figure 5.14 Typical behavior of casting process for polypropylene

Figure 5.15 shows the typical behavior of the casting process for low density polyethylene. Although the draw ratio increased, the necking width at the film edge is almost constant. Unlike the characteristics of polypropylene, the necking width at the film edge for low density polyethylene was independent of the draw ratio.

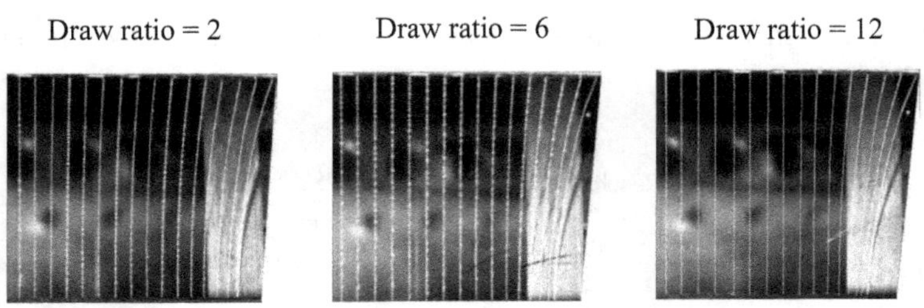

Figure 5.15 Typical behavior of casting process for low density polyethylene

There was a big difference in the behavior of the necking phenomenon. For polypropylene, the necking width at the film edge increased with the increase of the draw ratio. For low density polyethylene that had the notable increase of elongational viscosity at high strain, the necking width at the film edge was almost constant and independent of the draw ratio. Thus it was experimentally estimated that the necking phenomenon in the film casting process depends on the elongational properties.

From this theoretical analysis, we can understand that a polymer with long chain branching such as LDPE or a high molecular weight portion can produce wide film and shows little necking phenomenon [22, 25, 28]. The uniaxial elongational deformation is more dominant at the edge than the center position. This means that the necking phenomenon at the edge is much more dominant than at the center portion, so the wide die reduces the necking ratio.

As this analysis can predict the thickness distribution and velocity distribution along the width direction, it can predict increased thickness at the edge [28] (Fig. 5.16), low deformation speed at the edge, stress, and birefringence distributions. That is to say, the necking causes unbalanced shrinkage, a nonuniform optical axis, and a bad yield ratio. To reduce necking, a high cooling speed, a short air gap in the process conditions, and a strain hardening property as a rheological aspect are required.

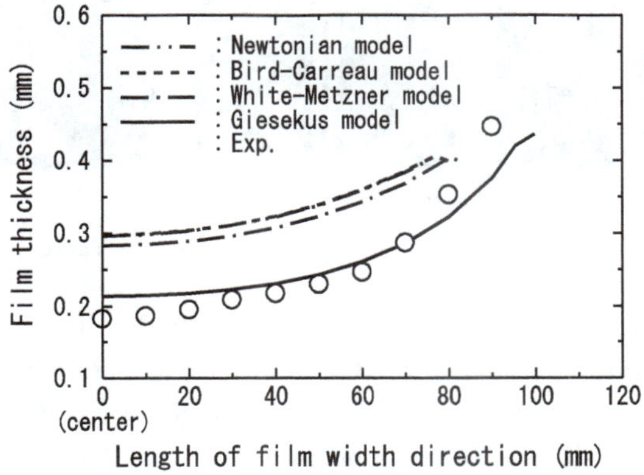

Figure 5.16 Experimental result and simulation results of film thickness in casting process for polypropylene under draw ratio of 6

■ 5.8 Surface Roughness Caused by Shark Skin and Melt Fracture

Two typical flow instabilities of "shark skin" and "melt fracture" sometimes occur and cause inferior surface roughness of film as shown in Figs. 5.17 and 5.18 [29].

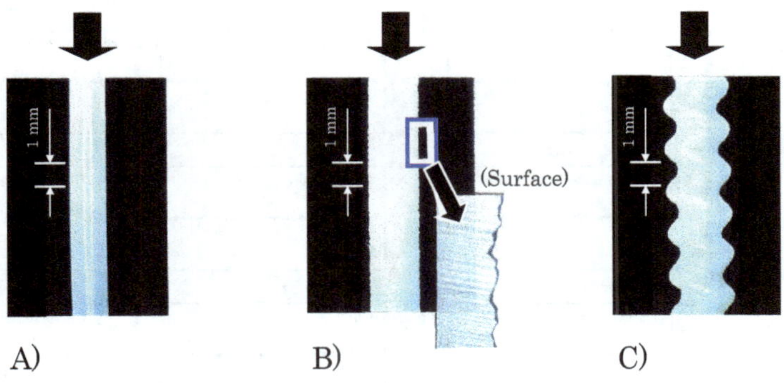

Figure 5.17 Photograph of typical extrudate of (A) stable flow, (B) shark skin, and (C) melt fracture

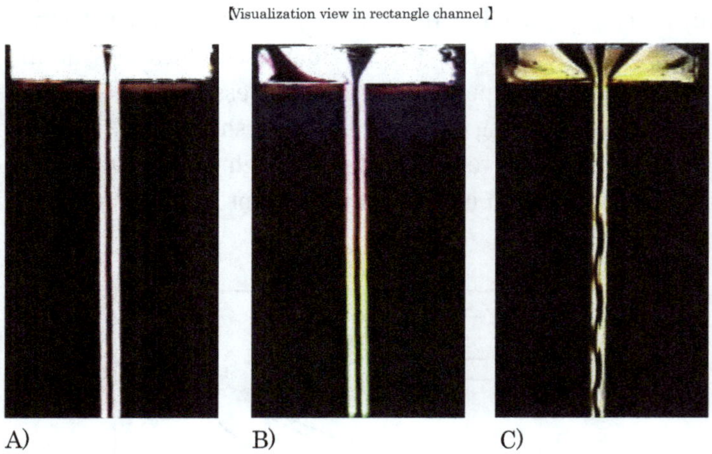

Figure 5.18 Streamline of (A) stable flow, (B) shark skin, (C) melt fracture in rectangle channel

Experimental measurements were carried out methodically in a wide flow region from the reservoir (entry region) to the die land with a specific slit die. In particular, the flow pattern was visualized with a high-speed camera and the frequency of the flow instability was analyzed. Corresponding to the visualization analysis, the pressure fluctuation in the slit die and the surface roughness at the die exit were measured simultaneously. The data are shown in Fig. 5.19. These frequency characteristics were analyzed with an FFT (Fast Fourier Transform) analyzer as shown in Fig. 5.20 [29].

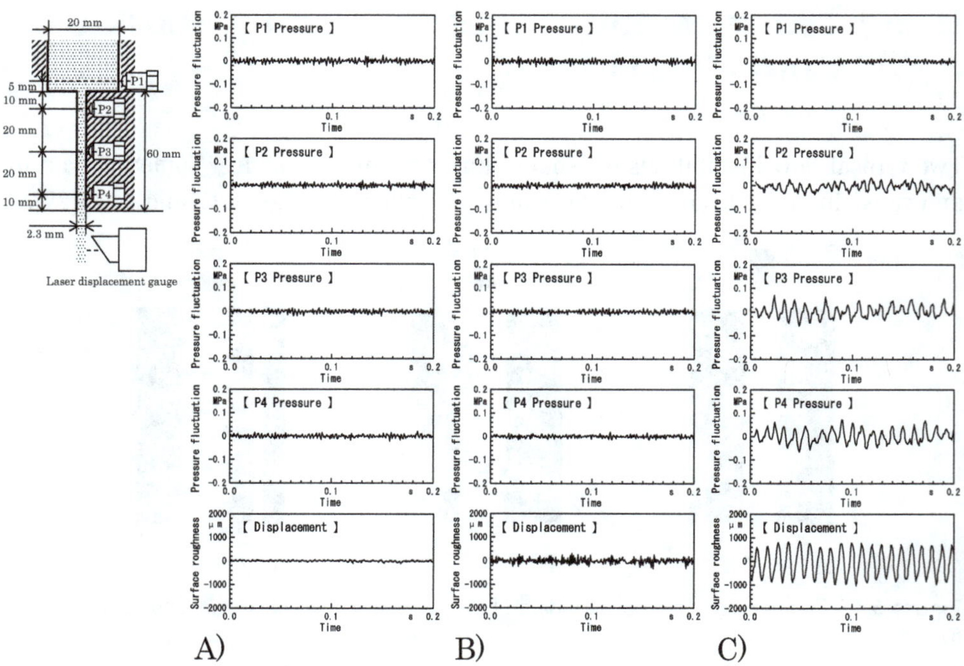

Figure 5.19 Pressure fluctuation and surface roughness of (A) stable flow, (B) shark skin, and (C) melt fracture

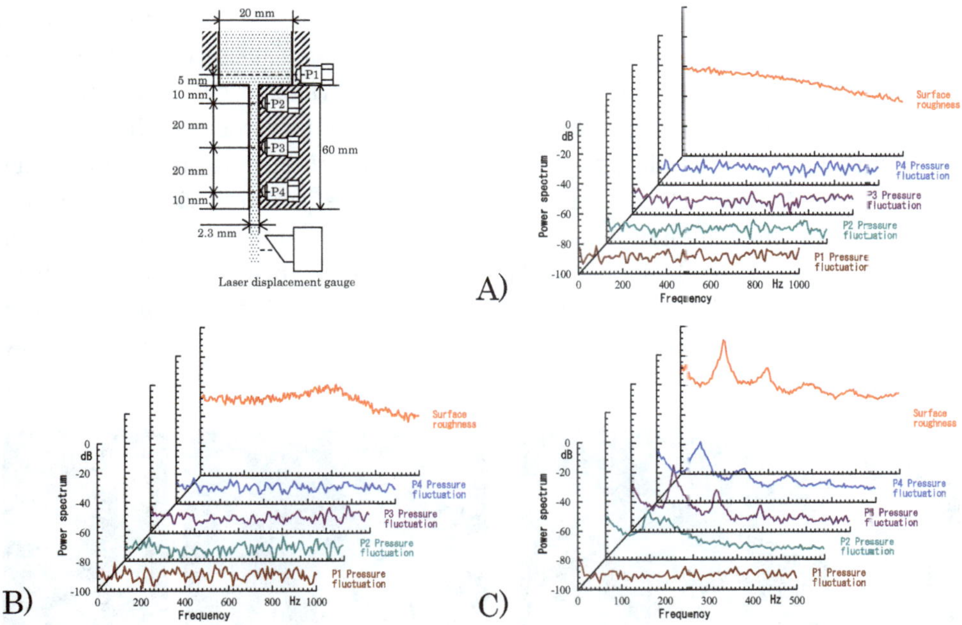

Figure 5.20 Relationship between frequency and power spectrum of pressure fluctuation and surface roughness of (A) stable flow, (B) shark skin, and (C) melt fracture

As for the "shark skin" region [29, 30], the irregularity of the surface roughness in the broad frequency band was observed at the die exit. In spite of this surface roughness, the pressure fluctuation and the fluctuation of the flow pattern at the wall vicinity in the die land were not observed. Shark skin occurs at the die exit. Figure 5.21 shows that the occurrence of shark skin (surface roughness) depends on the wall shear stress. The fluctuation becomes the surface distortion of extrudate because of the removal of shear stress at the exit of the flow channel.

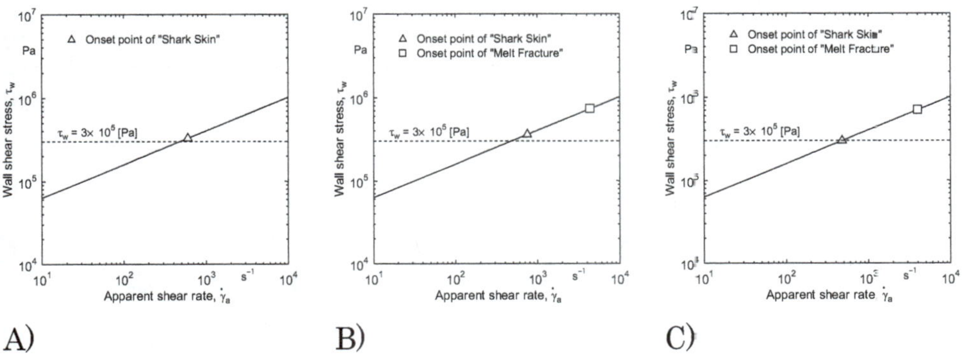

Figure 5.21 Normal stress distribution in circular channel with entrance angle of (A) 15°, (B) 90°, and (C) 180° under the same flow

As for the "melt fracture" region, the periodic oscillation of the flow pattern and the periodic pressure fluctuation that correspond to the periodic distortion of the extrudate were recognized in the die land. However, the periodic oscillation of the flow pattern and the periodic pressure fluctuation could not be recognized in the reservoir (entry region of the die land). "Melt fracture" is initiated at the die entry and occurs in the die land. The occurrence of melt fracture (severe periodic distortion) depends on the maximum normal stress of flow direction at the point of entrance of the circular flow channel. Figures 5.22 and 5.23 show the rheological data and theoretical data predicted by using a viscoelastic model [29]. The mechanism of melt fracture relates to the elongational behavior at the point of entrance of the flow channel.

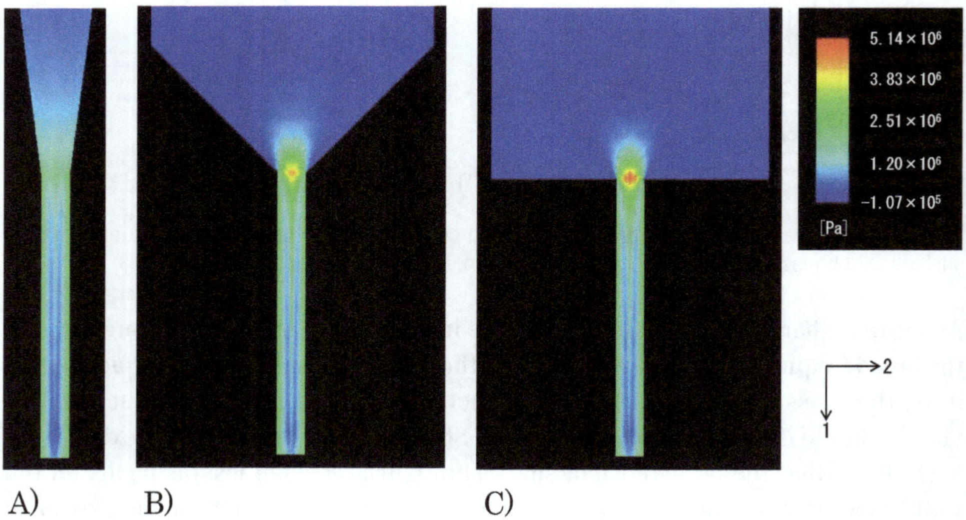

Figure 5.22 Normal stress distribution in a circular channel with entrance angle of (A) 15°, (B) 90°, and (C) 180° under the same flow

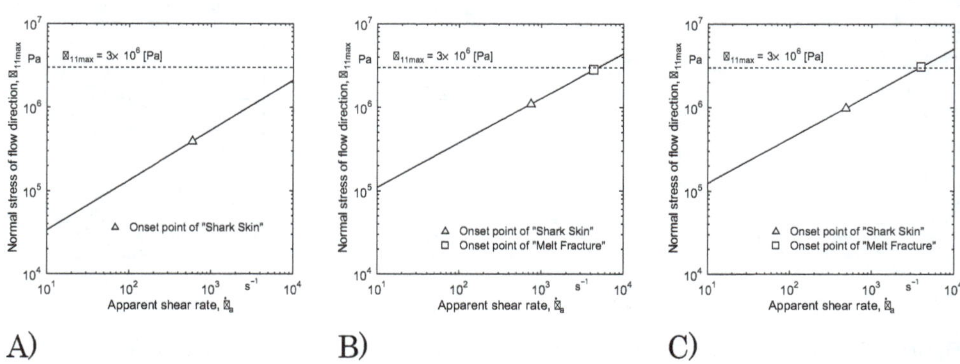

Figure 5.23 Relationship between apparent shear rate and maximum normal stress of flow direction in a circular channel with entrance angle of (A) 15°, (B) 90°, and (C) 180°

5.9 Film Physical Properties

5.9.1 Influence of Process Conditions

The influence of process conditions on the film physical properties were studied [25], and the results are listed in Table 5.1. Generally, it can be seen from Table 5.1 that compared to the processing factors in tubular film operations, those in T-die casting exert little influence on the mechanical properties of film, but they exert some influence on the heat sealing temperature of film, optical properties and the coefficient of static fiction. In Fig. 5.24, the relationships between (a) the film thickness and the coefficient of static fiction, (b) the polymer temperature and the haze, and (c) the chill roll temperature and the haze are shown.

Table 5.1 Influence of Process Conditions on the Film Physical Properties

	Direction	Tensile Modulus		Tensile Strength		Tear Strength		Heat Seal	Gloss	Impact Strength
		MD	TD	MD	TD	MD	TD	Temperature		
Resin Temperature	↑	↓	↓	↓	↓	↑	↓	↓	↑	↑
Out-Put Rate	↑	→	→	→	→	↓	→	→	→	→
Chill Roll Temperature	↑	↑	↑	↑	↑	↓	↓	↑	↓	↓
Film Thickness	↑	↓	↓	↓	↓	↑	↓	↑	↓	↑

The higher cooling temperature of the film causes higher crystallinity. As a result, the heat sealing temperature becomes higher and the optical properties deteriorate. Then the coefficient of static friction becomes lower because the surface becomes uneven due to crystallization, and the modulus becomes higher because of the high crystallinity.

The higher polymer temperature creates an improvement in the optical properties because it leads to the lower shear stress applied to polymer in the die and the higher relaxation time of polymer at the die slit. This, in turn, leads to a higher impact strength but lower modulus because of the lower crystallinity.

The film thickness affects the optical properties of film, and the coefficient of static friction sharply decreases with a greater thickness.

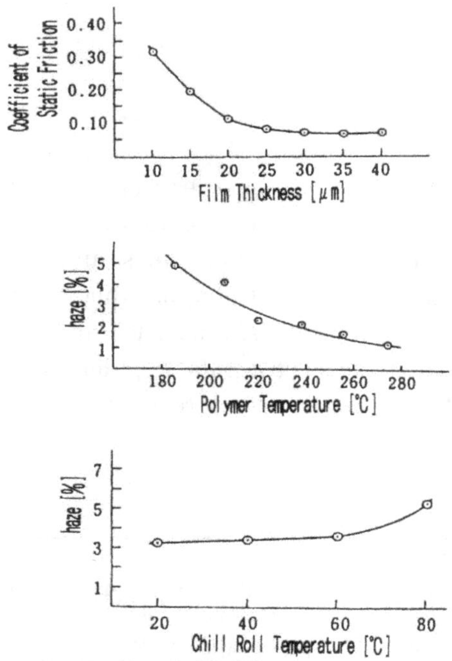

Figure 5.24 Relationship between process conditions and film physical properties

5.9.2 Influence of Polymer Design

The influence of polymer design on the film physical properties are introduced via an example of a polyolefin that is widely used for the T-die casting process. Take LLDPE for example. Table 5.2 show the relationship among the molecular structure and superstructure of LLDPE, processability, and film physical properties.

Table 5.2 Relationship among the Molecular Structure, Super Structure, and Film Properties of LLDPE

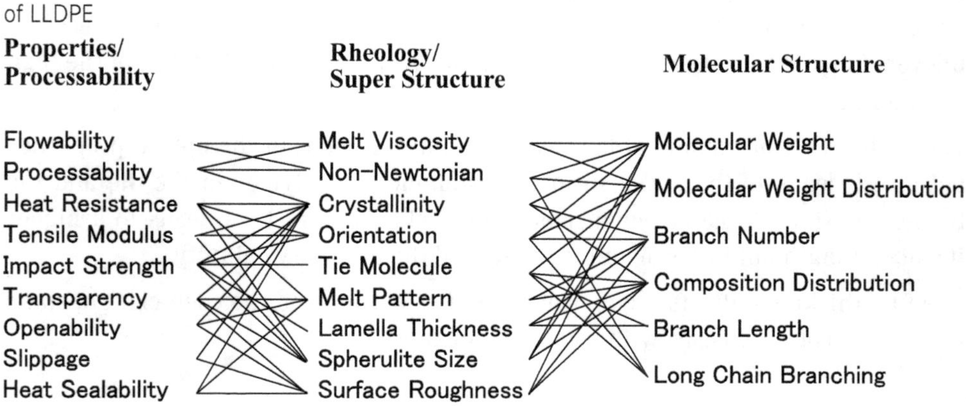

5.9.2.1 Branching and Properties of LLDPE

The film physical properties of LLDPE such as impact strength and tensile strength are influenced by composition distribution and branching length. In particular, branch length influences crystallization speed and crystallinity. The branching number per 1000 carbons of density 0.920 is as follows for different comonomers:

C3: 25–30/1000C

C4 15–20/1000C

C5 10–20/1000C

For example, if the branching number 20/1000C is introduced uniformly in the polymer, the continuous ethylene length is 50 CH_2 units, and the planar zigzag structure of PE crystalline reaches 60 Å, which is much thinner than HDPE lamella thickness of 100 Å.

5.9.2.2 Impact Strength

Film strength is related to stress acquired from the deformation of lamella to the fibril and cutting the tie molecule chain between lamellas. Because the long branch such as hexyl cannot exist in a lamella, the lamella thickness becomes thin. LLDPE having many long branches, which cannot exist in a lamella, has a structure of random reentry type rather than adjacent reentry type in Fig. 5.25. For this reason, LLDPE using high α olefin as a comonomer has many tie molecules and high strength.

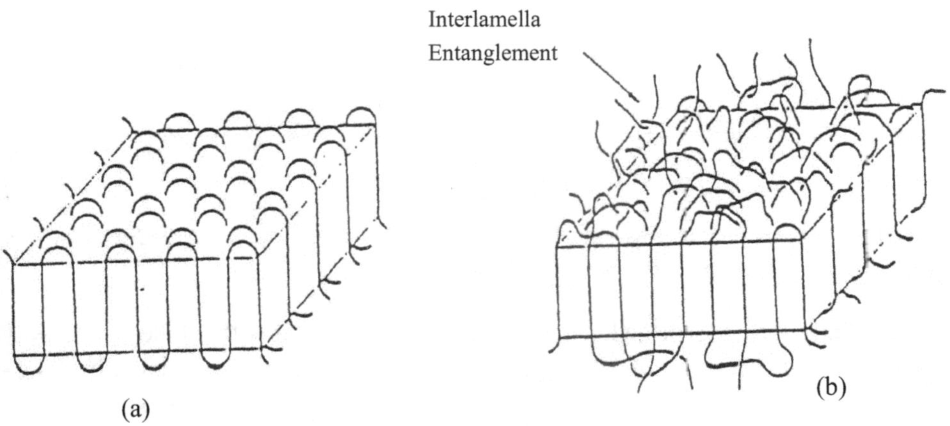

Figure 5.25 Lamella folding structure; (a) adjacent re-entry type; (b) random re-entry type

Table 5.3 shows the characteristics of a single-site catalyst LLDPE. As branches are introduced at uniform intervals and its lamella thickness becomes thin and the tie molecule number between lamellas increases, the impact strength increases. As shown in Fig. 5.26, low-density LLDPE having many branches has high elongation, and further C8 branches introduced uniformly by a single-site catalyst leads to high impact strength.

Table 5.3 Characteristics of a Single Site Catalyst LLDPE

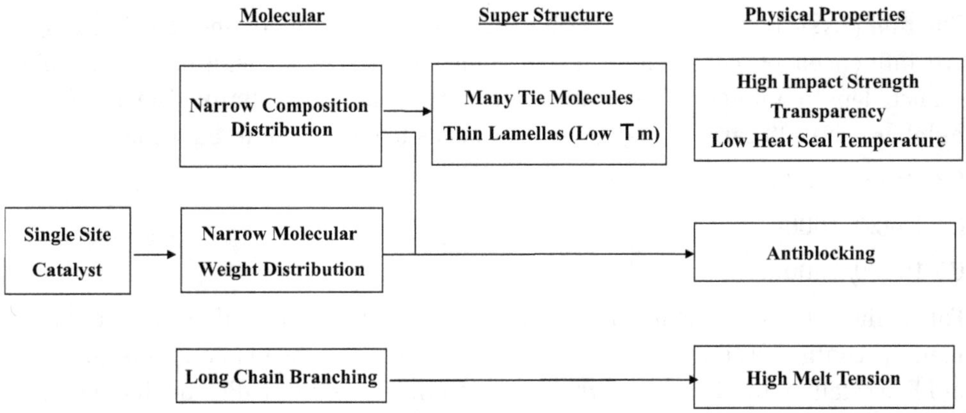

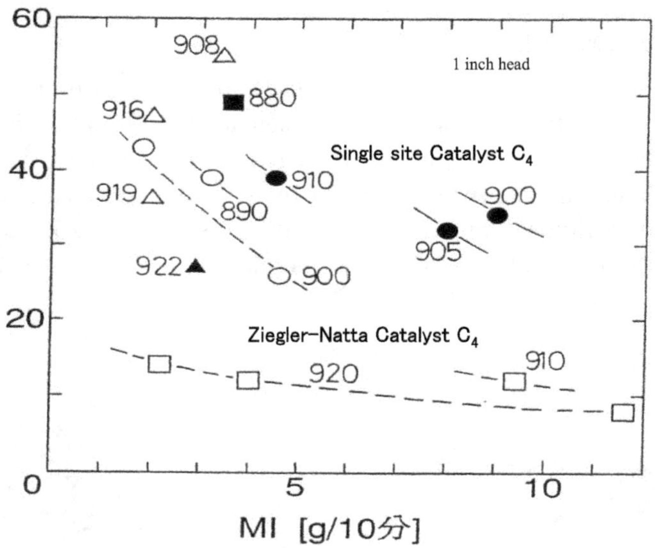

Figure 5.26 Comparison of single-site catalyzed PE film impact strength with Ziegler–Natta catalyzed PE film impact strength (Number: Density kg/m³)

The line speed of polypropylene has been increasing, and the take-up velocity reaches over 200 m/min, causing high orientation in MD. For this reason, film orientation between MD and TD is unbalanced, and this reduces the impact strength.

To restrain the film's impact reduction, PP having a narrower molecular weight distribution and a shorter relaxation time is used. The relationship between relaxation time λ and film impact strength is shown in Fig. 5.27, and it is found that the factors are closely related to one another.

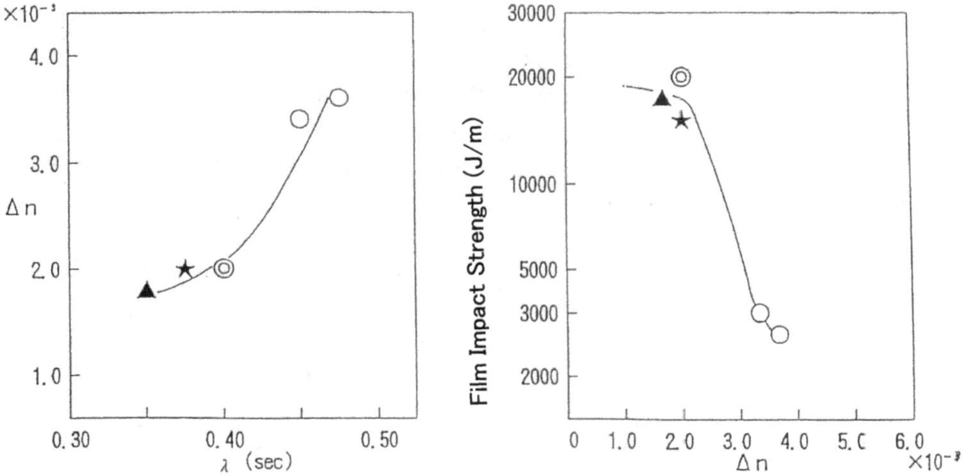

Figure 5.27 Relationship among relaxation time λ, birefringence Δn, and film impact strength. Process conditions: resin temperature 255°C, chill roll temperature 40, take-up velocity 125 m/min

5.9.2.3 Heat Seal Temperature

Heat seal temperature is one of the important film properties. Low heat seal temperature has merit for the secondary process, which can give a high production rate. In the case of LLDPE, heat seal temperature is influenced by the hexane-soluble portion and the low melt temperature portion. Because the hexane-soluble portion, which is shown in Fig. 5.28, is oily and not crystallized at room temperature, it bleeds out and remains on the film surface, and thus it reduces the heat seal strength. The low melt portion contributes to the heat seal at low temperature, and is molten between films and molecular chains entangled each other. Because the data from TREF (temperature rising elution fractionation) is usually 30°C lower than the corresponding melt data (heat seal temperature -30°C), the composition below 60°C is thus evaluated. An example of this type of analysis is proposed by the following equation.

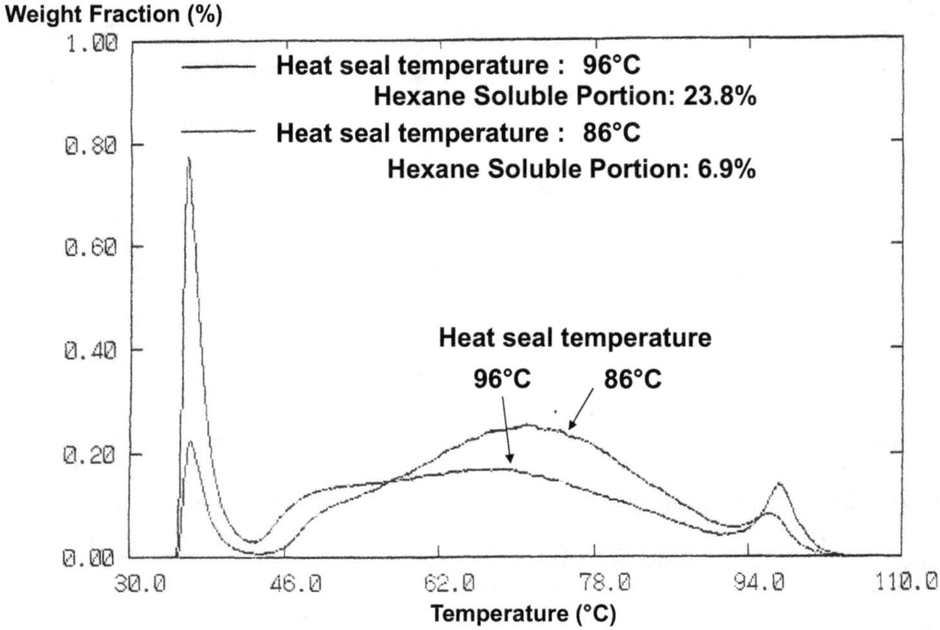

Figure 5.28 Relationship between TREF and heat seal temperature

$$\begin{aligned}&\text{Heat seal temperature} = -0.287 \times (\text{TREF component below } 35^\circ\text{C} \\ &-\text{C6 soluble portion}) \\ &-0.454 \times (\text{TREF component between } 35^\circ\text{C and } 60^\circ\text{C}) \\ &+8.6 \times (\text{hexane-soluble portion}) - 52.1 \times \text{MFR} + 138.6\end{aligned} \tag{5.9}$$

This equation suggests that the low heat seal temperature of this LLDPE is a consequence of reducing the low hexane soluble portion (noncrystalline portion at room temperature) and increasing the low melt temperature portion as indicated by TREF. Multisite catalysts increase the hexane-soluble portion more than the single-site catalyst as LLDPE density is below 0.910. In general, the single-site catalyst more readily produces LLDPE with a low heat seal temperature than a multisite catalyst.

Low heat seal temperature of polypropylene is also desirable. The idea of random copolypropylene heat seal temperature is similar to LLDPE and is influenced by the solvent-soluble portion and melt portion at a temperature below 100°C. A solvent-soluble portion increases heat seal temperature, and a melt portion below 100°C decreases heat seal temperature.

5.9.2.4 Blocking and Slippage

The film openability is influenced by the film surface roughness, bleeding on the surface, and elasticity. The surface roughness is dependent on HDPE composition, bleed on the surface depends on the hexane-soluble portion, and elasticity is dependent on density. For this reason, for the purpose of achieving both easy openability and antiblocking properties, the low temperature soluble portion is slightly shifted to a higher temperature, and at the same time the content of the hexane-soluble portion, which means a large part of the branching and the low molecular weight, should be decreased.

5.9.2.5 Transparency

In general, high transparency is preferable for film usage except for HDPE film requiring high haze. In order to increase transparency, it is important to reduce both extrusion haze and internal haze. Low extrusion haze is obtained by reducing the molecular weight distribution or the high molecular weight portion to get short relaxation during extrusion. Low internal haze is obtained by reducing the number of large spherulites produced by the HDPE portion, namely low density, narrow composition distribution, and narrow molecular weight distribution. In terms of high temperature resistance and high transparency, the film grade of LLDPE has a small amount of high density portion as a nucleation agent and a low temperature soluble portion.

Random copolypropylene that has high comonomer content and narrow composition distribution has small spherulites, so it has high transparency.

5.9.2.6 Summary of Film Physical Properties

Film physical properties are very much influenced by molecular structure and film process conditions. Tables 5.4 and 5.5 show the relationship among film physical properties, molecular structure, and film process conditions for polypropylene and linear low density polyethylene, respectively.

Table 5.4 Relationship among Film Physical Properties, Molecular Structure, and Film Process Conditions for Polypropylene

	PP Molecular Design				Process Condition		
Film Physical Properties	Tacticity	Molecular Weight	Molecular Weight Distribution	Low Molecular Weight Portion	Resin Temperature	Chill Roll Temperature	Take up Velocity
Direction	high	low	narrow	large	high	high	fast
Impact Strength	↓	↑	↑↑	↓	↑	↓	↓
Transparency	→	↑↑	↑↑	→	↑	↓	↓
Slippage	→	↓	↓↓	↑↑	↓	↑	↑
Antiblocking	↑↑	↓	↓↓	↓	↓	↑	↑
Tensile Modulus	↑↑	→	↓	→	↓	↑	↑

Table 5.5 Relationship among Film Physical Properties, Molecular Structure, and Film Process Conditions for Linear Low Density Polyethylene

		Number of Branch	Composition Distribution	Length of Branch	Molecular Weight	Molecular Weight Distribution	Long Chain Branching
Film Physical Properties	Transparency	↑	↑	→	↓	↑	→
	Heat sealability	↑	↑	→	→	→	→
	Antiblocking	↓	↑	→	→	→	→
	Impact Strength	↑	→	↑	↑	↑	↓
	Tear Strength	∩	↓	↑	→	↑	↓
	Tensile Modulus	↓	↓	→	↑	↓	→
	Heat Resistance	↓	↓	→	→	→	→
Processability	Productivity	→	→	→	↓	↓	↑
	Necking	→	→	→	↑	↓	↑
	Draw Resonance	→	→	→	↑	→	↑
	Melt Fracture	→	→	→	↓	↓	↑
	Stability	→	↓	→	↑	↓	↑

∩: Optimum density exists

■ 5.10 Bleeding of Additives in a Polypropylene Film

Slip agents, an antistatic additive, and other additives provide performance by bleeding on a film, and on the other hand, antioxidants, a nucleating agent, and so on act inside a film as internal additives. The bleeding process, by which the additive in a film comes out on the surface, will be considered to be effective in design development of additive prescription, if it can be predicted.

A new bleeding model of additives in a polypropylene film under atmospheric pressure was established [31–33]. Solubility and diffusion are found to be important for explaining this bleeding process. The experimental results were explained more precisely by assuming two transport processes between the amorphous regions and crystalline regions.

The solubility and diffusion coefficients of higher fatty acid amides and higher fatty acids such as erucamide (13-*cis*-docosenamide), behenamide (docosanamide), stearamide (octadecanamide), and behenic acid (docosanoic acid) were determined between 40 and 70 °C. The difference between the solubilities and the diffusion coefficients was discussed, with the size of these additives providing aggregates by hydrogen bonding in the polypropylene film.

The solubility and diffusion coefficients of UV stabilizers such as 2-(2H-benzotriazol-2-yl)-4-methylphenol and 2-(2H-benzotriazol-2-yl)-4-(1,1,3,3-tetramethylbutyl)phenol were determined at 40 °C. The difference between the solubility and the diffusion coefficients was discussed with the solubility parameters and the bulky structures of methyl group and *tert*-octyl group.

The additive-free isotactic polypropylene (MFR 8.0) was used. It has 47% crystallinity. The bleeding process of the additives under normal pressure is considered as follows. The additive in a PP film dissolves in an amorphous region first, and if it reaches saturation solubility it becomes impossible to dissolve more. The excess ingredient beyond this saturation solubility migrates to the film surface. Furthermore, an additive has a certain bleeding speed and migrates due to the excess amount beyond saturation solubility to the film surface. Therefore, in order to explain a bleeding process under the atmospheric pressure quantitatively, it is important to find saturation solubility and bleeding speed.

In order to find saturation solubility and bleeding speed under the atmospheric pressure, the following model, Equations 5.10 to 5.12, was proposed by considering an amorphous region and a crystalline region.

$$y(t) = \left(C_{0,i} - C_s\right)\left\{\alpha_i + (1-\alpha_i)\left(1-\exp(-kt)\right)\right\}\left(1 - \frac{1}{4l}\left(\int_{-l}^{l} c(x,t)dx\right)\right) \tag{5.10}$$

$$c(x,t) = erf\left(\frac{l-x}{2\sqrt{Dt}}\right) + erf\left(\frac{l+x}{2\sqrt{Dt}}\right) \tag{5.11}$$

$$erf(x) = \frac{1}{\sqrt{\pi}}\int_0^x \exp\left(-t^2\right)dt \tag{5.12}$$

where $y(t)$ is the amount of bleeding additive on the film surface at time t, C_0 is the added amount of an additive, C_s is the saturation solubility, α_i is a diffusion ratio of added amount i, k is the constant of first-order kinetics, l is the half thickness of film, $c(x,t)$ is the concentration at time t and distance x, erf(z) is the error function, and D is the diffusion coefficient.

Figure 5.29 shows the bleeding profiles of erucamide by changing the added amounts of erucamide at 40°C. It turned out that a bleeding process can be explained well. Figure 5.30 shows an example that applied the modified model to the bleeding process of benzotriazole stabilizer. It also turned out that these bleeding processes can be explained well.

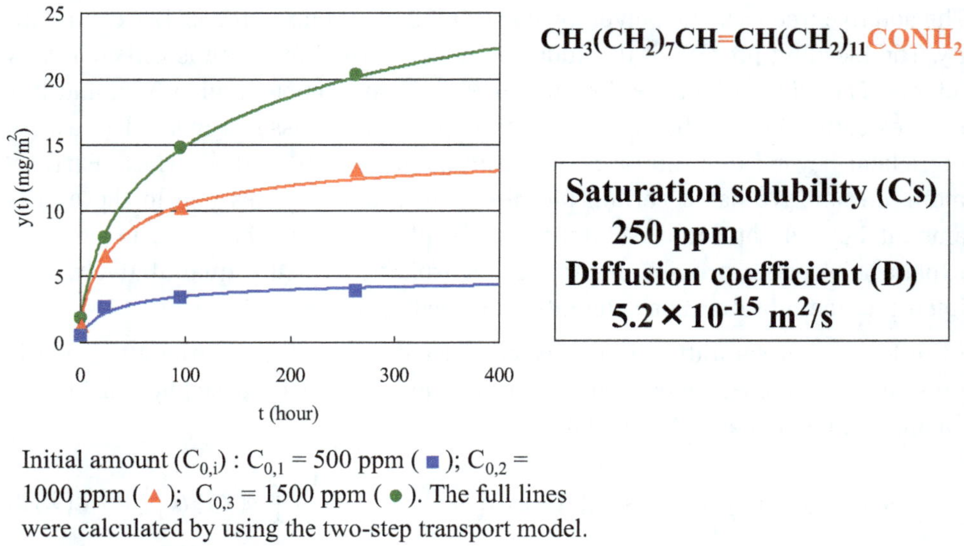

Figure 5.29 Bleeding profiles of erucamide(13-*cis*-docosenamide) at 40°C (saturation solubility C_s = 250 ppm; diffusion coefficient D = 5.2 × 10^{-15} m^2/s)

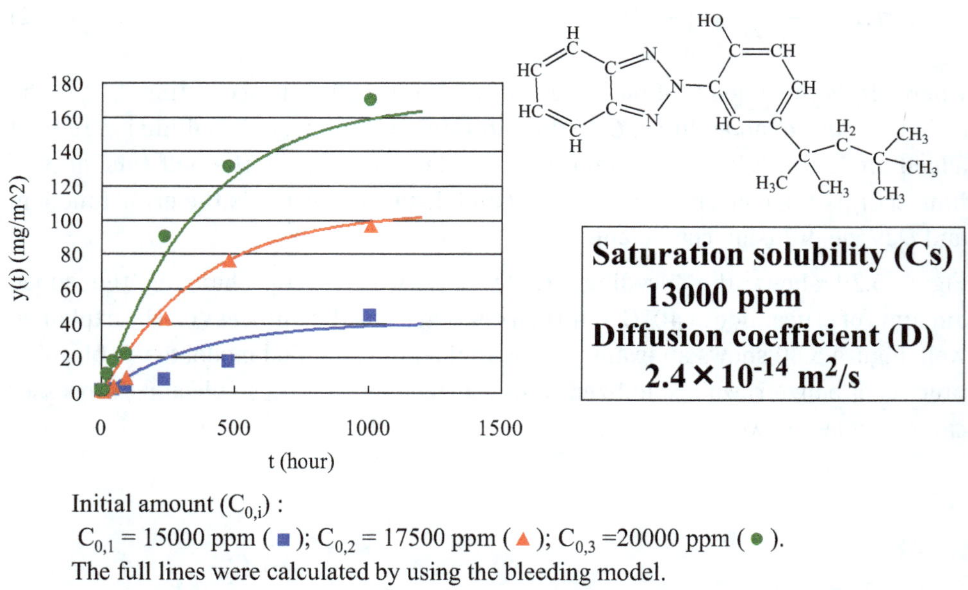

Figure 5.30 Bleeding profiles of 2-(2H-benzotriazol-2-yl)-4-(1,1,3,3-tetramethylbutyl)phenol at 40°C (saturation solubility C_s = 13,000 ppm; diffusion coefficient D = 2.4 × 10^{-14} m^2/s)

References

1. Yeow, Y. L., *J. Fluid Mech.* (1974) 66, p. 613
2. Aird, G. R., Yeow, Y. L., *Ind. Eng. Chem. Fundam.* (1983) 22, p. 7
3. Anturkar, N. R., Co, A., *J. Non-Newtonian Fluid Mech.* (1988) 28, p. 287
4. Iyenger, V. R., Co, A., *Chem. Eng. Sci.* (1996) 51, p. 1417
5. Cotto, D., Duffo, P., Haudin, J. M., *Int. Polym. Process.* (1989) 4, p. 103
6. Billon, N., Barq, P., Haudin, J. M., *Int. Polym. Process.* (1991) 6, p. 348
7. Barq, P., Haudin, J. M., Agassant, J. F., *Int. Polym. Process.* (1992) 7, p. 334
8. Kanai, T., *Seni-Gakkaishi* (1985) 41, p. 409
9. Smith, S., Stolle, D., *Polym. Eng. Sci.* (2000) 40, p. 1870
10. d'Halewyn, S., Aggasant, J. F., *Polym. Eng. Sci.* (1990) 30, p. 335
11. Sakaki, K., Kajiwara, T., Funatsu, K., 10th Annual Meeting Polymer Processing Society (1994) p. 321
12. Debbaut, B., Marchal, J. M., Crochet, M. J., *Z. Angew. Math. Phys.* (1995) 46 (special issue) p. 679
13. Silagy, D., Demay, Y., Agassant, J. F., *Polym. Eng. Sci.* (1996) 36, p. 2614
14. Silagy, D., Demay, Y., Agassant, J. F., *J. Non-Newtonian Fluid Mech.* (1998) 79, p. 563
15. Sakaki, K., Kajiwara, T., Funatsu, K., 11th Annual Meeting Polymer Processing Society (1995) p. 183
16. Kase, S., *J. Appl. Polym. Sci.* (1974) 18, p. 3279
17. Kanai, T., Funaki, A., *Seni-Gakkaishi* (1985) 41, p. 521
18. Kanai, T., Funaki, A., *Seni-Gakkaishi* (1986) 42, p. 1
19. Dobroth, T., Erwin, L., *Polym. Eng. Sci.* (1986) 26, p. 462
20. Ito, H., Doi, M., Isaki, T., Takeo, M., Yagi, K., *Nihon Reoroji Gakkaishi* (2003) 31, p. 149
21. Agassant, J. F., Demay, Y., Sollogoub, C., Silagy, D., *Int. Polym. Process.* (2005) 20, p. 136
22. Sakaki, K., Katsumoto, R., Kajiwara, T., Funatsu, K., *Polym. Eng. Sci.* (1996) 36, p. 1821
23. Fisher, R. J., Denn, M. M., *Chem. Eng. Sci.* (1975) 30, p. 1129
24. Ishihara, H., Kase, S., *J. Appl. Polym. Sci.* (1976) 20, p. 169
25. Kanai, T., *Plastics Age* (1986) 32, no. 10, p. 168
26. Giesekus, H., *Rheol. Acta* (1966) 5, p. 29
27. Giesekus, H., *J. Non-Newtonian Fluid Mech.* (1982) 11, p. 69
28. Kometani, H., Matsumura, T., Suga, T., Kanai, T., *J. Polym. Eng.* (2007) 27, no. 1, pp. 1–28
29. Kometani, H., Kitajima, H., Matsumura, T., Suga, T., Kanai, T., *Int. Polym. Process.* (2006) 21, p. 32
30. Agassant, F., Arda, D.R., Combeaud, C., Merten, A., Münstedt, H., Mackley, M. R., Robert, L., Vergnes, B., *Int. Polym. Process.* (2006) 21, p. 239
31. Wakabayashi, M., Kohno, T., Tanaka, T., Kanai, T., *J. Appl. Polym. Sci.* (2007) 104, pp. 3751–3757

32. Wakabayashi, M., Kohno, T., Tanaka, T., Kanai, T., *J. Appl. Polym. Sci.* (2007) 106, pp. 1398–1404
33. Wakabayashi, M., Kanai, T., *Int. Polym. Process.* (2009) 24, no. 2, pp. 133–139

6 An Overview of Molten Polymer Drawing Instabilities

Jean-Francois Agassant and Yves Demay

Abstract
Drawing instabilities and rupture are a serious limitation in polymer fiber and film processing. Onset of these defects depends on the processing conditions, on heat transfer, and on the rheology of the polymer even if some may be encountered for Newtonian fluids. The main part of this chapter is devoted to the different forms of the draw resonance instability encountered in fiber spinning, cast film, and film blowing. The time-dependent equations are presented for the simplified situation of constant width cast film. Results of the two possible modeling strategies, linear stability analysis and direct simulation, applied to the different fiber and film processes are discussed.

6.1 Introduction

Numerous polymer processes (fiber spinning, cast film, or film blowing) involve a stretching step after the initial extrusion step. Drawing instability (draw resonance) is a typical defect encountered during stretching in air. The existence of an air–polymer interface allows the onset of a periodic instability that leads to very large variations in thickness, width, and diameter. This paper deals with the different forms of drawing instabilities encountered in classical processes such as fiber spinning, cast film, and blown film.

In fiber spinning, a polymer (generally polyester or polyamide) is first melted by heat transfer and viscous dissipation in a screw/barrel device and then forced into a spinning pack consisting of a metal plaque with several hundred holes (typical diameter between 0.3 and 1 mm). The several hundred filaments emerging from the spinning pack are then stretched in order to obtain the final required diameter of textile filaments (several tens of micrometers). This requires high draw ratios (ratio between the drawing wheel velocity and the die exit velocity), under severe cooling conditions (air jet or water bath) in order to obtain a solidified spun fiber at the drawing wheel. In addition, stretching the filament allows the development of appropriate mechanical properties. The stretching distance may vary from a few centimeters for water bath cooling systems to several meters for classical air jet cooling.

There are two families of processes for polymer film production:

- In the cast film process, the polymer is melted in an extruder and forced into a coat-hanger flat die to produce an initial web (1 or 2 m in width, 1 mm in thickness), which then needs to be stretched between the die and a chill roll in order to obtain the required film thickness (several tens of microns are required for packaging applications).

- In the blown film process, the molten polymer is forced within a tube die. This tube is then simultaneously stretched in the vertical direction by nip rolls, inflated by an internal pressure, and cooled by external air rings. A more precise description of these processes may be found in Agassant et al. [1].

Let us note that both the diameter of the holes of a spinning pack and the final gap of a coat-hanger or tube die need to be sufficiently large in order to limit the required extrusion pressure. It leads naturally to an important stretching rate (draw ratio, *Dr*). In fiber and film production, two kinds of defects may be observed at high draw ratios, periodic fluctuations of the diameter (fiber spinning), of both the film thickness and width (cast film), of the film thickness and the bubble diameter (film blowing), but also filament or web rupture. All these phenomena depend on the stretching distance, the cooling conditions, and the rheology of the molten polymer, but it is important to mention that some of these instabilities may be encountered even for Newtonian fluids (in glass fiber processing, for example). These instabilities are the more drastic limitation for process productivity, and at present it is a real challenge to master and then to postpone their occurrence.

Several extensive review papers have been devoted to the experimental and theoretical analyses of these instabilities: Co [2] for cast film, Jung and Hyun [3, 4] for fiber spinning and film blowing.

In the present chapter, various drawing instability experiments performed at CEMEF-Mines ParisTech or in cooperation with Ecole Polytechnique-Montreal are presented in the next section. The time-dependent stretching equations for the constant width cast film process are presented as a reference case in the third paragraph. The different strategies used to explain draw resonance instabilities as well as fiber or web breakage at high draw ratios are discussed in the last paragraphs.

6.2 Experiments

Most of the published experiments have been devoted to filament stretching instabilities. The basic experiment may be done with purely viscous fluids (for example, corn syrup [5]) or viscoelastic (molten polymer) fluids [6–8]. The experiments consist of extruding a fluid from a capillary die at a constant flow rate (velocity u_0) and increasing progressively the take-up velocity u_L imposed at a short distance from the die, which corresponds to a quasi-isothermal situation. The fiber diameter is continuously recorded along the stretching path and after solidification (for molten polymers). Stretching step intensity is evaluated by the so-called draw ratio (a dimensionless number), defined as $Dr = u_L / u_0$. At low draw ratio, the stretching flow is stable, which means that the filament diameter is constant in time, and any perturbation decreases and goes to zero with time (Fig. 6.1(a)). When increasing the draw ratio above a critical value Dr^*, which is of the order of magnitude of one to several tens, the flow becomes unstable. This means that a sustained periodic diameter fluctuation develops (Fig. 6.1(b)). It appears just at the die exit and then propagates up to the extremity of the stretching path [9]. When the draw ratio is increased again, the amplitude of the periodic instabilities increases up to very large values (a ratio of 10 between the maximum and the minimum diameter may be reached) (Fig. 6.1(c)).

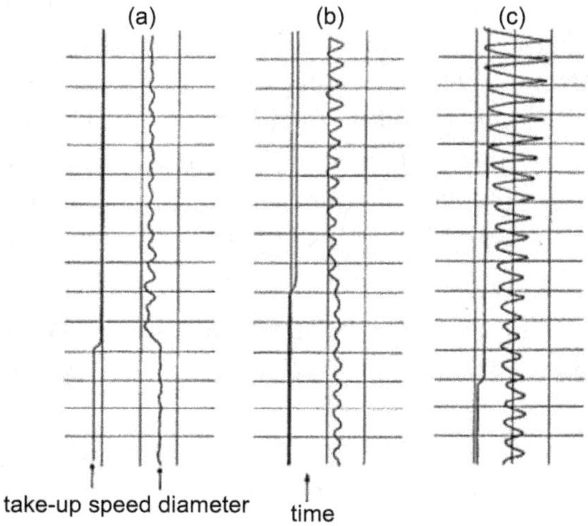

Figure 6.1 Recording of diameter vs. time showing the response to an increase of take-up speed. The stretching ratio goes from 46.6 to 57 for (a), from 57 to 65.6 for (b), and from 65.6 to 75.3 for (c) [16]

Ghiljels and Ente [10] identified directly the occurrence of these instabilities by recording continuously the stretching force, which is obviously very small (of the order of several milli-Newton). When increasing the stretching distance, the process becomes progressively nonisothermal, and the critical draw ratio Dr^* increases too [8, 11, 12]. When the stretching distance L is sufficiently large, more complex phenomena occur as stress-induced crystallization involving changes in the temperature field. This may result in a strong necking phenomenon in the crystallization zone, which prevents drawing instability from developing [13, 14]. When stretching polymers of increasing elasticity in quasi-isothermal conditions, the critical draw ratio remains first more or less constant, and then the periodic instability disappears and frequent filament breakages are then observed [5]. Spinning experiments have been performed with polyesters of various macromolecular architecture, which results in different viscoelastic properties [15]. As pointed out by Demay and Agassant [16], different instabilities may be observed depending on the viscoelastic properties (Fig. 6.2).

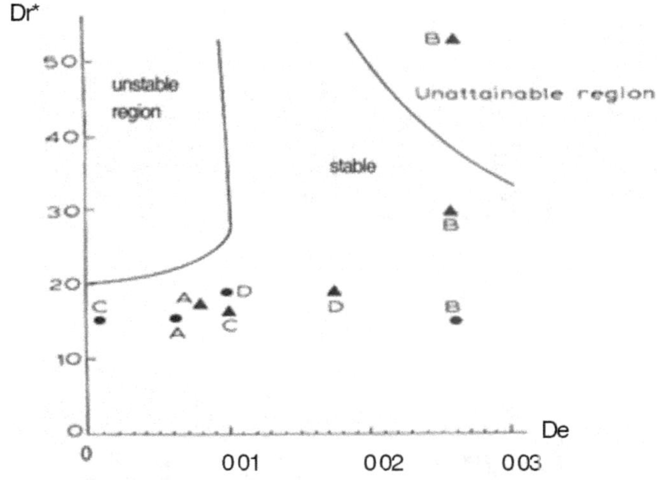

Figure 6.2 Theoretical and experimental critical draw ratio *Dr** vs. Deborah number *De.* A, B, C, and D refer to different polyesters. (▲) corresponds to u_0 = 2.09 cm/s. (●) corresponds to u_0 = 4.18 cm/s [16]

The Deborah number *De* weights the mean relaxation time of the polymer μ to a characteristic time of the process L / u_0. For polyesters A and C, whatever the value of the extrusion velocity u_0, the spinning process becomes unstable for critical draw ratios Dr^* between 15 and 20. For polyester B, the periodic instability is never encountered, but filament breakage is. The behavior of polyester D is in between the preceding ones: periodic instabilities at low extrusion velocity (low Deborah number) and filament breakage at high extrusion velocity (high Deborah number).

In the cast film process, both thickness and width periodic fluctuations are observed above a critical draw ratio of the same order of magnitude as in fiber spinning. But the phenomenon is more complex because film stretching also induces width reduction (called neck-in) and related lateral over-thicknesses (called dog bone defects). This is the reason why very few experimental results are available. Barq et al. [17] performed careful and tedious width and thickness fluctuation measurements (in the central part of a PET film and along the edges) on very long distances (Fig. 6.3).

They point out periodic fluctuations with the same period for width and thickness. The maximum thickness in the central part of the film corresponds more or less to the minimum width, but the two periodic fluctuations do not balance one another, which means that the flow rate at take-up is also a function of time. Shin et al. [18] recorded the film thickness in the central part of HDPE and LDPE films as a function of time.

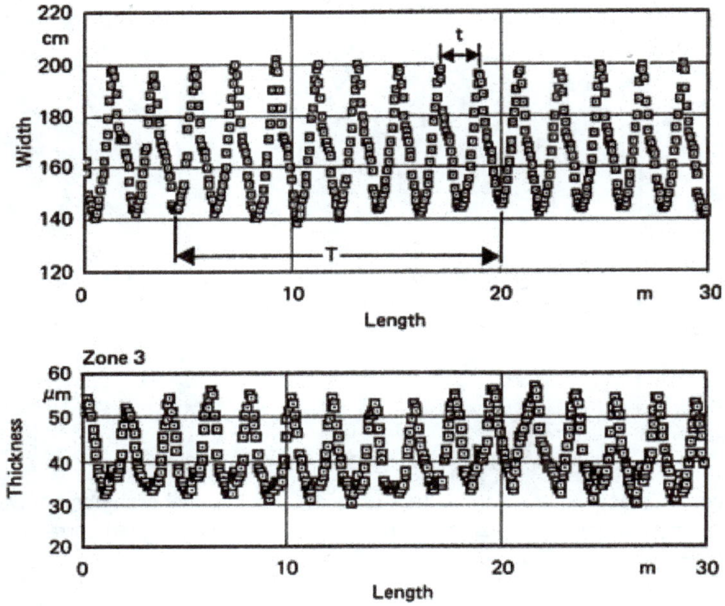

Figure 6.3 Experimental measurement of thickness (a) and width (b) instabilities in the polyester cast film process; *Dr* = 28.4 [17]

Agassant et al. [19] were able to identify the occurrence of the cast film instability by recording continuously the stretching force as a function of time for various LLDPE films (Fig. 6.4). This is easier than in fiber spinning because the stretching force is much higher (several Newtons). Bourrigaud et al. [20] have shown for a series of LDPE of various viscoelastic properties that web rupture may be observed at a draw ratio that decreases when the Deborah number increases (Fig. 6.5).

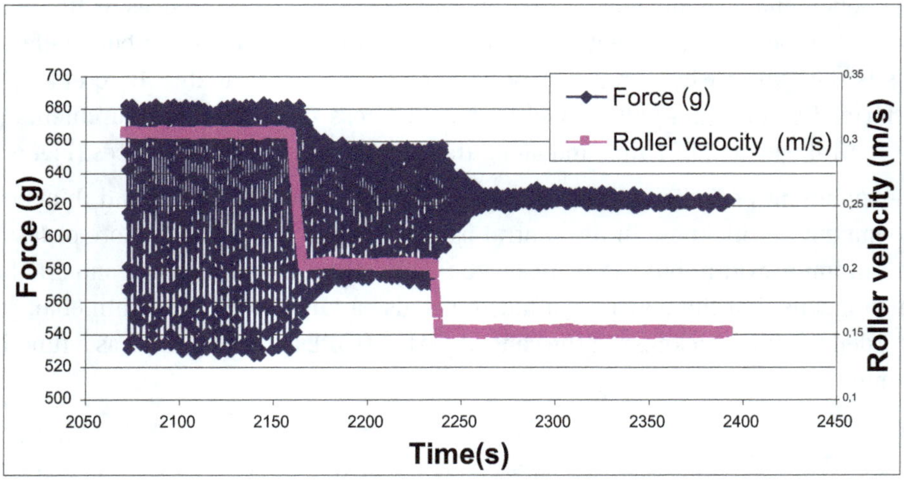

Figure 6.4 Stretching force as a function of time for various take-up velocities [19]

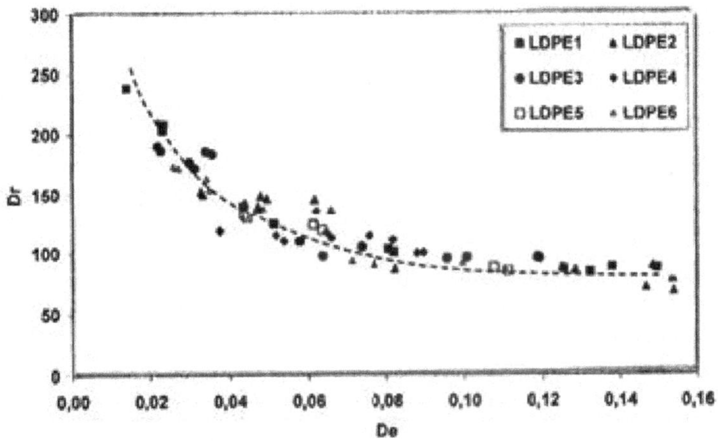

Figure 6.5 Experimental unattainable zone for different LDPE films produced by the cast film process [20]

In film blowing, stretching intensity is evaluated by the draw ratio but also by the blow-up ratio ($Bur = R_L / R_0$; R_L is the final bubble radius; and R_0 the tube die radius). The same type of phenomenon as in cast film may be expected, which means a periodic fluctuation of thickness and diameter (Fig. 6.6) [21–25]. In fact a nonaxisymmetric instability may also be observed that corresponds to a helical shape of the bubble between the die exit and the film freezing line (Fig. 6.7).

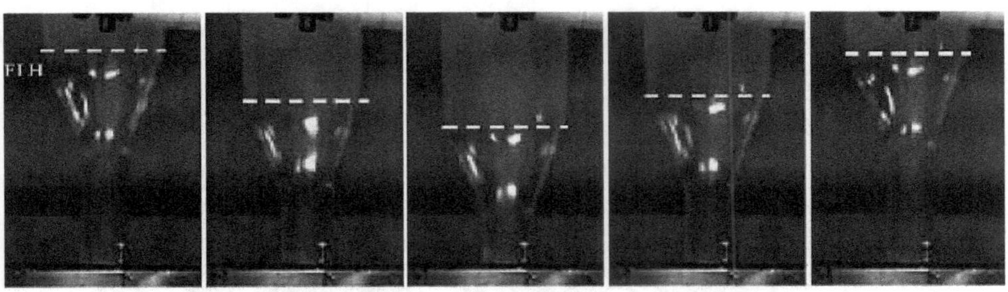

t = 0 s t = 1 s t = 2 s t = 3 s t = 4 s

Figure 6.6 Axisymmetric film blowing instability

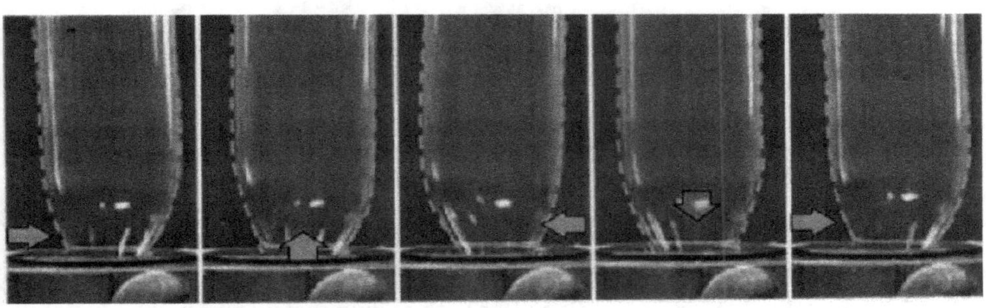

t = 0 s t = 1 s t = 2 s t = 3 s t = 4 s

Figure 6.7 Helical film blowing instability

Ghaned-Fard et al. [23] developed a sophisticated optical device to capture quantitatively and simultaneously the diameter of the bubble and the location of its center at each position between the die exit and the freezing line. Figure 6.8 shows, for a blow-up ratio (*Bur*) of 2.75, that the bubble radius remains constant for a draw ratio (*Dr*) of 9.4. When the *Dr* is increased to 9.9, pronounced periodic radius variations are observed. When the *Dr* is decreased again to 9.4, the bubble radius stabilizes. At the same time, the center of the bubble remains stable. For a *Bur* of 3.25, the bubble radius remains constant when the *Dr* is increased from 6.6 to 7.1, but at the same time the center of the bubble rotates (Fig. 6.9) in a horizontal plane, giving a helical shape to the bubble.

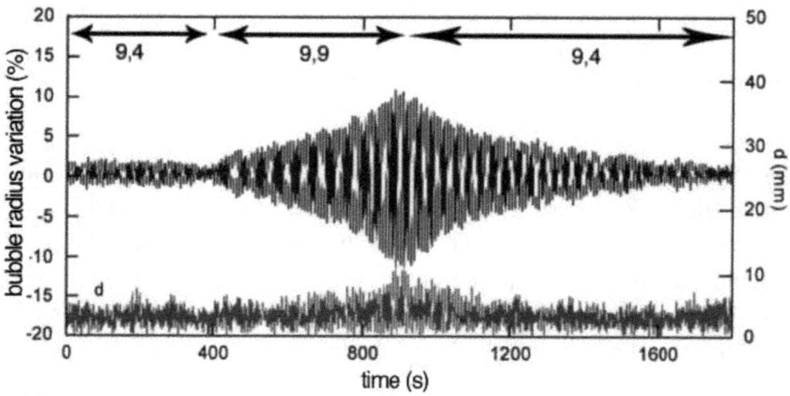

Figure 6.8 Bubble radius variation and location of the center of the bubble as a function of time for a varying draw ratio and a constant blow-up ratio (*Bur* = 2.75) [24]

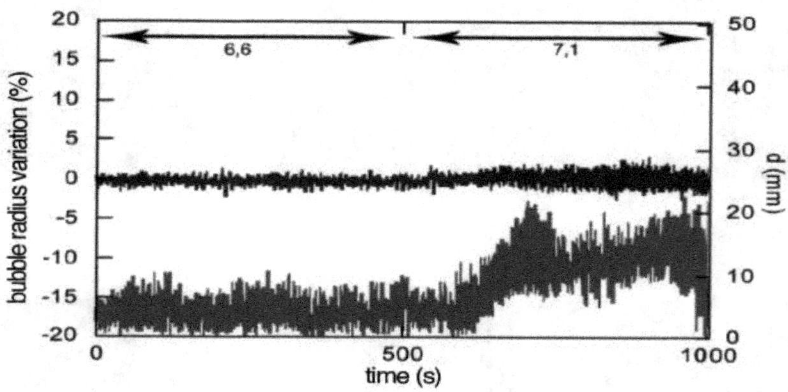

Figure 6.9 Bubble radius variation and location of the center of the bubble as a function of time for a varying draw ratio and a constant blow-up ratio (*Bur* = 3.25) [24]

Laffargue et al. [25] performed film blowing experiments on various extrusion lines but with the same LLDPE. They point out the existence of a stability hill splitting stable and unstable blowing conditions (for both axisymmetric and helical instabilities (Fig. 6.10)).

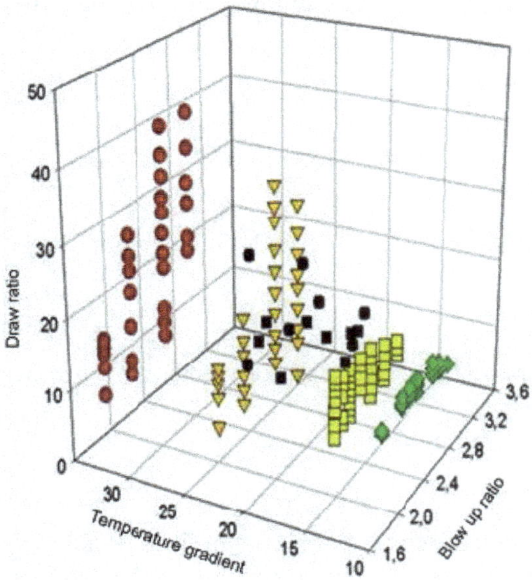

Figure 6.10 Experimental stability map with the same LLDPE and various film blowing machines [25]

■ 6.3 Modeling Strategy: Constant Width Cast Film Process

The physics that governs the appearance of the drawing instability may be explained qualitatively. Having in mind that the requirement of the process is to accelerate the molten polymer from the die exit (velocity u_0) to the take-up wheel (velocity u_L), one may consider that this may be done in two different ways:

- At low draw ratio the velocity increases progressively from $x = 0$ to $x = L$ (x = distance from the extrusion die), the flow is steady, and the dissipated energy is mainly due to stretching.
- At higher draw ratio, a time-dependent flow will require less energy than the steady one: first flow packets are formed at the die exit, and then these packets are accelerated from u_0 to u_L at low deformation rate and then at low energy expense because inertia is negligible.

Models will be built to account for the onset and the development of the draw resonance instability but "without necessarily revealing the fundamental cause of the instability itself" [9]. The first step consists in obtaining a time-dependent model for the stretching flow. A web with an initial thickness $e_0 \approx 1$ mm is stretched between a die and a chill roll (or a water bath) over a distance $L \approx 10$ cm. The width W of the web is assumed to be constant (the neck-in phenomenon is neglected), and the velocity u and the current thickness e are assumed to be uniform in each cross section (Fig. 6.11).

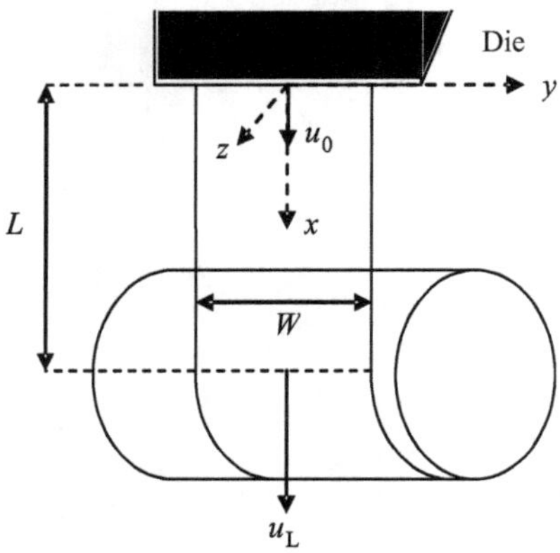

Figure 6.11 Sketch of a constant width cast film model

Because the aspect ratio ($\varepsilon = e_0 / L$) of the stretching flow is small, it is possible to build a simple but realistic membrane model [17, 26–28]. Let us note x the distance to the extrusion die ($0 < x < L$) and t the time.

- Mass conservation writes:

$$\frac{\partial}{\partial t}(e) + \frac{\partial}{\partial x}(eu) = 0 \tag{6.1}$$

- Assuming a Newtonian behavior (viscosity η) and neglecting the shear terms in front of the elongation terms leads to the Trouton equation for film stretching:

$$\sigma_{xx} = 4\eta \frac{du}{dx} \tag{6.2}$$

- The force balance becomes, neglecting mass forces and inertia:

$$F = eW\sigma_{xx} = \text{constant} \tag{6.3}$$

- Boundary conditions imposed by the process are the following:

$$u(t;\,0) = u_0;\ u(t;\,L) = u_L;\ e(t;\,0) = e_0 \tag{6.4}$$

Let us recall that u_0 and u_L are respectively velocity at die exit and take-up. Notice that the final thickness is imposed only if the process is steady. It is then possible (and in this case very easy) to compute a steady (not time-dependent) flow. It may be written as:

$$\bar{u}(x) = u_0 (Dr)^{\frac{x}{L}} \qquad \bar{e}(x) = e_0 (Dr)^{-\frac{x}{L}} \tag{6.5}$$

The key point is to know if this steady flow is stable (really observed in the experiments). It means that a small perturbation of the steady state solution (Eq. 6.5) introduced in Eqs. 6.1 to 6.4 decreases with time until the perturbations vanish. Two different strategies (giving in fact complementary information) may be used.

- The first one is direct numerical simulation. Equations 6.1 to 6.4 are numerically solved according to variables t and x using a suitable time and space scheme. The initial data at $t = 0$ is the stationary solution slightly disturbed. If this initial perturbation decreases and goes to zero, the process is stable. If the initial perturbation develops, the process is unstable. This method is relatively costly because of the time and space discretization but is efficient for determining the periodic flow far away from transition (at relatively high draw ratio). Some examples will be presented in the next paragraph.
- The other one is linear stability analysis, which allows rapid determination of the critical parameters (critical draw ratio) for onset of the instability. In this case a perturbed solution of the following form is introduced.

$$u(x)=\overline{u}(x)+\exp(\lambda t)\hat{u}(x) \qquad e(x)=\overline{e}(x)+\exp(\lambda t)\hat{e}(x) \tag{6.6}$$

In this equation, λ is an eigenvalue (a complex number), and $\hat{u}(x)$ and $\hat{e}(x)$ are eigenfunctions. Linearized equations are obtained by substituting the perturbed values of velocity and thickness (Eq. 6.6) in Eqs. 6.1 to 6.4 and neglecting terms of order 2 according to $\hat{u}(x)$ and $\hat{e}(x)$. This system allows computation of the dominant eigenvalues λ (there are theoretically an infinite number of eigenvalues) and associated eigenfunctions. The steady flow is stable if all the eigenvalues have a negative real part. The steady flow becomes unstable if the real part of a pair of complex conjugate dominant eigenvalues vanishes for $Dr = Dr^*$ ($\lambda = \pm i\omega$) and becomes positive for $Dr > Dr^*$. This is the general framework of the Hopf bifurcation theory. For $Dr > Dr^*$ the steady flow becomes unstable, and generally a stable periodic flow appears. At the transition between stable and unstable flow, the period is determined by the corresponding imaginary part ω of the eigenvalue ($T = 2\pi / \omega$). This simple model predicts the onset of the instability for a critical draw ratio $Dr^* = 20.2$, and it will be seen in the following that mechanical and thermal processing conditions as well as polymer rheology can influence significantly the value of the critical draw ratio. However, this order of magnitude is in good agreement with experimental measurements.

These mathematical approaches described in a simple case have to be extended to account for more realistic situations:

- Geometry and kinematics: neck-in and varying thickness for the cast film process, axisymmetric situation in fiber spinning, nonuniform velocity profile through the thickness when the stretching distance decreases (the aspect ratio ε is not small), nonaxisymmetric bubble shape to account for the helical defect in film blowing.

- Heat transfer: as soon as the stretching distance is significant, the polymer temperature decreases, and the energy equation as well as the temperature dependence of the viscosity have to be accounted for.
- Constitutive equations: The Newtonian behavior is a crude approximation for polymer stretching. Viscoelastic constitutive equations need to be accounted for and, surprisingly, even an upper-convected Maxwell constitutive equation provides qualitative results.

Jung and Hyun [3] proposed a more physical understanding of the development of these drawing instabilities based on different kinematic waves travelling along the stretching line.

In the following sections, the stability of the different stretching processes will be studied, taking progressively into account heat transfer, enriched kinematics, and constitutive equations.

6.4 Cast Film Process

6.4.1 Influence of Cooling

In the cast film process, the stretching distance is generally as small as possible in order to limit neck-in and the related dog bone defect [29], and so temperature remains more or less uniform. For some coating applications, this stretching distance is more important (for geometrical constraints) and temperature decreases, but crystallization temperature is generally not reached before the contact point with the chill roll. The temperature balance equation reduces to:

$$\rho ce(\frac{\partial T}{\partial t}+u\frac{\partial T}{\partial x})=-h_c(T-T_\infty) \tag{6.7}$$

In this equation ρ denotes the polymer density (function of the temperature), c the heat capacity, T the mean temperature in the thickness (which is consistent with the shell approximation for the kinematics), T_∞ the surrounding temperature (a complex compromise between air and tool temperature), and h_c the heat transfer convection coefficient (free or forced convection depending on the web velocity). A system of equations [6.1–6.4, 6.7] accounts for the influence of temperature on viscosity. Using the same perturbation procedure as in the preceding paragraph, the stability of the process is determined as a function of the Stanton number *St*, which represents a nondimensional value of the heat transfer coefficient. It is found that cooling largely stabilizes the process for any values of the draw ratio at large enough values of the Stanton number.

6.4.2 Influence of the Neck-In Phenomenon

In order to account for the width reduction, the thin shell membrane model is enriched in two successive steps:

- First, both thickness and velocity in the stretching direction $u(x)$ are assumed to remain constant through the width of the web $W(x)$. The velocity component $v(x,y)$ in the transverse direction has to be introduced, but the continuity equation is introduced to express it as a function of x: $v(x,y) = yf(x)$. The linear stability analysis is then used to compute the value of the critical draw ratio Dr^* as a function of the shape factor of the film $A = L/W$ [30, 31]. The case $A = 0$ corresponds to a very short stretching distance, and in this situation the constant width cast film analysis developed in the preceding paragraph is valid (the critical draw ratio is $Dr^* = 20.2$). Increasing the A value means increasing the drawing distance, and the critical draw ratio Dr^* increases too. For example, for $A = 1$, the value of the critical draw ratio is increased by 50% as compared to $A = 0$ (Fig. 6.12).

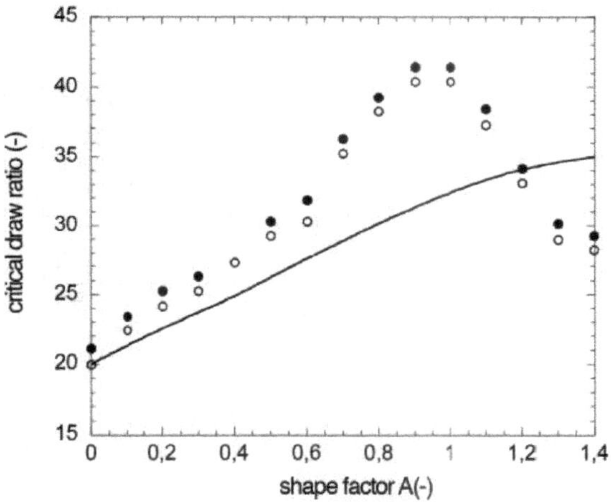

Figure 6.12 Influence of the shape factor A on the critical draw ratic Dr^*: (–) 1-D model; (○) 2-D stable situation; (●) 2-D unstable situation [31]

- The membrane model is then extended to capture both development of the dog bone defect and neck-in by considering thickness and mean velocities in stretching and transverse directions to be functions of x and y ($e(x,y)$, $u(x,y)$, and $v(x,y)$). Direct numerical simulation was used by Silagy et al. [31]. Figure 6.13 shows the thickness as a function of time for increasing values of the draw ratio after introducing at $t = 0$ a small arbitrary perturbation of the steady flow ($A = 0.5$). For $Dr < 21$, the initial perturbation vanishes rapidly. For $Dr = 27$, a small but nonperiodic perturbation is observed. For $Dr = 31.5$ a periodic well-marked perturbation develops and grows slowly in amplitude. It becomes very important for $Dr = 36$, and saturation of the amplitude is rapidly reached (with a ratio of eight between the maximum and the minimum thickness). As compared to the linear stability analysis, the border between the stable and unstable drawing regime is less precisely defined but it allows prediction of the shape of the film in unstable conditions. Figure 6.14 shows that width and thickness fluctuations (in the central part and at the periphery of the film) as a function of time are in phase opposition, which is in agreement with the experiments of Fig. 6.3. Figure 6.12 compares the 2-D direct numerical stability simulation and the 1-D linear stability analysis. The development of the dog bone defect in the 2-D model could contribute to increasing the stability of the process. For A = 1, the critical draw ratio is around 40 for the 2-D model; in contrast it was around 30 for the 1-D model.

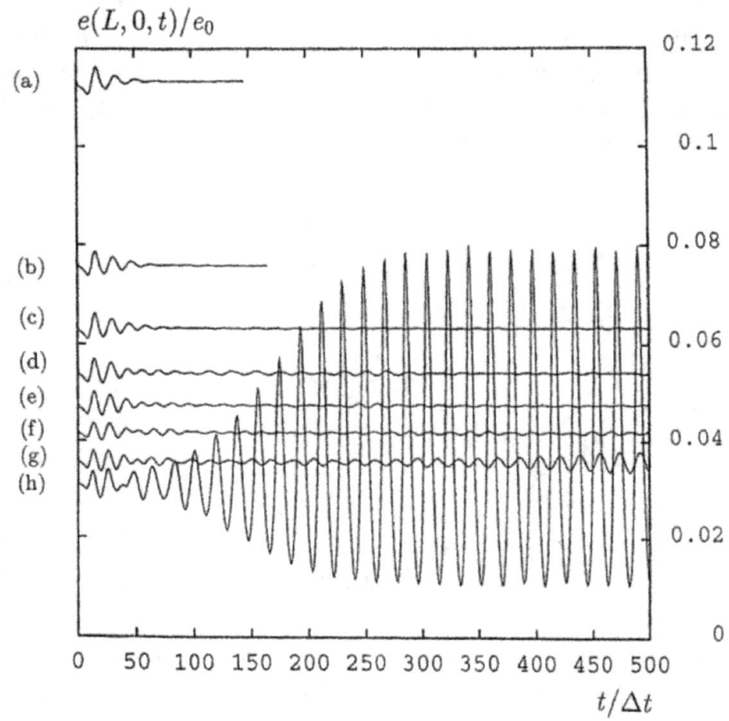

Figure 6.13 2-D cast film model, film thickness as a function of time after imposing a small initial perturbation. The film shape factor is A = 0.6. (a) Dr = 10; (b) Dr = 15; (c) Dr = 18; (d) Dr = 21; (e) Dr = 24; (f) Dr = 27; (g) Dr = 31.5; (h) Dr = 36 [31]

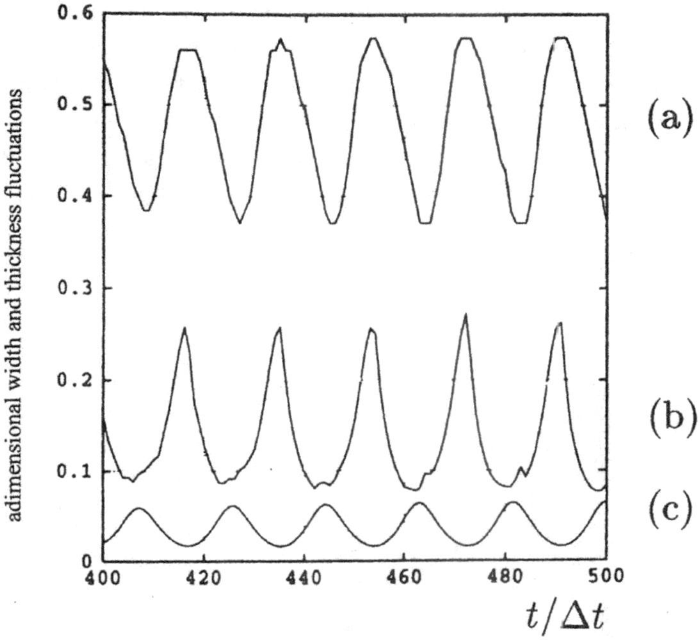

Figure 6.14 2-D Newtonian unsteady cast film computation: Dr = 33; A = 0.44; (a) width fluctuations; (b) film thickness fluctuations at the periphery of the film; (c) film thickness fluctuations in the center of the film [31]

6.4.3 Validity of the Membrane Model: 2-D Transverse Simulation

The membrane model previously described assumes a constant velocity profile through the thickness. This hypothesis is clearly valid in the stretching part of the flow, that is, at a distance x larger than one or several thicknesses of the die gap. To describe precisely the rearrangement from a shear flow in the slit die (the extremity of the coat-hanger die) to a stretching flow, it is necessary to use a 2-D model in the (x, z) plane (coordinates of Fig. 6.11). For the sake of simplicity it is assumed as previously that the fluid has a Newtonian behavior and that mass forces, inertia, and surface tension are neglected. A 2-D Stokes problem has to be solved on an unknown domain, and determination of the polymer–air interface is a part of the solution. It is a so-called free interface problem, and two strategies are then possible:

- If a surface tracking method is used the kinematic and dynamic interface boundary conditions are explicitly enforced. The Stokes problem is solved on a computation domain that is restricted to the polymer flow. The interface with air is determined by successive iterations (by adjusting the position of the nodes on the interface) to satisfy the kinematic interface equation. This first strategy is relatively precise but unable to describe the transient evolution of the interface. The linear stability method is then used to predict onset of draw resonance [32].

- If a surface-capturing method is used the computation domain is extended to the neighborhood of the air interface, and the Stokes problem is coupled with a transport equation. Classical approaches are the level set and the volume of fluid method. A complex mesh refinement is required in the vicinity of the interface, but it allows capture of severe topological changes of the flow domain. Then a time and space numerical scheme (direct numerical simulation) is used to predict onset of draw resonance. In this case the periodic solution is obtained even for high values of the draw ratio [33].

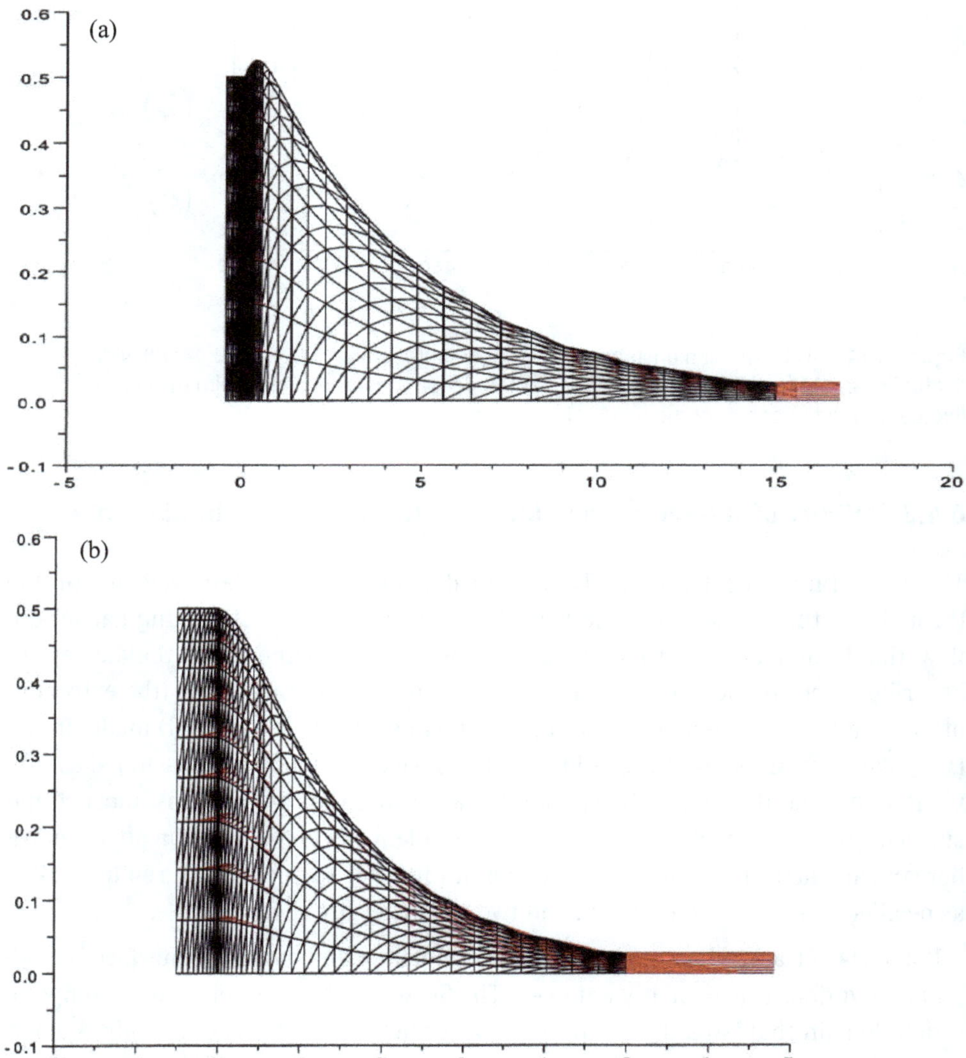

Figure 6.15 Coupled die and stretching flow computation: (a) $e_0 / L = 1/15$; (b) $e_0 / L = 1/5$

If this represents an important increase in the computation complexity, it also brings an important precision that is illustrated in the following two examples obtained at a draw ratio $Dr = 18$ using the tracking method and modelling both extrusion and stretching.

- Figure 6.15(a) presents the mesh and the flow domain computed for an aspect ratio $\varepsilon = 1/15$. A significant swelling at the die exit is observed as it would be expected that drawing will eliminate die swell. This explains why the experimental critical draw ratios are always less than the theoretical ones (Fig. 6.2). The dominant eigenvalue ($\lambda = -0.084 + i11.99$) is very close to the limit of stability (real (λ) = 0). It is interesting to notice that this critical draw ratio $Dr^* = 18$ accounting for the transition between the slit die and the stretching flow is near the critical draw ratio of the 1-D membrane model $Dr^* = 20.2$.
- Figure 6.15(b) presents the mesh and the flow domain computed for an aspect ratio $\varepsilon = 1/5$. There is no visible swelling in this case, and the flow is more stable because the real part of the eigenvalue ($\lambda = -0.38 + i12.149$) is much more negative. It points out that 2-D simulation is required to determine stability conditions when the stretching distance decreases. Figure 6.16 illustrates the variation of the real part of the dominant eigenvalues as a function of the stretching distance.

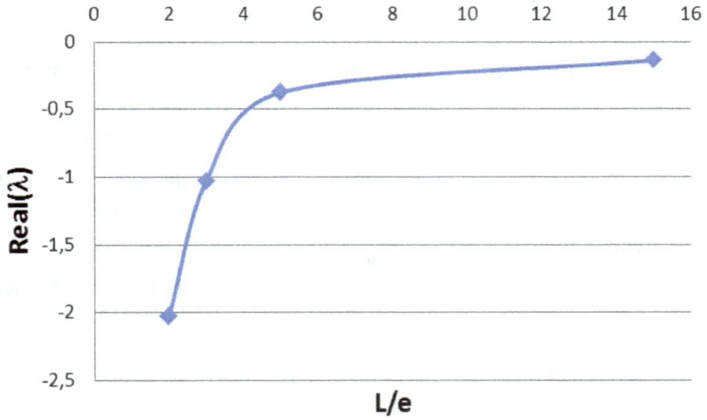

Figure 6.16 Real part of the leading eigenvalue as a function of the stretching distance (a dimensionless value is obtained using the residence time L / u_0)

6.4.4 Influence of Rheology

As a first approach, the viscoelastic behavior of the polymer may be introduced through the simple upper-convected Maxwell (UCM) model, which for the constant width model leads to the following equations:

$$\sigma'_{xx} + \mu\left(\frac{\partial \sigma'_{xx}}{\partial t} + u\frac{\partial \sigma'_{xx}}{\partial x} - 2\sigma'_{xx}\frac{du}{dx}\right) = 2\eta\frac{du}{dx} \tag{6.8a}$$

$$\sigma'_{zz} + \mu\left(\frac{\partial \sigma'_{zz}}{\partial t} + u\frac{\partial \sigma'_{zz}}{\partial x} + 2\sigma'_{zz}\frac{du}{dx}\right) = -2\eta\frac{du}{dx} \tag{6.8b}$$

where μ is the relaxation time and σ'_{xx} and σ'_{zz} are the extra-stress components of the stress tensor. The thin shell approximation implies:

$$\sigma_{xx} = \sigma'_{xx} - \sigma'_{zz} \tag{6.8c}$$

Solving stationary equations is not possible for every value of the draw ratio *Dr* and the Deborah number *De* (it is the nondimensional value of the relaxation time μ: $De = \mu\frac{u_0}{L}$). This has been mentioned by Denn et al. [34] and Antukar and Co [27], who pointed out the existence of an unattainable drawing region in the vicinity of which the stretching force increases by several orders of magnitude (Fig. 6.17). This is related to the dramatic increase of the elongation viscosity observed for the upper-convected Maxwell model. Even the use of more realistic viscoelastic models, limiting the increase of the elongation viscosity, leads to nonphysical values of the stretching force [31]. This could explain the web breakage experimentally observed by Bourrigaud et al. [20] for a collection of LDPE films of varying viscoelasticity (Fig. 6.5). It is interesting to notice that the shape of the experimental film stretching limit is similar to the theoretical one, even if the experimental draw ratios at rupture are more important than the ones theoretically predicted with the upper-convected Maxwell model.

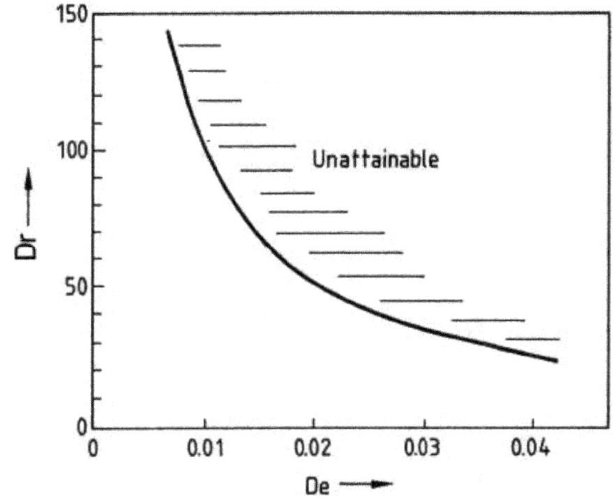

Figure 6.17 Unattainable stretching zone as a function of the draw ratio and the Deborah number

For stability analysis, Eqs. 6.1, 6.3, and 6.8 are linearized in the same way as in the preceding paragraph. Figure 6.18 discriminates stable and unstable drawing conditions in the (*Dr, De*) plane for different values of the film shape factor *A*. At low Deborah number, which corresponds to a weak elasticity of the polymer, a low extrusion velocity, or a long stretching distance, the critical draw ratio *Dr** is slightly dependent on the film shape factor *A*. Above a critical Deborah number *De**, which decreases markedly when the film shape factor increases, the periodic instability disappears. It is noteworthy to observe another stable region, for small Deborah numbers, at high draw ratios. This is not a mathematical artifact, and a patent has been granted [35] to draw films in these superstable conditions. Silagy et al. [31] and Kim et al. [36] coupled the UCM model with a 2-D membrane model accounting for width and thickness reduction. Kim et al. [36] computed the transient thickness and width responses to small perturbations using a time-space numerical scheme. Shin et al. [18] developed the same approach but with a Phan-Thien Tanner model, which limits the increase of the elongation viscosity.

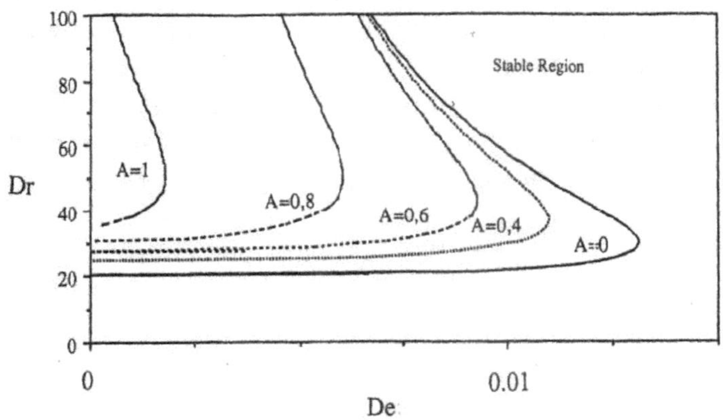

Figure 6.18 1-D viscoelastic cast film model: influence of the film shape factor *A* on the drawing stability [31]

These models lead to the following interesting conclusions about the stability of the cast film process:

- For short stretching distances, periodic thickness and width instabilities may be observed for a draw ratio higher than a critical value around 20.
- Increasing the stretching distance will stabilize the process, due to both neck-in and cooling phenomena, but then the dog bone defect develops, which necessitates trimming lateral sides of the film.
- Curiously, decreasing the stretching length to a distance of the order of magnitude of the die gap dimension will also stabilize the process.
- Using polymers of increasing viscoelasticity only slightly affects the web stability until a critical value above which the process becomes much more stable. The development of periodic instabilities is then no more a limitation, but web rupture

may be observed. Let us point out that swelling at the die exit will increase too, and it is expected that it modifies the critical draw ratio computed from the membrane model. Accounting for the coupled flow in the die and along the stretching distance for a viscoelastic fluid remains a challenge for a viscoelastic behavior because it is difficult in this case to deal with the stress singularity at the die exit.

6.5 Fiber Spinning

The basic equations of the fiber spinning process are essentially the same as for the constant width cast film process. As a consequence, the critical draw ratio in isothermal Newtonian fiber spinning is equivalent ($Dr^* = 20.2$) [6, 37–41], and this is in agreement with experiments presented in the second section.

In the high velocity fiber spinning process (up to 5000 m/min for polyamide or polyester) the stretching distance is important (several meters), and an additional air jet is applied in order to cool the filament, which makes the process highly nonisothermal [37]. The first linear stability analysis shows [41–42] that the critical draw ratio increases rapidly with the nondimensional heat transfer coefficient when neglecting crystallization. This is in agreement with PET experiments (Fig. 6.19).

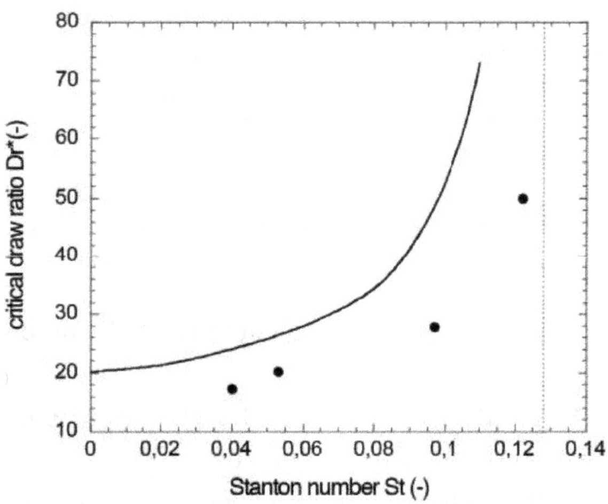

Figure 6.19 Critical draw ratio as a function of the Stanton number; (–) theoretical result [41]; (●) measurements [16]

When introducing a viscoelastic UCM constitutive equation, the same phenomena are obtained as in the constant width cast film situation [8, 43]: an unstable region with the development of periodic instabilities at moderate Deborah numbers and intermediate draw ratios, and an unattainable region at high Deborah number and high draw ratios where the stretching force becomes infinite, which would explain filament breakage. This is consistent with experiments (Fig. 6.2). More realistic viscoelastic models have been introduced by Lee et al. [44], while Jung et al. [45] coupled the UCM model and heat transfer. Lee et al. [46], Shin et al. [13], as well as Kohler and McHugh [14, 47] introduced stress-induced crystallization and the corresponding crystallization energy in the temperature balance equation. Kohler and McHugh [14, 47] used a linear stability analysis and both Giesekus and Pom-Pom viscoelastic models to describe the response of the spinning line to a small variation of a spinning parameter in stable processing conditions. Interestingly, Shin et al. [13] followed the real part of the dominant eigenvalue as a function of the take-up speed and pointed out the stabilizing effect of the necking phenomenon, which occurs in the crystallization temperature range.

The following conclusions may be drawn for the fiber spinning process:

- As soon as the stretching distance and the cooling conditions are important, the periodic instability is not a true limitation.
- When increasing the extrusion velocity or the relaxation time at constant draw ratio, the stretching force increases dramatically, which induces breakage.
- The rheology of the polymer during the crystallization step, especially at the high cooling rates encountered in the process, is not well mastered, and this is the main limitation of the existing models [48].

6.6 Film Blowing

At a first glance, one may consider that film blowing is simply a nonisothermal rolled-up cast film process, with a uniform thickness and with an increasing width instead of a width reduction (neck-in). Heat transfer would stabilize the process, and viscoelasticity would cause web breakage at high draw ratios. However, experiments listed in the second section showed unexpected phenomena: the existence of axisymmetric instabilities (which after unrolling correspond to both thickness and width variations) but also of nonaxisymmetric instabilities. The reason is that the molten polymer tube is subjected to a draw ratio, but also to a pressure differential (between the air inside the bubble and the atmospheric pressure), which induces a blow-up ratio *(Bur)*, and the combination of these two orthogonal drawings makes the stability of the process much more complex.

It is surprising that modeling such a widely used process presents so many mathematical difficulties. All models use a membrane approximation leading to a system of partial differential equations according to the drawing distance z and time t. Computation of the steady-state solution remains a numerical challenge for some operating conditions: both nonexistence or multiplicity of numerical solutions are found according to processing parameter values. The early steady-state Newtonian isothermal model [49] has been extended to account for nonisothermal effects [50] coupled with crystallization [21, 51]. Introducing an isothermal Maxwell (UCM) viscoelastic constitutive equation may lead to multiple solutions depending on the boundary conditions [52]. Andre et al. [53] studied the Maxwell nonisothermal film blowing process and also pointed out that no or multiple solutions may be obtained depending on the processing conditions. Hyun et al. [54] introduced a viscoelastic Phan-Thien Tanner (PTT) nonisothermal model. Doufas and McHugh [55] and Pirkle and Braatz [56] introduced flow-induced crystallization with a two-phase model.

Even though the time-dependent equations for the axisymmetric model were introduced for a long time, the transient studies remain difficult because the boundary condition inside the bubble is not clear (fixed air volume or fixed pressure as a function of time). Yeow [26], Cain and Denn [52], and Yoon and Park [57] assume a Newtonian isothermal behavior and developed the same kind of linear stability analysis as in fiber spinning or cast film. They point out the big influence of the boundary condition inside the bubble on the instability onset. Pirkle and Braatz [58] performed time-dependent simulations and obtained the steady-state result by integrating transient equations for a sufficiently long time. Shin et al. [59] studied the stability of a PTT nonisothermal model, while Henrichsen and McHugh [60] used a viscoelastic Giesekus constitutive equation and compared their results to LDPE film blowing experiments. Both used a constant air volume condition inside the bubble.

All of these models are only focused on the axisymmetric instabilities. Following Housiadas and Tsamopoulos [61], Laffargue et al. [25] wrote and solved the film blowing equations in the frame affixed to the die geometry. This leads to complex developments that are detailed in Laffargue [62] and Laffargue et al. [25], but this allows capture of both axisymmetric and helical instabilities. Each variable (bubble radius, bubble thickness, velocity, vertical stress, pressure, temperature, and so on) is expressed as the sum of the stationary solution and a perturbation:

$$X(t, z, \varphi) = \overline{X}(t, z) + \exp(\lambda t + im\,\varphi)\hat{X}(z) \tag{6.9}$$

In this equation X corresponds to any geometrical, kinematics, stress, or temperature variable listed before. $\overline{X}$ is the corresponding stationary solution, and $\hat{X}$ is an eigenfunction. The complex number λ is, as previously, an eigenvalue describing growth or decay of perturbations. The positive integer m is the azimuthal wave number: $m = 0$ corresponds to an axisymmetric instability as $m = 1$ corresponds to a helical instability (higher order instabilities ($m > 1$) have not been investigated yet). Figure 6.20 shows the real part of the most important eigenvalue as a function of the draw ratio for $m = 0$ (axisymmetric case) and for $m = 1$ (helical case) when the blow-up ratio (*Bur*) is 2. For $m = 0$, the eigenvalue becomes positive for a draw ratio of 29, as the eigenvalue corresponding to $m = 1$ is negative. This means that the axisymmetric instability will be activated. In contrast, Fig. 6.21 shows the real part of the eigenvalue corresponding to $m = 1$ as a function of *Bur* for a fixed $Dr = 6$.

The real part of the eigenvalue remains negative until *Bur* = 3.5 and then becomes positive. In that blowing situation the eigenvalue corresponding to $m = 0$ is negative, which means that the helical instability will be activated. Figure 6.22 summarizes the results in the (*Dr*, *Bur*) plane: when varying the temperature gradient along the bubble (which means varying the air flow in the surrounding air ring) it is possible to build an envelope delimiting stable and unstable film blowing conditions (Fig. 6.23). This figure allows one to draw interesting conclusions for the stability of the film blowing process:

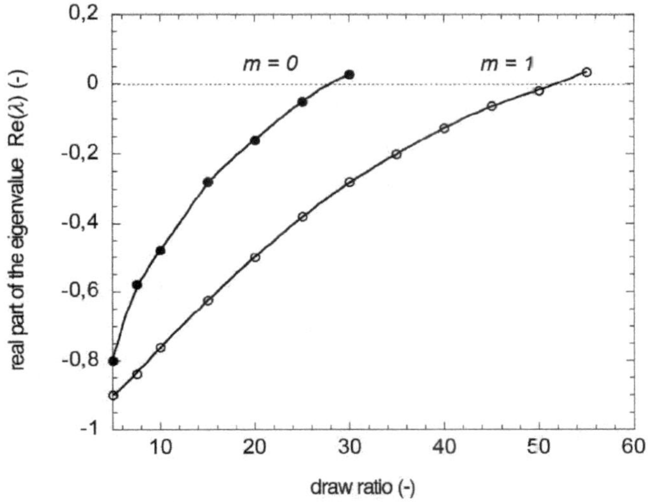

Figure 6.20 Real part of the dominant eigenvalue as a function of the draw ratio for the axisymmetric mode ($m = 0$) and the helical mode ($m = 1$) (*Bur* = 2) [25]

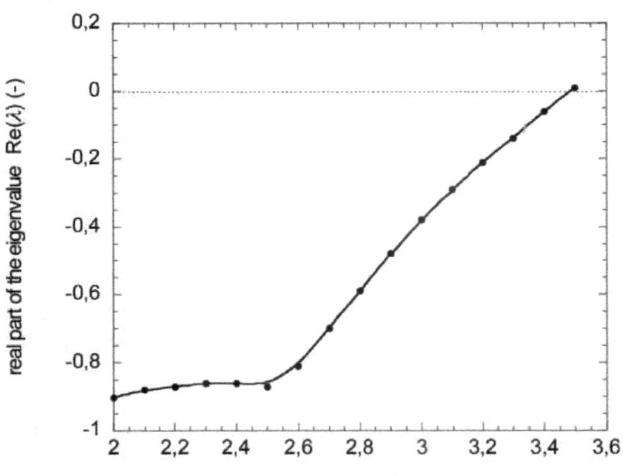

Figure 6.21 Real part of the dominant eigenvalue as a function of the blow-up ratio for the helical mode ($m = 1$) (*Dr* = 6) [25]

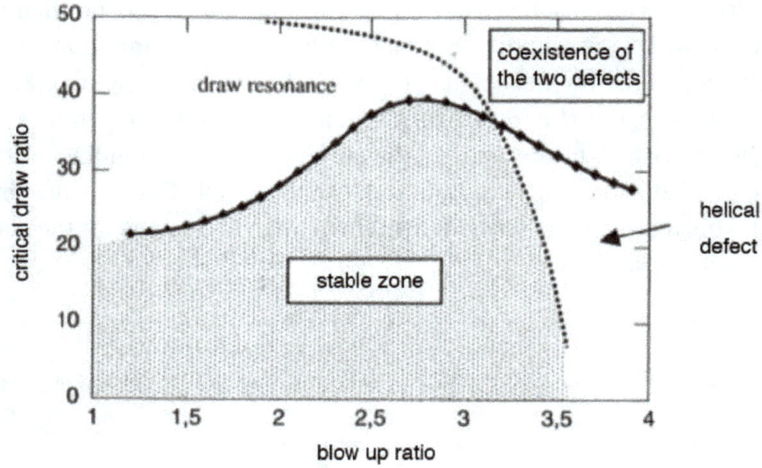

Figure 6.22 Stability curve in the (*Dr*, *Bur*) frame [25]

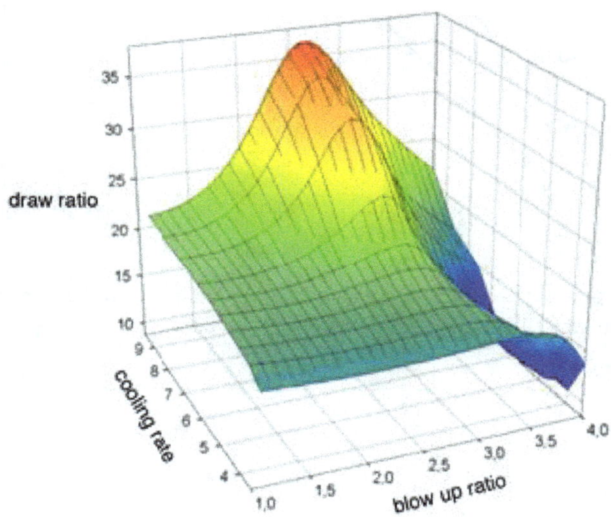

Figure 6.23 3-D stability map as a function of the draw ratio, the blow-up ratio, and the temperature gradient [62]

- For given heat transfer conditions, increasing the blow-up ratio first stabilizes the process (which means it is possible to impose more important draw ratios and so to decrease significantly the film thickness in stable conditions) as it was likely to wait for the contrary.
- Above a critical *Bur*, *Dr* needs to decrease in order to remain in stable conditions.
- This “stability hill” is less pronounced when the temperature gradient (the air flow in the surrounding ring) increases.

Recently, Lee et al. [63] showed that crystallization may stabilize the film blowing process as in fiber spinning. Using a constant blowing pressure inside the bubble in a viscoelastic nonisothermal model, Lee et al. [64] showed very different results than with the classical constant air volume boundary condition.

■ 6.7 Conclusions

Instabilities are encountered in most polymer forming processes. In extrusion or coextrusion, these instabilities are directly linked to the viscoelastic behavior of the polymer. In fiber and film stretching processes, drawing instabilities are present even for purely Newtonian fluids, which means that their occurrence is, at the first order, governed by dissipation minimization mechanisms. However, heat transfer and viscoelasticity will influence the onset and the development of the defect and even suppress it. In shear-dominant flows, as encountered in extrusion dies, sophisticated constitutive equations have to be used in order to account for instabilities. In fluid stretching operations, even a simple constitutive equation, as for example the upper-convected Maxwell model, is able to capture the main features. The reason is that in confined flows a wide distribution of shear rates is present, as in fluid stretching the elongation rates remain within a narrow range, which induces activation of a dominant relaxation time.

References

1. Agassant, J. F., Avenas, P., Sergent, J. Ph., Vergnes, B., Vincent, M., "La Mise en Forme des Matières Plastiques," Techniques et Documentation (Lavoisier) Paris (1996)
2. Co, A., "Draw Resonance in Film Casting," Chapter 10 in *Polymer Processing Instabilities: Control and Understanding*, Hatzikiriakos, S. V., Migler, K. B. (Eds.) (2005) Marcel Dekker, New York
3. Jung, H. W., Hyun, J. C., "Fiber Spinning and Film Blowing Instabilities," Chapter 11 in *Polymer Processing Instabilities: Control and Understanding*, Hatzikiriakos, S. V., Migler, K. B. (Eds.) (2005) Marcel Dekker, New York
4. Jung, H. W., Hyun, J. C., "Instabilities in Extensional Deformation Polymer Processing," *Rheol. Rev.* (2006) pp. 131-164
5. Chang, J. C., Denn, M. M., "An Experimental Study of Isothermal Spinning of Newtonian and Viscoelastic Liquid," *J. Non-Newtonian Fluid Mech.* (1979) 5, pp. 369-385
6. Kase, S., "Studies on Melt Spinning IV: On the Stability of Melt Spinning," *J. Appl. Polym. Sci.* (1974) 18, pp. 3278-3304
7. Ishihara, H., Kase, S., "Studies on Melt Spinning. V. Draw Resonance as a Limit Cycle," *J. Appl. Polym. Sci.* (1975) 19, pp. 557-566

8. Demay, Y., "Instabilité d'Etirage et Bifurcation de Hopf," Thèse de Doctorat d'Etat, Université de Nice (1983)
9. Hyun, J.C., "Draw Resonance in Polymer Processing: A Short Chronology and a New Approach," *Korea-Australia Rheol. J.* (1999) 11, pp. 279–285
10. Ghiljels, A., Ente, J.J.S.M., "Draw Resonance Studies on Polypropylene Melts," Proceedings of the International Conference on Rheology, Naples (1980)
11. Blyler, L.L., Gieniewski, C., "Melt Spinning and Draw Resonance of a Poly(α-methylstyrene/silicone) Block Copolymer," *Polym. Eng. Sci.* (1980) 28, pp. 140–148
12. Gupta, R.K., Drechsel, P.D., "Draw Resonance in the Non-isothermal Spinning of Polypropylene," 2nd World Congress of Chemical Engineering, Montréal, Canada (1981)
13. Shin, D.M., Lee, J.S., Jung, H.W., Hyun, J.C., "High-Speed Fiber Spinning Process with Spinline Flow-Induced Crystallization and Neck-like Deformation," *Rheol. Acta* (2006) 45, pp. 575–582
14. Kohler, W.H., McHugh, A.J., "Prediction of the Influence of Flow-Enhanced Crystallization on the Dynamics of Fiber Spinning," *Polym. Eng. Sci.* (2008) 48, pp. 88–96
15. White, J.L., Yamane, H., "A Collaborative Study of the Rheological Properties and Unstable Melt Spinning Characteristics of Linear and Branched Polyethylene Terephthalates," *Pure Appl. Chem.* (1985) 57, pp. 1441–1452
16. Demay, Y., Agassant, J.F., "Experimental Study of the Draw Resonance in Fiber Spinning," *J. Non-Newtonian Fluid Mech.* (1985) 18, pp. 187–198
17. Barq, P., Haudin, J.M., Agassant, J.F., Roth, H., Bourgin, P., "Instability Phenomena in Film Casting Process. Experimental and Numerical Approaches," *Int. Polym. Process.* (1990) 5, pp. 264–271
18. Shin, D.M., Lee, J.S., Jung, H.W., Hyun, J.C., "Transient and Steady-State Solutions of 2-D Viscoelastic Non-isothermal Simulation Model of Film Casting Process via Finite Element Method," *J. Rheol.* (2007) 51, pp. 393–407
19. Agassant, J.F., Bouamra, R., Peiti, C., Haudin J.M., "Cast Film Extrusion, Geometrical Defects and Stability," Proceedings of the Polymer Processing Society Meeting, Yamagata (Japan) (2006)
20. Bourrigaud, S., Marin, G., Dabas, V., Dupuy, C., Silagy, D., "The Draw Ratio-Deborah Number Diagram: A Useful Tool for Coating Applications," *Polym. Eng. Sci.* (2006) 46, pp. 372–380
21. Kanai, T., White, J.L., "Kinematics, Dynamics and Stability of the Tubular Film Extrusion of Various Polyethylenes," *Polym. Eng. Sci.* (1984) 24, pp. 1185–1201
22. Minoshima, W., White, J.L., "Instability Phenomena in Tubular Film and Melt Spinning of Rheologically Characterized High Density, Low Density and Linear Low Density Polyethylenes," *J. Non-Newtonian Fluid Mech.* (1986) 19, pp. 275–302
23. Ghaneh-Fard, A., Carreau, P.J., Lafleur, P.G., "Study of Instabilities in Film Blowing," *AIChE J.* (1996) 42, pp. 1388–1396

24. Laffargue, J., Parent, L., Lafleur, P. G., Carreau, P. J., Demay, Y., Agassant, J. F., "Investigation of Bubble Instabilities in Film Blowing Process," *Int. Polym. Process.* (2002) 17, pp. 347-353

25. Laffargue, J., Demay, Y., Agassant, J. F., "Investigation of Polymer Stretching Instabilities: Application to Film Blowing," *Int. Polym. Process.* (2010) 25, pp. 356-371

26. Yeow, Y. L., "Stability of Tubular Film Flow: A Model of the Film-Blowing Process," *J. Fluid Mech.* (1976) 75, pp. 577-591

27. Anturkar, N. R., Co, A., "Draw Resonance in Film Casting of Viscoelastic Fluids: A Linear Stability Analysis," *J. Non-Newtonian Fluid Mech.* (1988) 28, pp. 287-307

28. Barq, P., Haudin, J. M., Agassant, J. F., Bourgin, P., "Stationary and Dynamic Analysis of Film Casting Process," *Int. Polym. Process.* (1994) 9, pp. 350-358

29. Silagy, D., "Modélisation du Procédé d'Extrusion de Film à Plat de Polymère," Thèse de Doctorat, Ecole des Mines de Paris (1996)

30. Silagy, D., Demay, Y., Agassant, J. F., "Study of the Stability of the Film Casting Process," *Polym. Eng. Sci.* (1996) 36, pp. 2614-2625

31. Silagy, D., Demay, Y., Agassant, J. F., "Stationary and Stability Analysis of the Film Casting Process," *J. Non-Newtonian Fluid Mech.* (1998) 79, pp. 563-583

32. Souli, M., Demay, Y., Habbal, A., "Finite Element Study of the Draw Resonance Instability," *Eur. J. Mech., B: Fluids* (1993) 12, pp. 1-13

33. Fortin, A., Carrier, P., Demay, Y., "Numerical Simulation of Coextrusion and Film Casting," *Int. J. Numer. Methods Fluids* (1995) 20, pp. 31-57

34. Denn, M. M., Petrie, C. J. S., Avenas, P., "Mechanics of Steady Spinning of a Viscoelastic Liquid," *AIChE J.* (1975) 21, pp. 791-799

35. Chambon, F., Agassant, J. F., Demay, Y., Silagy, D., Ohlsson, S., "A Method to Reduce Draw Resonance," WIPO patent WO9631348 (1996)

36. Kim, J. M., Lee, J. S., Jung, H. W., Hyun, J. C., "Transient Solutions of the Dynamics of Film Casting Process Using a 2-D Viscoelastic Model," *J. Non-Newtonian Fluid Mech.* (2005) 132, pp. 53-60

37. Kase, S., Matsuo, T., "Studies on Melt Spinning I: Fundamental Equations on the Dynamics of Melt Spinning," *J. Polym. Sci. A* (1965) 3, pp. 2541-2554

38. Pearson, J. R. A., Matovich, M. A., "Spinning of a Molten Threadline Stability," *Ind. Eng. Chem. Fund.* (1969) 8, pp. 605-609

39. Ishihara, H., Kase, S., "Studies on Melt Spinning. VI. Simulation of Draw Resonance Using Newtonian and Power Law Viscosities," *J. Appl. Polym. Sci.* (1976) 20, pp. 169-191

40. Fischer, R. J., Denn, M. M., "Finite-Amplitude Stability and Draw Resonance in Isothermal Melt Spinning," *Chem. Eng. Sci.* (1975) 30, pp. 1129-1134

41. Demay, Y., Agassant, J. F., "Application de la Stabilité Linéaire à l'Etude du Filage Textile Isotherme et Non-isotherme," *J. Mec. Theor. Appl.* (1982) 1, pp. 763-782

42. Shah, Y.T., Pearson, J.R.A., "On the Stability of Non-isothermal Fiber Spinning," *Ind. Eng. Chem. Fund.* (1972) 46, pp. 145–149
43. Fischer, R.J., Denn, M.M., "A Theory of Isothermal Melt Spinning and Draw Resonance," *AIChE J.* (1976) 22, pp. 236–246
44. Lee, J.S., Jung, H.W., Kim, S.H., Hyun, J.C., "Effect of Fluid Viscoelasticity on the Draw Resonance Dynamics of Melt Spinning," *J. Non-Newtonian Fluid Mech.* (2001) 99, pp. 159–166
45. Jung, H.W., Song, H.S., Hyun, J.C., "Analysis of the Stabilizing Effect of Spinline Cooling in Melt Spinning," *J. Non-Newtonian Fluid Mech.* (1999) 87, pp. 165–174
46. Lee, J.S., Shin, D.M., Jung, H.W., Hyun, J.C., "Transient Solutions of the Dynamics in Fiber Spinning Process Accompagnied by Flow-Induced Crystallization," *J. Non-Newtonian Fluid Mech.* (2005) 130, pp. 110–116
47. Kohler, W.H., McHugh, A.J., "Sensitivity Analysis of Low-Speed Fiber Spinning of Isotactic Polypropylene," *Chem. Eng. Sci.* (2007) 62, pp. 2690–2697
48. Jarecki, L., Ziabicki, A., "Viscosity Effects in Computer Modeling of Fiber Spinning from Crystallizing Polymer Melts," *Polimery* (2004) 49, pp. 101–109
49. Pearson, J.R.A., Petrie, C.J.S., "The Flow of a Tubular Film, Part 1: Formal Mathematical Representation," *J. Fluid Mech.* (1970) 40, pp. 1–19
50. Han, C.D., Park, J.M., "Studies on Blow Film Extrusion. I. Experimental Determination of Elongational Viscosity," *J. Appl. Polym. Sci.* (1975) 19, pp. 3257–3276
51. Kanai, T., White, J.L., "Dynamics, Heat Transfer and Structure Development in Tubular Film Extrusion of Polymer Melts: A Mathematical Model and Predictions," *J. Polym. Eng.* (1985) 5, pp. 135–157
52. Cain, J.J., Denn, M.M., "Multiplicities and Instabilities in Film Blowing," *Polym. Eng. Sci.* (1988) 28, pp. 1527–1541
53. Andre, J.M., Agassant, J.F., Demay, Y., Haudin, J.M., Monasse, B., "Numerical Modelling of the Polymer Film Blowing Process," *Int. J. Form. Processes* (1998) 1, pp. 197–210
54. Hyun, J.C., Kim, H., Lee, J.S., Song, H.S., Jung, H.W., "Transient Solutions of the Dynamics in Film Blowing Processes," *J. Non-Newtonian Fluid Mech.* (2004) 121, pp. 157–162
55. Doufas, A.K., McHugh, A.J., "Simulation of Film Blowing Including Flow-Induced Crystallization," *J. Rheol.* (2001) 45, pp. 1085–1104
56. Pirkle, J.C., Braatz, R.D., "A Thin-Shell Two-Phase Microstructural Model for Blown Film Extrusion," *J. Rheol.* (2010) 54, pp. 471–505
57. Yoon, K.S., Park, C.W., "Stability of a Blown Film Extrusion Process," *Int. Polym. Process.* (1999) 14, pp. 342–349
58. Pirkle, J.C., Braatz, R.D., "Instabilities and Multiplicities in Non-isothermal Blown Film Extrusion Including the Effects of Crystallization," *J. Process Control* (2003) 21, pp. 405–414
59. Shin, D.M., Lee, J.S., Jung, H.W., Hyun, J.C., "Multiplicity, Bifurcation, Stability and Hysteresis in Dynamic Solutions of Film-Blowing Process," *J. Rheol.* (2007) 51, pp. 605–621

60. Henrichsen, L.K., McHugh, A.J., “Analysis of Film Blowing with Flow enhanced Crystallization, Part 2: Linearized Sensitivity and Stability Behavior,” *Int. Polym. Process.* (2007) 22, pp. 190–197

61. Housiadas, K., Tsamopoulos, J., “Unsteady Flow of an Axisymmetric Annular Film under Gravity,” *Phys. Fluids* (1998) 10, pp. 2500–2516

62. Laffargue, J., “Etude et Modelisation du Procédé de Soufflage de Gaine,” Thèse de doctorat, Mines-ParisTech (2003)

63. Lee, J.S., Jung, H.W., Hyun, J.C., “Transient Solutions of Non-linear Dynamics in Film-Blowing Accompanied by On-line Crystallization,” *J. Rheol.* (2011) 55, pp. 257–271

64. Lee, J.S., Jung, H.W., Hyun, J.C., “Non-linear Dynamic Behavior of Non-isothermal Film Blowing Process under the Condition of Constant Bubble Pressure,” *AIChE J.* (2011) 57, pp. 3565–3570

7 Biaxial Oriented Film Technology

J. Breil

7.1 Introduction

The stretching of films constitutes a finishing process in which the mechanical properties, optical characteristics, and barrier properties are increased significantly. The improvements of the properties result from the orientation of the molecular chains due to the stretching as well as from the increase in the degree of crystallinity in the case of semicrystalline plastics. These effects have been known since the 1930s where stretching processes were already used for polystyrene and PVC (polyvinyl chloride). However, there was not a commercial breakthrough until the mid-1950s, when ICI and DuPont developed the biaxial stretching of polyester (BOPET). This technology initially had proved of value due to the achievable property profiles for technical applications and was spread around the world in a short time due to licensing. The biaxial stretching of polypropylene (BOPP) followed in the mid-1960s, which gained a large market share mainly in the field of packaging and substituted for the previously dominant cellophane [1].

The stretching technologies can distinguished in terms of the orientation of the stretching and the stretching process itself. Longitudinal stretching, transverse stretching, sequential biaxial stretching, and simultaneous biaxial stretching as well as the double-bubble process do not really represent competing technologies but rather complete each other in order to achieve specific film characteristics, each of which is suitable for certain product groups (Fig. 7.1).

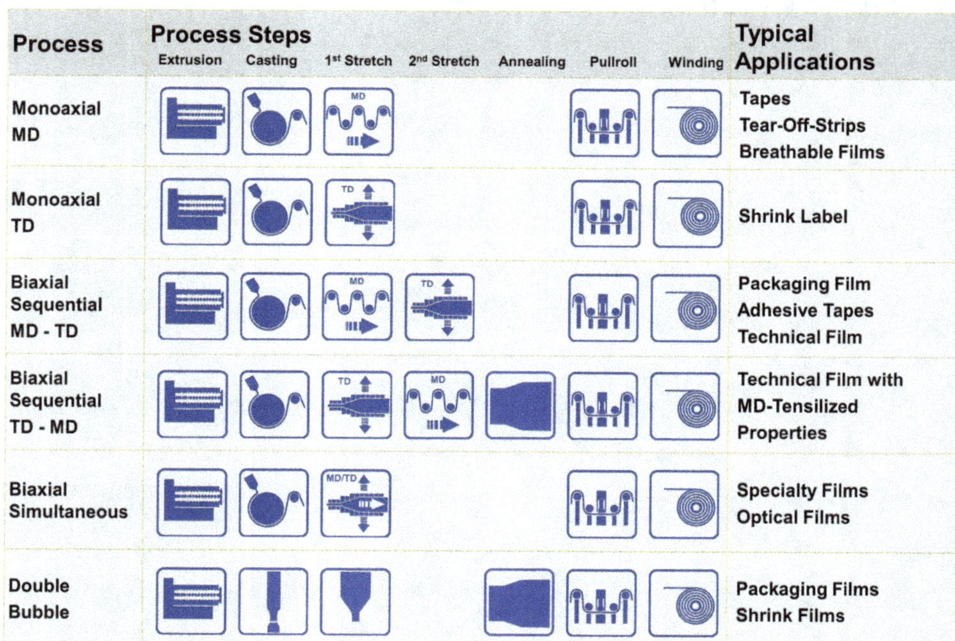

Figure 7.1 Outline of stretching technologies

In the case of monoaxial stretching in the machine direction (MD), the stretching is realized by means of rollers with increasing speeds, which in turn results in an orientation of the molecules in the machine direction. This leads to mechanical properties that are mainly enhanced in the machine direction and are thus suitable for high-strength packaging straps and tear-off strips. Furthermore, this process is also applicable to breathable films, where polymers with a high content of inorganic filling materials are employed.

Tenters are used in the case of monoaxially stretched film in the transverse direction (TD), with the most common application being shrink labels with high shrink values in the transverse direction and low shrinkage in the machine direction.

By far the most commonly used stretching process is the biaxial sequential technology. In this case, the film is usually first stretched in the longitudinal direction and then in the transverse direction. This method prevails for most of the packaging and also technical applications because it makes it possible to combine the highest productivity and very good quality. Another variant is biaxial sequential stretching in which the film is first stretched in the transverse direction and then in the machine direction. In this case an additional annealing oven is required in order to reduce the shrink values to acceptable limits. This process is restricted with respect to the maximum working width and speed, with the result that the productivity of sequential MD-TD machines cannot be attained. For this reason this method is only used for a few applications in which a very high strength in the machine direction is decisive.

Another long-established process is simultaneous biaxial stretching, where the film is stretched in the longitudinal and transverse directions at the same time. In this case, the clips that hold the film move on diverging rails so that the film is stretched in the transverse direction while the distance of the clips is increased at the same time [2]. There are different technical solutions available to control the clip distance: spindle, pentagraph, and LISIM (linear motor simultaneous stretching technology), which will be explained in detail in Section 7.2.2.

The so-called double-bubble process is another simultaneous biaxial stretching method. In this case, first a tube is extruded and cooled down. It is then heated to stretching temperature and stretched afterwards. This is simultaneously done by increasing the haul-off speed and the effect of the internal pressure, which forms a bubble. This process usually serves to realize lower outputs, with the result that the productivity data of state-of-the-art tenter stretching machines cannot be attained. The prevailing application for this method is shrink film on PE-basis (polyethylene-basis) as well as, to a minor degree, BOPA, BOPP, BOPET (biaxially oriented polyethylene terephthalate), and multilayer films.

7.2 Biaxial Oriented Film Lines

In the following, sequential and simultaneous film stretching machines are explained in detail in order to illustrate the current state-of-the-art configuration of the systems and system components.

7.2.1 Sequential Film Lines

BOPP production lines represent the majority of the sequential film stretching systems. Therefore the typical components can be explained by taking the example of state-of-the-art BOPP machines. With net film widths up to 10.4 m, speeds up to 525 m/min, machine lengths up to 150 m, and output capacities up to 7.5 t/h (metric tons per hour), these types of biaxial stretching machines rank among the largest plastic processing systems ever. Dimensions and design of the individual components depend on the film types to be produced, the layer structure, and the output. In general, the layout of the systems for different film types with its components of raw material supply, extrusion, casting unit, longitudinal stretching machine, transverse stretching machine, pull roll stand, and winder are basically similar (Fig. 7.2). In detail, however, the system components must be adapted to the specific requirements of each raw material.

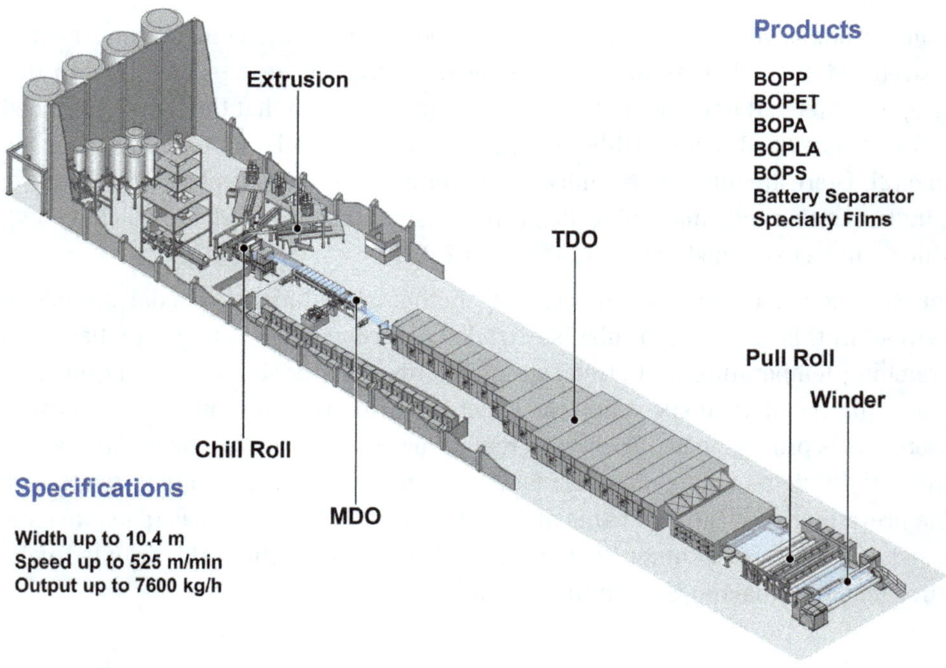

Figure 7.2 Sequential biaxial stretching line

The outline of typical state-of-the-art line data for different film types is given in Table 7.1.

Table 7.1 Outline of Typical Thickness Ranges and Line Data

Line Types		PP		PET				PA
		Capacitor	Packaging	Capacitor	Packaging	Industrial/Optical		Packaging
						Medium	Thick	
Max. Line Width	m	5.8	10.4	5.7	8.7	5.3	5.8	6.6
Thickness Range	µm	3–12	4–60	3–12	8–125	20–250	50–400	12–30
Max. Production Speed	m/min	280	525	330	500	325	150	200
Max. Output	Kg/h	600	7600	1100	4250	3600	3600	1350

The dimensioning of the widths of all components results from the required net film widths taking into consideration the TD stretching ratio, edge trim, and neck-in effects. The chill roll diameter and roll number for the longitudinal stretching machine as well as the zone lengths of the transverse stretching machine are calculated from the heating and cooling time as well as the required dwell time in the individual temperature zones. For this purpose, suitable calculation programs are available that calculate both the film temperature in the machine direction and the temperature profile along the film cross section. Figure 7.3 depicts an example of the typical temperature profile for BOPP production. The maximum temperature differences along the cross section can be determined from the temperatures of the chill roll, center, and air-knife side, which are illustrated separately. In particular, the cooling process on the chill roll is decisive for the molecular structure along the cross section. Here a symmetric cooling is required, which is achieved by placing the cooling roller into a water bath.

Exact temperature control in all process steps is essential in order to attain the required characteristics of the stretched film. This must also be ensured in all system components along the complete system width.

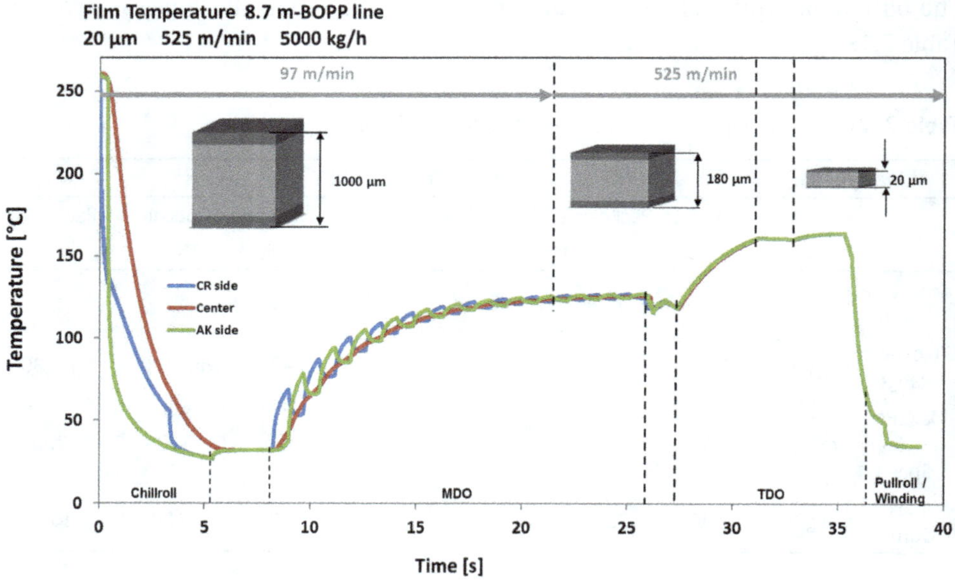

Figure 7.3 Calculated film temperatures for BOPP stretching process

7.2.1.1 Extrusion

The extrusion unit of the stretching machine is usually equipped with a main extruder and several coextruders in order to meet the requirements for a multilayer structure with different raw materials by means of coextrusion. The most common variant is the three-layer structure with one extruder for each layer. For BOPP systems, however, five-layer coextrusion with five extruders is also used (Fig. 7.4); in rare cases, there are seven or nine layers, which is primarily for barrier film applications.

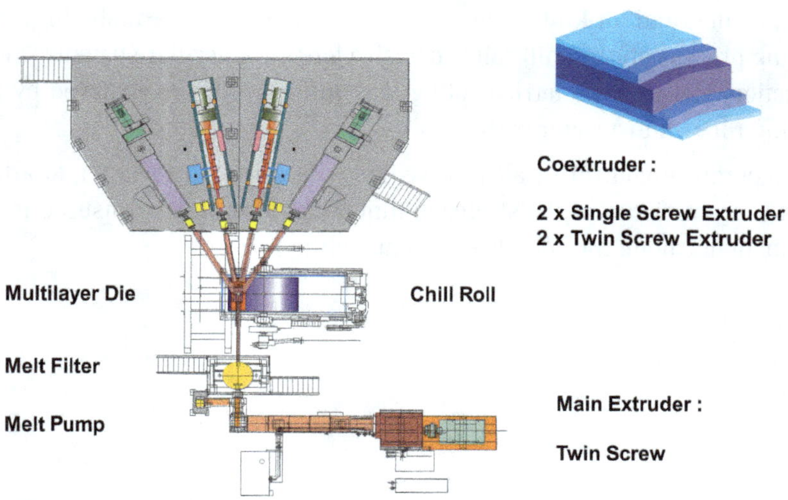

Figure 7.4 Extrusion configuration for five-layer structures

Corotating twin-screw extruders have become common for the main extrusion (Fig. 7.5). The advantages over single-screw and cascade extruders are

- low specific-energy consumption,
- compact machine design,
- continuous vacuum degassing,
- adaptability by modular design of screw and cylinder,
- direct additive compounding of powder and liquid components,
- adjustable melt temperature,
- good homogenization and mixing, and
- availability for maximum output of up to 8.2 t/h.

For coextrusion, single-screw and twin-screw extruders are used. In twin-screw extruders, melt pumps are installed both for the main extrusion and the coextrusion in order to realize the required pressurization for filtration and nozzle as well as to make the output as constant as possible. The filtration is done by means of large-area filters that, in BOPP systems, are usually equipped with plain or pleated candle filters of up to 6 m^2 filter surface in order to achieve a service life of greater than four weeks. For BOPET, large-area filtration is also required, but in this case disk filters are used. Figure 7.6 shows both types of filter systems.

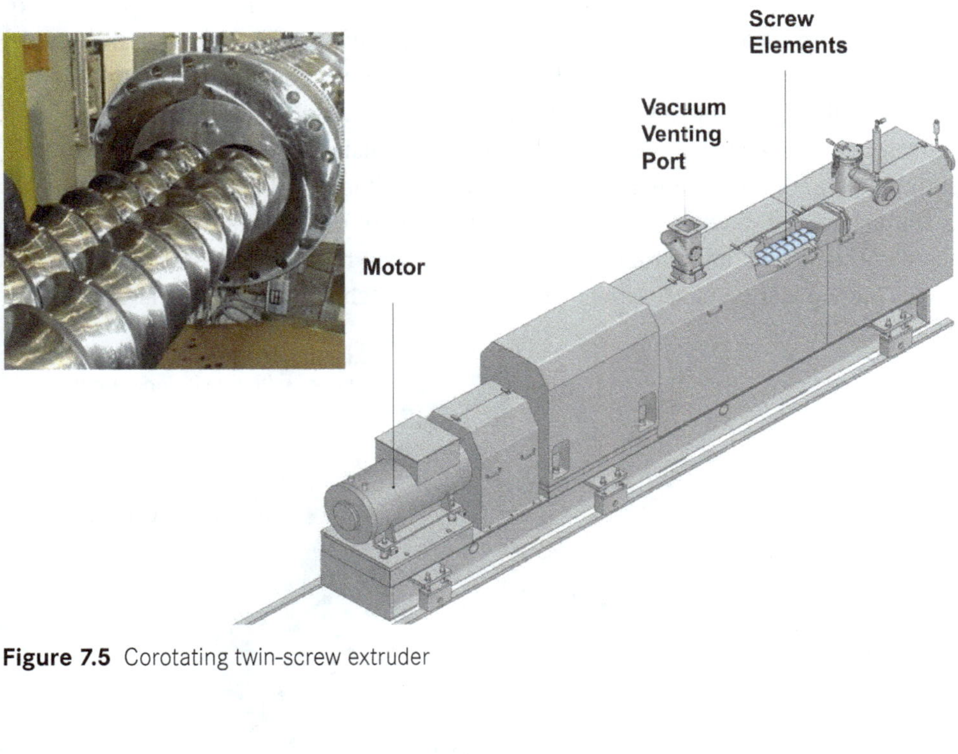

Figure 7.5 Corotating twin-screw extruder

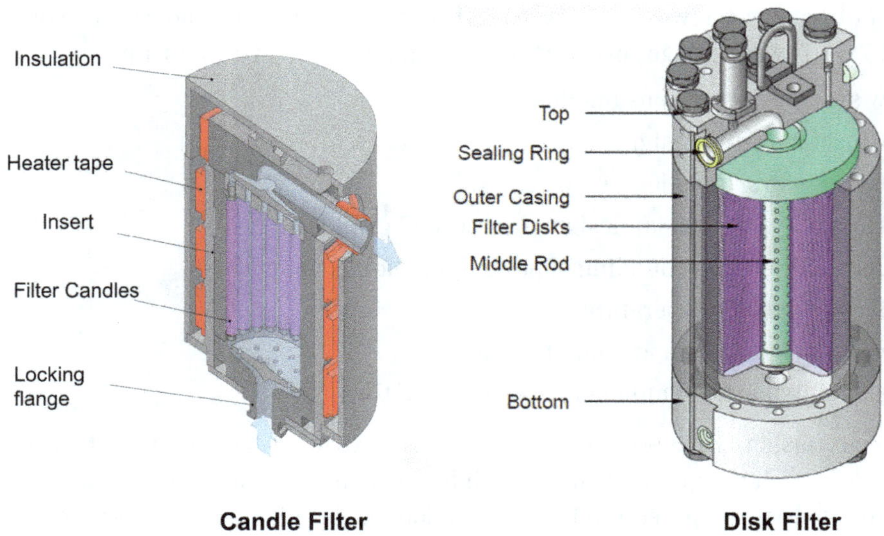

Figure 7.6 Large area melt filtration

After the filtration the melt is led to the extrusion die, which is designed as a multi-channel coat-hanger die in BOPP systems. This has been proven for obtaining a very uniform thickness of the individual layers across the width in the desired thickness and output range even for different viscosities. In general the extrusion dies are equipped with an automatic die bolt adjustment, which is operated in a closed loop with the thickness gauge in the pull roll stand in order to be able to control the film thickness along the entire working width. The automatic die bolt systems have been optimized in such a way that, on the one hand, the distance of the actuators could be diminished and, on the other hand, the response time could be reduced. Figure 7.7 is the schematic diagram illustrating that the individual bolts are heated by rod heaters and cooled down from the outside. The good heat contact and the low mass allows for the fast response time of 20 s, and the pitch of the die bolts is 10 mm.

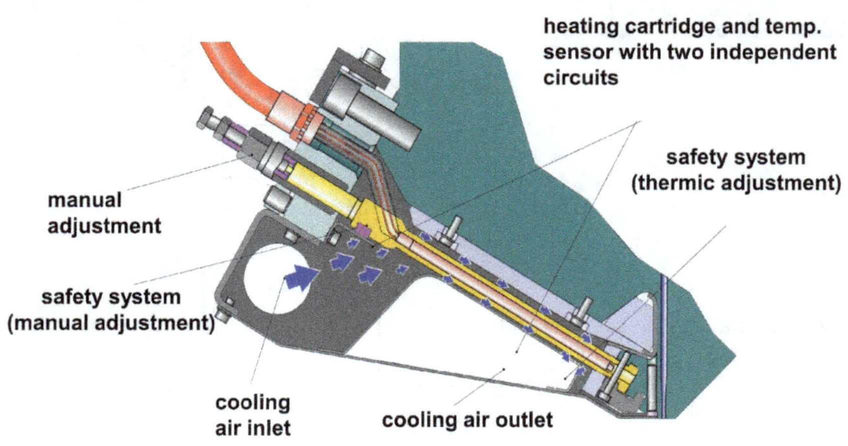

Figure 7.7 Automatic die bolt system

7.2.1.2 Casting Machine

When the melt exits the extrusion die it must be cooled down rapidly, even across the width, and a homogenous cast film must be formed as a base for the subsequent stretching process. This process significantly affects the output capacity and film quality obtainable with the system. Figure 7.8 depicts a typical configuration for a chill roll unit for BOPP systems. Typical is the arrangement of the chill roll in a water bath in order to achieve a symmetric cooling and the subsequently required water removal, which is realized by a high-pressure air blowing nozzle. The cooling in a water bath has the advantage that a high heat transfer is realized on both film sides and that the cooling happens as fast and as symmetrically as possible in this way. Furthermore, it is critical that the same cooling conditions are achieved along the entire working width.

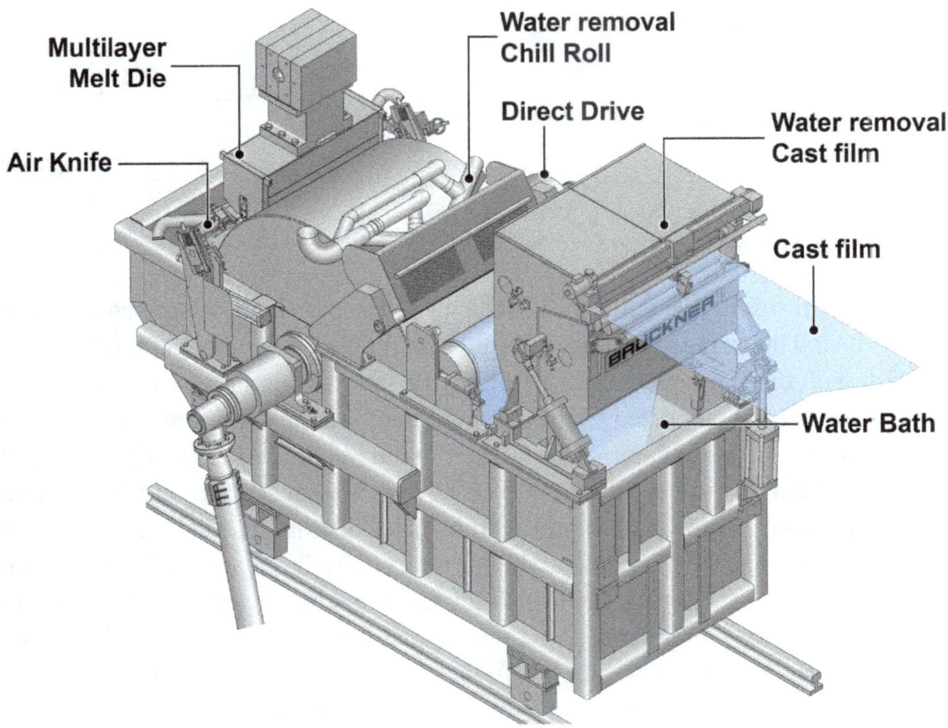

Figure 7.8 Casting unit with chill roll

This is, on the one hand, made possible by efficient water circulation; on the other hand, the internal layout of the chill roll is of vital importance in achieving temperature equality. Figure 7.9 illustrates the principle of internal cooling in which the demand for the preferably uniform chill roll surface temperature is taken into consideration by use of a degressive design of the cooling ducts. The heat transfer depends on the temperature difference between melt and chill roll surface as well as on the heat transfer coefficient. This, in turn, is defined by the speed of the water in the cooling duct. The fact that the cooling medium warms up in the spiral winding is compensated here by the increased speed due to the diminishing design of the duct cross sections in the spiral, which leads to a higher heat transfer. The resulting uniform cooling of the film along the working width makes it possible to attain correspondingly uniform film properties along the working width.

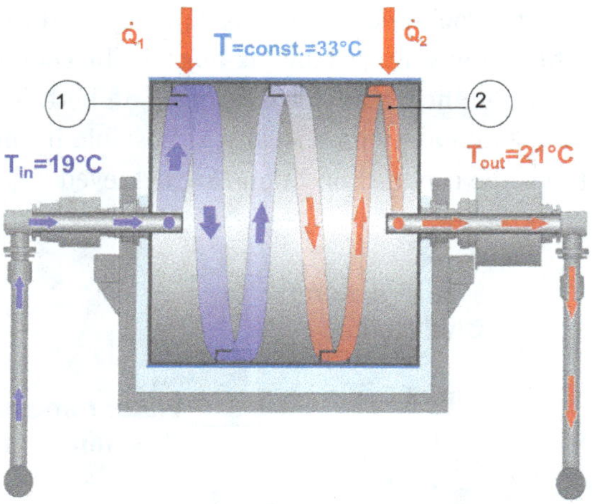

$\Rightarrow v_2 > v_1$

$\Rightarrow \alpha_2 > \alpha_1;\ \alpha = f(v)$

$\Rightarrow \dot{Q}_2 = \dot{Q}_1$

$$\dot{Q} = \alpha \cdot A \cdot \Delta T$$

$$\alpha = \frac{\lambda}{L} \cdot Nu(Re, Pr)$$

Figure 7.9 Chill roll with degressive cooling channel

In cast film processes, the uniformity of the melt discharge and the speed of the chill roll are of vital importance for the achievable longitudinal tolerances of the film. The constant melt discharge is defined by the extrusion, with minimum pressure fluctuations having to be realized when the melt enters the die. The uniformity of the chill roll surface speed is determined by the run-out tolerance of the chill roll as well as the speed stability of the drive unit. Direct drives are particularly suitable for this purpose because faults by mechanical transmission units like belts or gears are avoided in this way. Here a high-resolution rotary encoder is attached to the same axis, and a corresponding control loop is implemented that is optimized in order to achieve the best speed tolerance. Another advantage is the fact that this drive type is generally maintenance-free and low in losses. Figure 7.10 describes the configuration for a direct-drive system for a chill roll using a torque motor.

Another important component of the casting unit is the pinning system. In the case of BOPP lines, air-knife units are preferentially used. For this purpose, the uniformity of the air discharge is decisive as well as the airflow after the discharge. In addition to the air-knife, either external-edge blow nozzles are used or this functionality is integrated into the air-knife. Furthermore, an exact adjustability of the position relative to the die and chill roll is required for all pinning devices, which is either realized by manually operated or automatic two-axis positioning units. Pinning technologies have to be variably adjusted and optimized for each raw material. Whereas the air-knife application is the standard for PP packaging film, an electrostatic pinning device either with a high-voltage wire or blade is typically used for BOPET lines. In BOPA lines an electrostatic needle pinning is the state of the art. As an alternative, an HPA (high-pressure air-knife) was developed that serves to realize approximately 20% higher line speed compared with the electrostatic pinning.

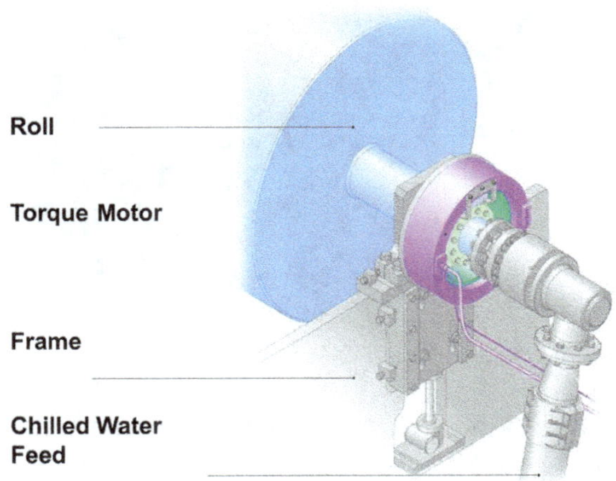

Figure 7.10 Chill roll with direct drive

In considering the pinning technology, it must be noted that the maximum achievable speed depends not only on the pinning device but also on the raw material used. In particular, the melt viscosity is important and, in the case of electrostatic pinning, also the electrical conductivity of the melt. There is a correlation that with increased melt conductivity higher pinning speeds can also be realized. In the case of polyester, pinning speeds up to 130 m/min can be realized, which lead to line speeds after stretching of more than 500 m/min.

7.2.1.3 Machine Direction Orienter (MDO)

The first stage of the sequential biaxial stretching process is uniaxial in the machine direction and is realized by rollers with increasing speeds. The typical stretching ratios are for PP 1:5, for PET 1:3.5–4.5, and for PA (polyamide) 1:3. According to their function, the rollers can be classified into three groups (Fig. 7.11): preheating zone, stretching zone, and annealing zone.

The preheating zone requires a homogenous preheating of the cast film to the target stretching temperature, which is realized by a high heat transfer coefficient and uniform heating of the rollers along the working width. The stretching zone is characterized by rollers with a small diameter that are arranged in such a way that a nip roller can be added to each roller and that the roller distance can be adjusted in order to be able to adapt the stretching gap to the product.

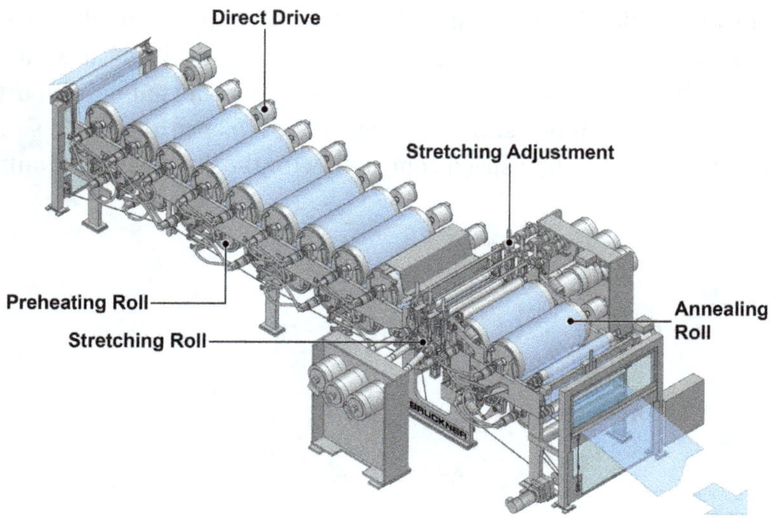

Figure 7.11 Machine direction orienter (MDO)

This arrangement makes it possible to minimize the neck-in effect, which causes a width reduction during the MD stretching process, to avoid slippage at the rollers and to control the stretching speed. In doing so it is advantageous to use direct drives in the entire MDO in order to control each roller separately, that is, to set the optimal speed and torque. It is also important for the preheating zone to ensure a good contact of the film to the roller by means of sufficient tension, thus ensuring a uniform heat transfer, and to compensate for the length variation that is caused by the thermal expansion during the heating process at the same time.

In the stretching system, the drive concept of direct drives has the advantage that the stretching gaps can be individually divided by adjusting the increase in speed between the individual rollers, which therefore leads to optimal settings for each product (Fig. 7.12).

In particular, sensitive skin layers, such as low-temperature sealing layers, can be shielded from surface damage in this way. By separating the total MD stretching ratio into several individual stretching gaps, a higher total stretching ratio can be realized, which leads to better product properties and process stability.

The rollers are heated using thermal oil or pressurized water. In the case of BOPET lines, additional infrared heating elements are employed in the stretching gap. Also in this case the total stretching ratio can be increased correspondingly by means of multigap stretching, which has the advantage for BOPET that this also allows for a higher total line speed because the bottleneck of the line is the limited pinning speed at the chill roll. Due to the speed increase in the MDO, a production speed over 500 m/min can therefore be realized for BOPET lines.

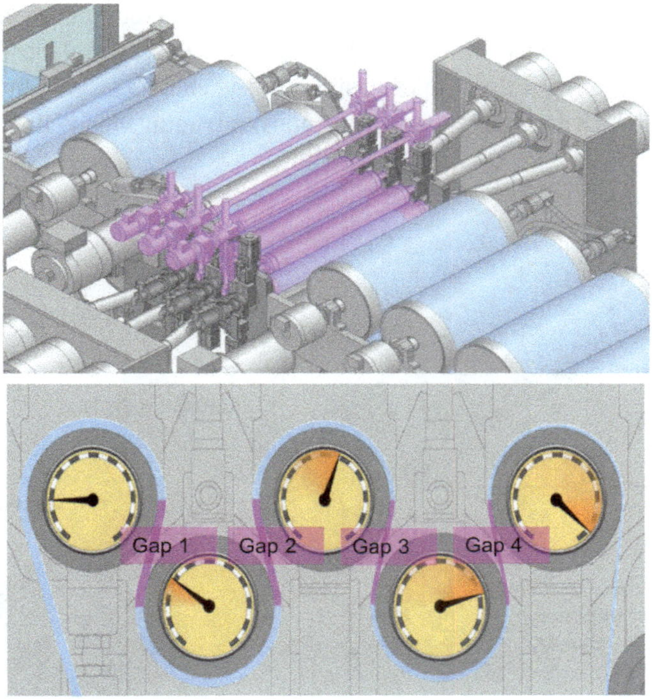

Figure 7.12 MDO multigap stretching

7.2.1.4 Transverse Direction Orienter (TDO)

In the transverse direction orienting machine, the MD stretched film is fixed with holding devices (clips) at the film edges and stretched along an adjustable rail in the transverse direction. This process takes place in an oven that is segmented into a preheating zone, a stretching zone, a thermosetting zone, and a cooling zone (Fig. 7.13). The machine component that ensures the transport and holding mechanism through the oven is the so-called chain-track system, which has to perform with high reliability, low wear, and long lifetime. For this purpose, roller chains or sliding chains with speeds over 550 m/min are available (Fig. 7.14). In the sliding systems the chain is guided on the rail with replaceable sliding elements, where a thin wetting of oil for lubrication has to be ensured on the sliding surfaces.

These systems are designed in such a way that the chain system is effectively shielded from the process environment in order to avoid the contamination of the film surfaces with oil spots. In the roller chains the support and guide of the chain track system is realized by means of roller bearings. In this case less oil is used for lubrication compared with a sliding chain. The roller bearings and track are exposed to wear, which can be reduced by a special chain geometry that ensures that the bearings have permanent contact with the rail. The clips must be closed at the beginning of the transverse stretching machine in order to hold the film and must be opened at the end in order to allow for the further transport of the film in the pull roll stand. This is realized contact-free by magnetic opening and closing bars. During the gripping process the proper position of the closing system must follow the film edge, for which hydraulic systems or, most advantageously, linear motors are used.

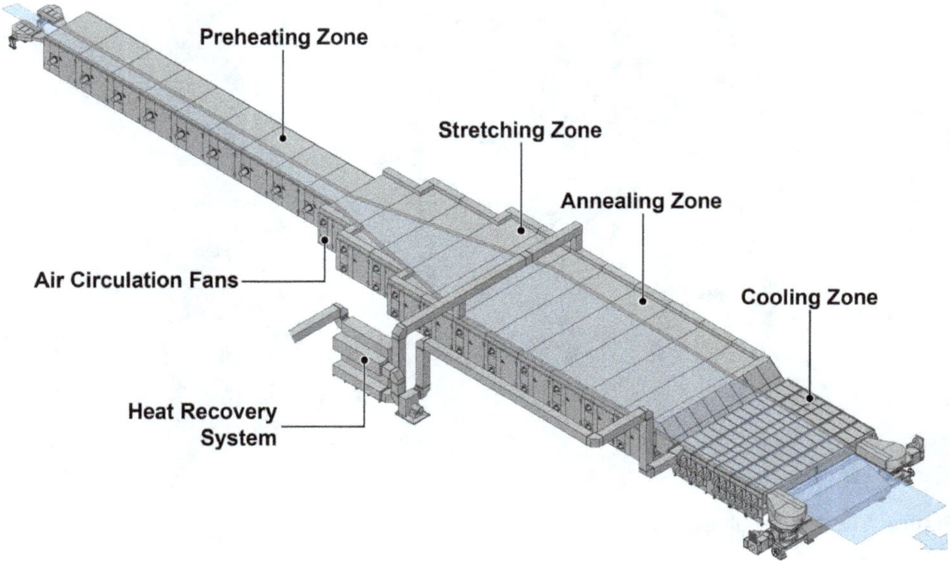

Figure 7.13 Transverse direction orienter (TDO)

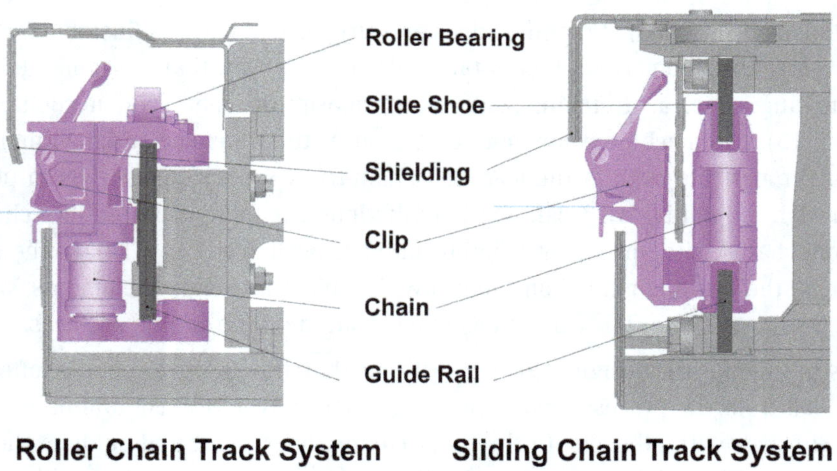

Figure 7.14 Roller chain- and sliding chain track systems

In the transverse stretching machine, the temperature control is decisive for achieving the desired film properties and their distribution along the working width. Therefore it is necessary that both the temperature and the heat transfer conditions are as uniform as possible along the working width.

Also, in a 10-m-wide oven, a temperature accuracy of ±1 °C must be complied with over the entire working width. For this purpose a circulating air system is used (Fig. 7.15), which is designed in such a way that a very uniform air discharge is realized by means of slot or hole nozzle boxes. The air sucked back flows over heater exchangers, which are optionally heated using electric, oil, steam, or a direct gas heating. The circulating air system is designed with separate fans above and below the film surface, which are controlled by frequency controllers in order to adjust the fan speed individually for different products. The additive content of the films and the surface enlargement during the transverse stretching process leads to evaporation at the applied temperatures, which results in condensate formation in the oven. This must be reduced to a tolerable amount by means of suitable air exchange rates. Because high exchange quantities entail a corresponding energy loss, the application of heat recovery systems is useful at this stage. In doing so, the fresh air is heated by the exhaust air by a heat exchanger (Fig. 7.16) and added to the individual zones by a central fresh air channel.

Approximately 300 kW can be recovered in this way. For BOPET lines, the installation of catalysts and filters, which are integrated into the circulating air system, have proven necessary in order to reduce the concentration of oligomers in the oven. In optical applications, large-area HEPA filters are mounted on the oven roof (penthouse design), and thus much cleaner conditions are obtained.

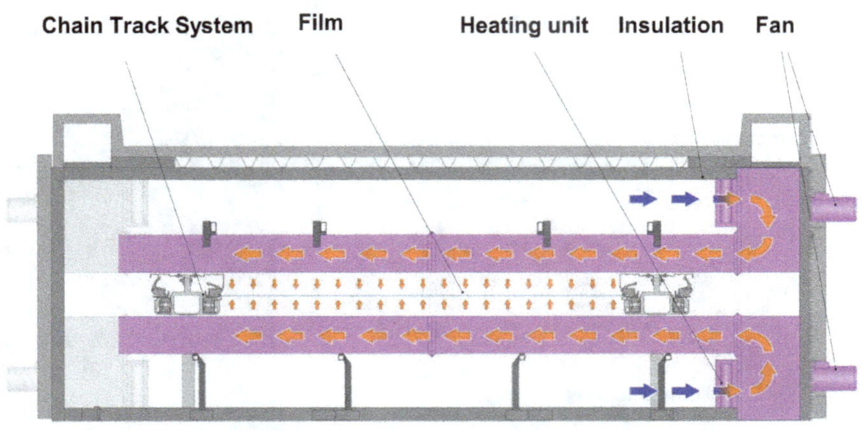

Figure 7.15 Cross section of a TDO oven zone

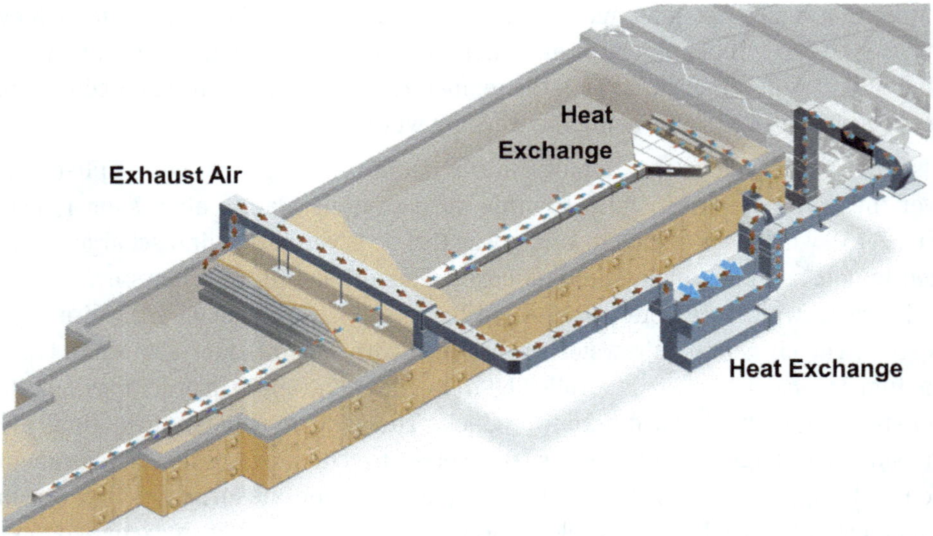

Figure 7.16 TDO heat recovery system

7.2.1.5 Pull Roll Stand

When the film exits the transverse stretching machine, it must first be cooled down before the edges are cut, the thickness is measured, and the surface is treated. These functions are realized in the so-called "pull roll stand," which is preferably realized in a C-frame design in order to allow for a simple and safe film feed from the operating side (Fig. 7.17).

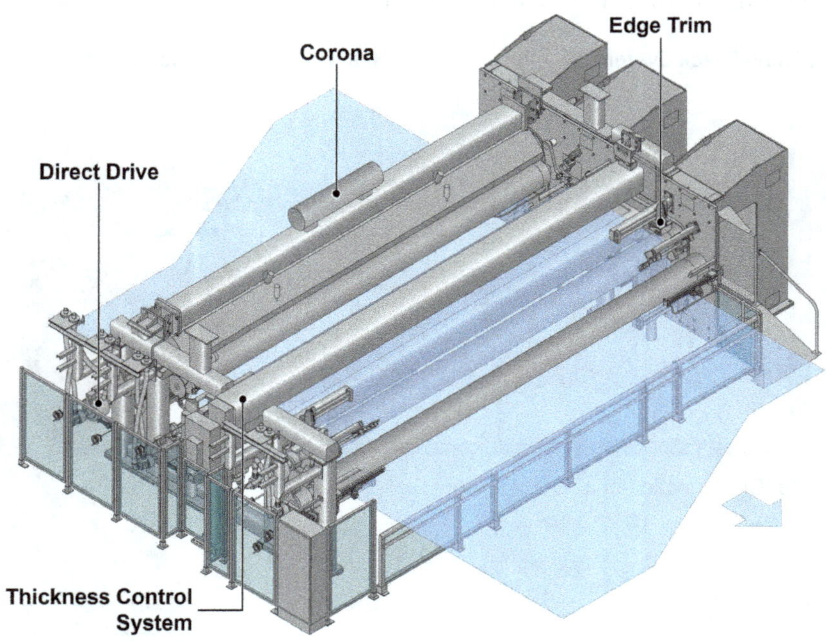

Figure 7.17 Pull roll stand

This requires a correspondingly robust structure for rollers of greater than 10 m working width and diameters up to 600 mm in order to avoid vibrations even for high speeds of over 500 m/min.

The edges are trimmed off by an appliance in which a blade and an automatic cutting device guarantee a safe cutting-off of the thicker edges, which are continuously sucked off on both sides. These edges are shredded to film fluff and then fed to the raw material supply for the extrusion via pipes, with the result that the edge trimming does not result in a loss of material.

After the edge trimming the thickness is measured continuously by a thickness gauge head that is traversing over the working width. Depending on the film type, different measuring procedures are employed. Beta-ray, X-ray, and infrared are the most common. The accuracy requirement for the measurement is typically 0.05 µm because the final film thickness and the thickness profile must be controlled on the basis of this signal.

The control is accomplished by activating the automatic die, with a special algorithm being required that allows for the correct alignment of the die bolt position to the corresponding film position of the biaxial stretched film. After the thickness measurement the surface treatment is realized in most cases by means of one or several corona stations. In some cases a flame treatment is applied, too. Its aim is to modify the surface tension in such a way that the film is suitable for subsequent processing (printing, lamination, metallization). The fact that the film is guided over several rollers with a large working width requires special focus on the tension control. For this purpose, each roller is driven by an individual torque motor, and a superimposed tension control ensures that in varying conditions the necessary tension is applied in each section of the pull roll stand.

7.2.1.6 Winder

The winding process in film stretching lines takes place on a winder with the entire working width, where winding weights of up to 7 metric tons are reached. According to the film type, the contact winding mode or gap winding mode can be selected.

The contact rollers, which are made of carbon fiber laminates, are dimensioned in such a way that a preferably low deflection along the working width as well as a good vibration damping are ensured. In order to adjust the contact rollers, a mechatronic system has been developed (LIWIND) with which the following functions are realized in one functional unit (Fig. 7.18): contact roll position, contact pressure, and damping function against vibrations. Linear motors in combination with precision linear scales are employed here, and a special control software is implemented that ensures all three functions. The programmed preselection of the winding tension and contact pressure characteristics is required for the ideal winding structure.

The information on film thickness, winding length, and contact roller position serves to calculate and control the winding density. The correct winding density setting, in turn, is a quality criterion for the subsequent storage process of the winding roll in which the postcrystallization takes place. Only this guarantees optimal prerequisites for the subsequent cutting process on primary cutting machines, where the customization to user-specific working widths and roll lengths from the primary roll takes place. The dimensions of the film winder in a 10.4 m BOPP line can be seen in Fig. 7.19.

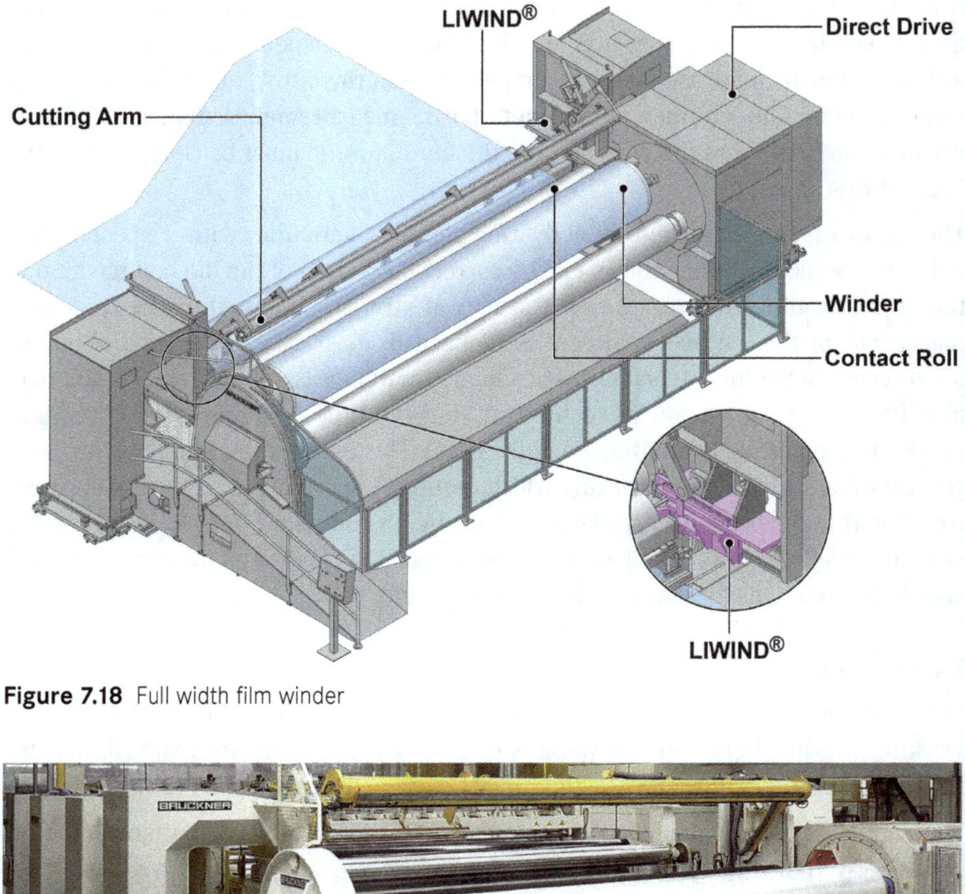

Figure 7.18 Full width film winder

Figure 7.19 Winder of a 10.4 m BOPP production line

7.2.2 Simultaneous Stretching Lines

As a contrast to sequential film stretching lines, in simultaneous stretching lines the film is not stretched in two separate steps in the MD and TD directions but simultaneously in both directions in one oven. In this case, the clips that hold the film move on diverging rails, so the film is stretched in the transverse direction while the distance of the clips is increased at the same time. There are different technical solutions for this process. In the so-called "pentagraph method" the clip distance is adjusted by a folding pentagraph geometry of the chain, whereas the distance of the clips is determined by the geometry of the guiding rails. In the spindle method, the clips moving along the rail lock into a spindle with a progressive notch and are separated in this way, which makes the longitudinal stretching possible. A third variant is the LISIM technology (linear motor simultaneous stretching) [3]. With this method the clips are driven by linear motors, which allows for a free adjustability of the clip distances along the entire machine and thus the local MD stretching ratio. This technology has the following advantages over the previously described mechanical solutions, which have a significant effect on both productivity and product quality:

- production speed up to 400 m/min,
- high flexibility of the stretching ratios in the longitudinal and transverse directions,
- variable setting of the relaxation in the longitudinal and transverse directions,
- low maintenance costs,
- high uptime,
- suitability for clean room conditions, and
- applicability for all stretchable polymers in a large thickness range.

This technology has been employed in production scale since 1998 and is available in adapted versions for different plastic film types (Fig. 7.20).

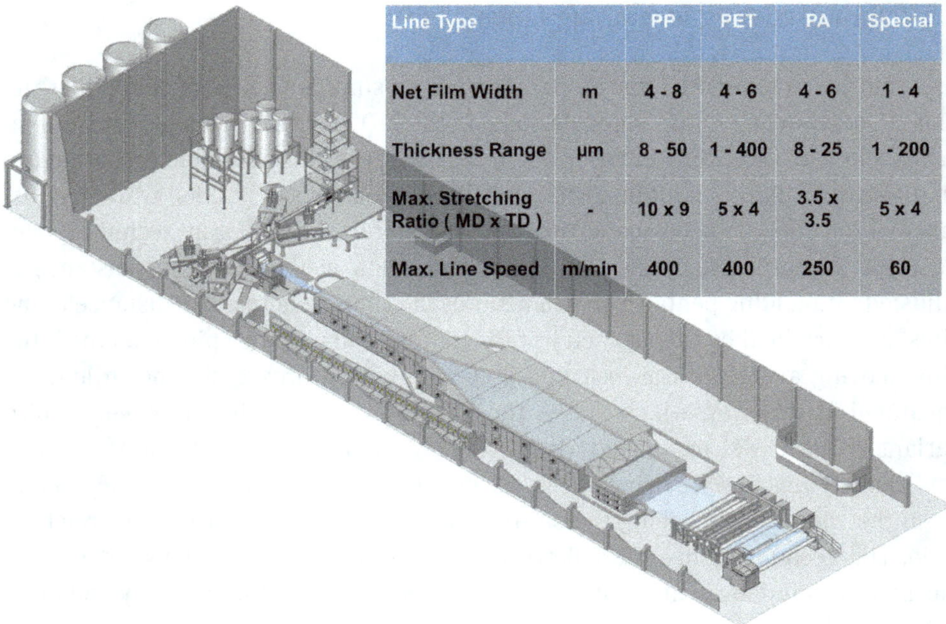

Line Type		PP	PET	PA	Special
Net Film Width	m	4 - 8	4 - 6	4 - 6	1 - 4
Thickness Range	µm	8 - 50	1 - 400	8 - 25	1 - 200
Max. Stretching Ratio (MD x TD)	-	10 x 9	5 x 4	3.5 x 3.5	5 x 4
Max. Line Speed	m/min	400	400	250	60

Figure 7.20 LISIM (linear motor simultaneous stretching) technology

The above-mentioned advantages are realized by a symmetric monorail track system (Fig. 7.21). The clip is guided by eight roller bearings, and permanent magnets are mounted on the top and bottom of the clip opposite to the linear motor stators that are fixed on the track system. The force that each clip needs for moving, acceleration, and film stretching is generated by the interaction of the magnetic fields of the permanent magnets and the stator, following the principle of the synchronous linear motor. In this case the moving magnetic wave of the linear motor stator is generated by the current supplied by adjustable frequency drives, whereas the current amplitude defines the force and the frequency defines the speed of the magnetic wave. The linear motor stators are cooled by integrated water pipes; therefore they can resist the severe conditions of a hot oven environment for a long lifetime. The system is designed for clean room conditions so a protective oil shield is mounted on the top and bottom of the system.

Due to the design feature that there are no mechanical links between the clips, there is an extreme flexibility regarding the speed patterns and MD stretching ratios, which are determined by the local distance of the clips throughout the whole machine. This flexibility in stretching patterns can be utilized to enhance the film properties significantly (Fig. 7.22).

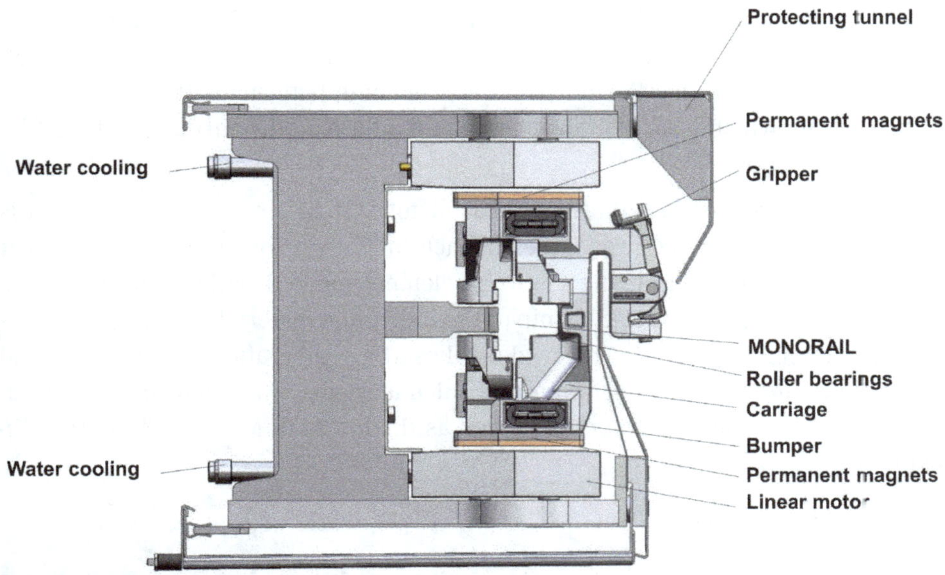

Figure 7.21 Cross section of a LISIM track system

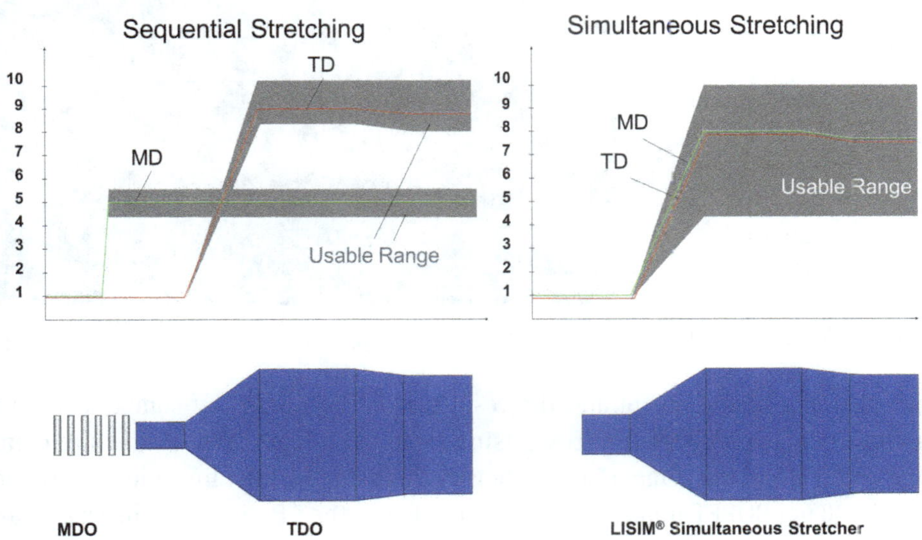

Figure 7.22 Comparison of stretching curves for BOPP with sequential and simultaneous stretching

An example is shown for the BOPP process comparing sequential and simultaneous stretching. Whereas in sequential stretching the MD and TD stretching ratios are determined by process limitations in the corresponding machine (MDO and TDO), the simultaneous stretching process allows a much wider range for the MD and TD stretching ratios. The advantage in this case that in contrast to sequential stretching, the MD ratio can be adjusted to higher values than TD, which results in higher mechanical properties in MD as well, or it is possible to adjust the stretching pattern for MD and TD in exactly the same way, which results in isotropic properties. Another advantage of this flexibility of the stretching patterns is the possibility of MD relaxation, which is an effective method for adjusting the shrinkage of the film in the MD direction.

For the product-related layout of the linear motor system for a production line it is necessary to know the required forces in each individual zone of the machine. For that purpose a simulation using the finite element method (FEM) is used. This is based on the stress-strain relationship of the individual materials during the simultaneous stretching process. The model takes the temperature, strain rate, and stretching ratio under consideration and calculates the two-dimensional distribution of the stress and the thickness as well as the forces in MD and TD on the clip positions (Fig. 7.23).

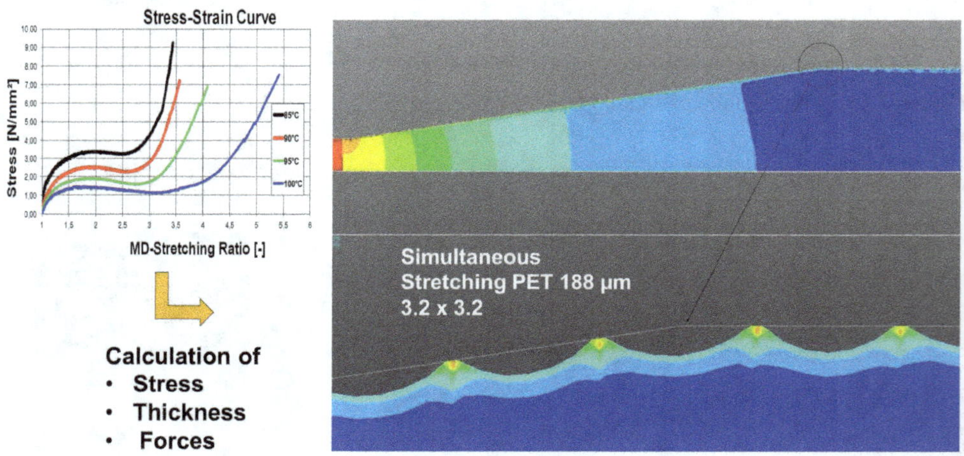

Figure 7.23 FEM simulation of the simultaneous stretching process

The reliability of the model and the results could be proven by comparison with measured data from a real process, using a clip with load cells for the stretching force (Fig. 7.24). This comparison is shown with data of a pilot line and the production of 188 µm BOPET film. The force calculation in the TD direction shows a steadily increasing force until the end of the stretching zone, whereas the force in the rail direction, which corresponds to the necessary motor force, shows positive and negative components during the path through the machine. Positive means the clip has to pull; a negative force means the clip has to hold back the maximum forces reached after the stretching zone, which is the basis for the layout of the linear motor driving system. The forces can be influenced in a wide range by adjusting the MD and TD stretching patterns as well as the temperature patterns of the individual zones.

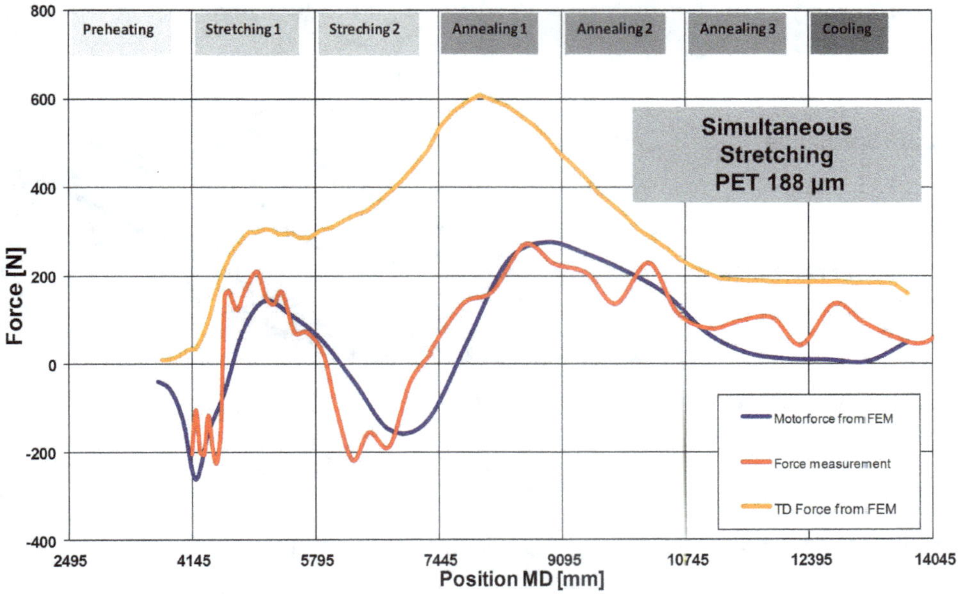

Figure 7.24 Comparison of simulated and measured forces during simultaneous stretching

The advantages of a high flexible simultaneous stretching technology are different and individually decisive for the different products due to the requirements and enhancement characteristics.

For BOPP, some products require high shrinkage in the MD direction and low in the TD direction, which can be adjusted by the individual stretching patterns in MD and TD. Such film is produced for MD shrink labels based on BOPP. Another example is the fact that different materials can be coextruded and stretched together in an appropriate process window (stretching ratio, temperature, strain rate). This effect can be used, for example, by stretching EVOH (ethylene vinyl alcohol) grades with low ethylene content with polypropylene in order to combine the good barrier characteristics of both materials for enhancing barrier values regarding OTR (oxygen transmission rate) and WVTR (water vapor transmission rate).

For BOPA, simultaneous stretching in combination with simultaneous relaxation allows very low and isotropic shrinkage characteristics, which is significant for the BOPA converting processes, that is, lamination with PE film. In this case the simultaneously oriented film has much fewer distortion characteristics when the laminates are under the temperature influence that is required during hot fill and sterilization processes.

For BOPET, two different cases have been proven in production scale; ultrathin film with high mechanical properties and isotropic film characteristics has been produced down to 0.5 micron, which is the lowest end for capacitor applications. For thick film for optical applications the advantages are summarized in Fig. 7.25.

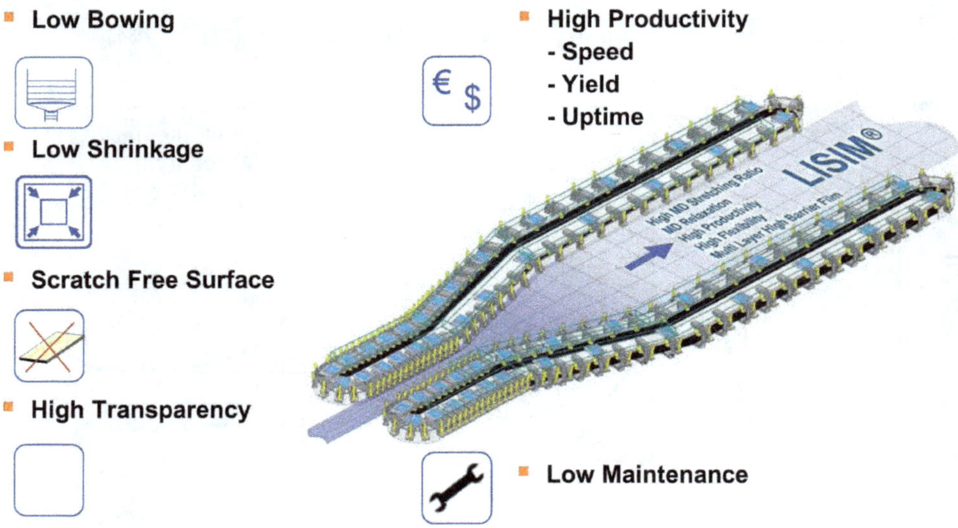

Figure 7.25 Advantages of LISIM technology for BOPET optical film

In flat-screen displays the angle of the molecular orientation, low and isotropic shrink values, a scratch-free surface, and a very high transparency are crucial for high-quality optical films. All of these properties can be set using the simultaneous technique by means of adapted stretching profiles. Such high-tech films are by now produced with a thickness up to 400 µm. The profitability is essentially determined by the high output capacity, high A-quality yield, and high availability that can be achieved with this technology.

Another economic advantage results if subsequent processing steps are not necessary due to the integration of all required functionalities into the simultaneous stretching process. Among them are off-line tempering processes in order to minimize the shrinkage values, as for some BOPET thick film with very low shrinkage requirements in the MD and TD directions. In this case the feature of MD relaxation at the end of the annealing zone is used to lower the MD shrinkage significantly (Fig. 7.26).

A relaxation rate of minimum 6% is needed in order to bring the shrinkage values close to zero, which is a requirement for some applications (for example, substrates for organic electronics).

Some optical films also require a special characteristic regarding the molecular orientation angle, which can be influenced in a wide range by the stretching patterns as well.

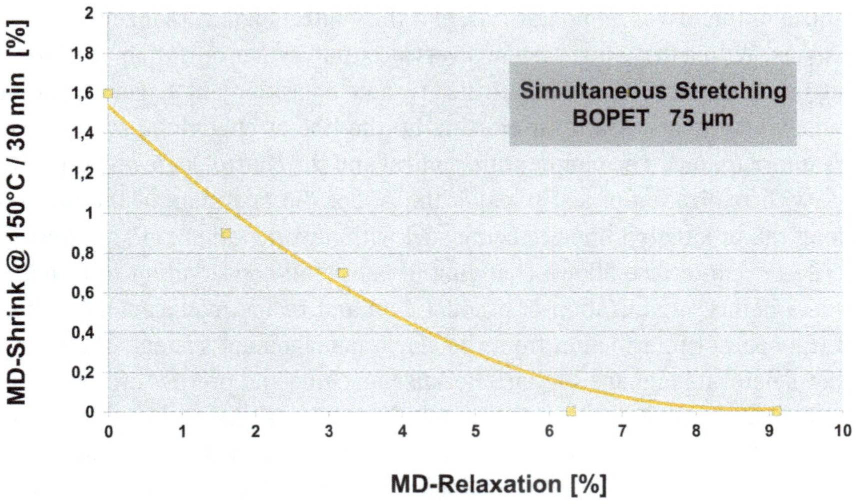

Figure 7.26 Influence of MD-Relaxation on MD-Shrink values

■ 7.3 Process Control

The proper function and synchronization of all components of a biaxial film orienting line as well as continuous quality control have to be secured by an integrated process control (IPC) system, which has a modular design as illustrated in Fig. 7.27.

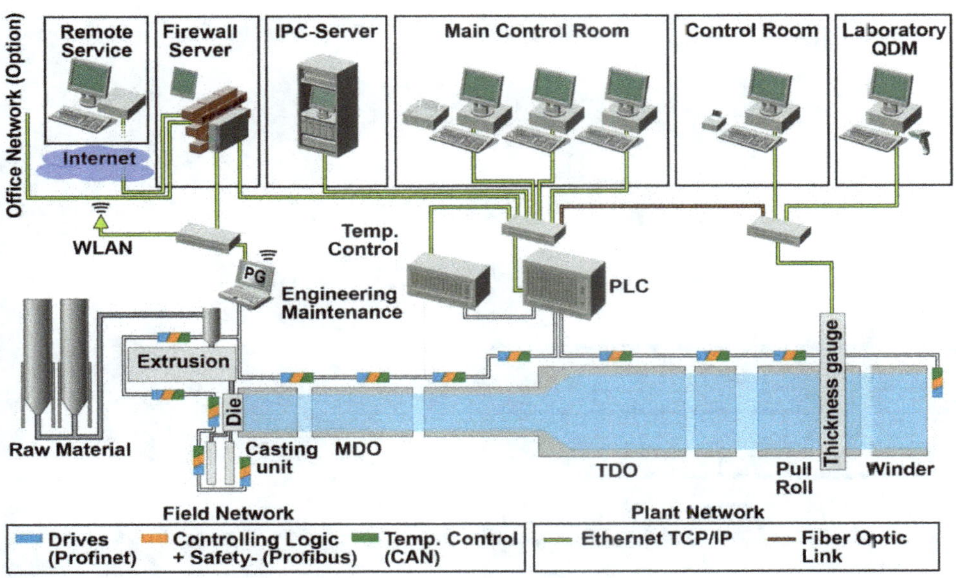

Figure 7.27 Integrated process control (IPC) for a biaxial stretching line

The control of the drives, temperatures, and the control logic is realized by dedicated bus systems. With a drive bus system a very accurate synchronization of all line components has to be secured not only in steady-state operation but also during ramping functions, which is necessary for starting up the line or changing products without process interruptions. The temperature control and the control logic are implemented via PLC with realization of fast reaction times. For the operation of the line several PCs along the production line are connected with a workstation via an Ethernet network. The user interface allows changing of set points, observation of trends of all major data points, organization of product data and recipe management, and depiction of transparent information from the alarm management system. The user interface allows natural language support. Because machine and process are very complex and the huge amount of data has to be controlled, it is necessary to give the operator a guide in order to avoid failures during operation. This is realized by a simple push-button operation, which brings the production line from one preconfigured parameter setting to the next in order to allow product changes or ramping functions without process interruptions. As a result, the uptime of the line can be maximized, which is an important factor for the overall production costs. If there are troubles at the line, in addition to the immediate messages from the alarm management system, there is also a remote service available, which provides fast support from specialists who have access to all parameters of the line by using the internet connection.

The thickness control is fully integrated into the IPC system and is based on the signal of the thickness gauge in the pull roll stand in order to control the automatic die (Fig. 7.28).

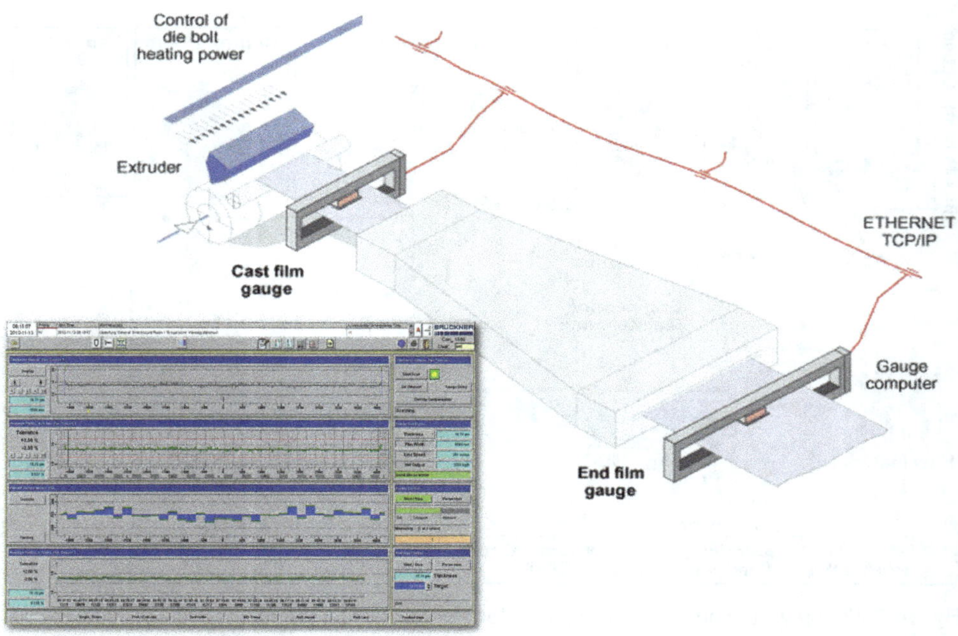

Figure 7.28 Thickness control system

The goal for the thickness-control system is to achieve a constant thickness over the full working width in the range of ±1%, which is only possible by a combination of a very precise measurement of the final film and a sophisticated control loop. In order to achieve this, a cascaded control loop is realized where the signal of the final film thickness gauge is based for the calculation of the set points for the die bolt temperatures in the extrusion die. The internal control loop controls the actual temperatures of the individual die bolts according to the corresponding set points. For that purpose also a correct allocation of the individual die bolts to the corresponding segments of the final film is necessary. This is realized by an automatic, self-learning, bolt-mapping function that incorporates the nonlinear assignment during casting, MD stretching, and TD stretching.

A precondition for the control is the precise and reliable measurement of the final film thickness, which can range between 1 micron and 500 micron, depending on the line type and product range. There are different methods available that fit in general for the requirements of the biaxial oriented film lines. Table 7.2 gives an overview of the advantages and disadvantages of the individual systems.

Table 7.2 Comparison of Thickness Gauges

Beta	X-ray	Infrared
Pro	**Pro**	**Pro**
+ easy to operate	+ easy to operate	+ very accurate
+ only one value to calibrate (density)	+ only one value to calibrate (density)	+ wide gap
+ good accuracy	+ good accuracy	+ insensitive to ambient conditions
	+ wide measurement range	+ mechanically tolerant
	+ high TD resolution	+ multilayer film measurement
	+ no license needed	+ no license needed
		+ low maintenace costs
Con	**Con**	**Con**
– sensitive to ambient temperature, air pressure	– sensitive to ambient temperature, air pressure	– calibration with two factors
– sensitive to vertical variation of gap	– big passline error	– sensitive to changes of material
– big passline error	– small gap	– no black film (dark colors)
– small gap	– medium maintenance costs	
– contaminated zone		
– radiation license required		
– high maintenance costs		

In the past, radiometric thickness-measuring devices using beta radiation were mostly used with sources PM 147 or KR 85, depending on the required thickness range. The advantage is the stability of the measurement, which is based on the absorption characteristics of the materials or the beta radiation. Because radiometric sources require special licenses for operation and safety training and have a limited lifetime, the beta-gauge system has been more and more replaced by other alternatives. One option is to use X-ray radiation with low beam energy because less than 5 kV does not require a license in most countries. X-ray is also more sensitive to changes in recipe, additives in film, and ambient air temperature. Another option is the infrared (IR) absorption method, where the absorption of the infrared light is either measured in specific frequency bands or in an analysis of the complete absorption spectrum. The different absorption characteristics of individual polymers also allow measurement of coextruded multilayer films where the individual layers can be distinguished if the IR spectra show a sufficient difference.

For continuous quality control it is most attractive to measure as much as possible quality data in line. Measurement devices that allow a spot measurement can be attached to the traversing head of the thickness measurement in order to also get data for the full width of the film. This method is possible especially for optical data like haze, gloss, and birefringence. Figure 7.29 shows as an example a method to measure the molecular orientation angle (MOA) of the final film over the working width by using a birefringence sensor with fast data processing [4]. With the information of the molecular orientation angle, uneven properties over the working width such as caused by the bowing effect can be optimized, which is important for specific optical films for flat-panel displays. For these applications it is necessary to keep the MOA below a specific limit value.

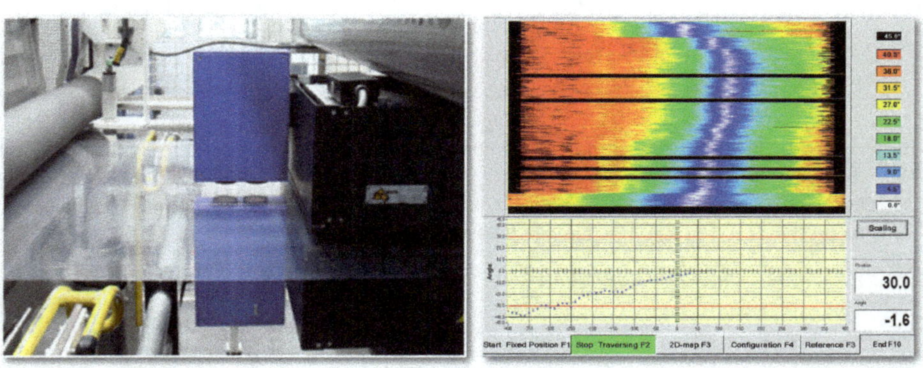

Set-Up

- **Measurement of MOA (Molecular Orientation Angle)**
- **Birefringence sensor**
- **Mounting at traversing thickness gauge**

Analysis

- **MOA over the full working width**
- **Influence of process parameters**
- **Trend analysis**
- **Inline quality monitoring**

Figure 7.29 Inline measurement of the MOA (molecular orientation angle)

Another example of in-line quality control are web inspection systems, which are used to detect defects in the film or on the surface. By means of high-resolution line scanners in combination with fast data processing, a classification of the defects (gels, scratches, inclusions, dust, and others) can be recorded and documented during production.

In addition to the control of the line and the in-line measurement of the quality, there is also another possibility for optimizing the production efficiency, that of adapted software tools. As with the integrated process control system there is access to all necessary process and quality data; an intelligent line management system (ILS) can utilize the data to optimize the production yield (Fig. 7.30). A roll data history module (RDH) collects all production data including the raw material recipes and process data for each produced roll and stores it for later data processing. With a quality data management system (QDM) the data that are measured in the laboratory (mechanical, optical, shrinkage, surface characteristics, and others) are also stored in an appropriate database. A production planning system (PPS) helps the production manager to minimize losses due to product changes. A slitting optimization system (CUT) is dedicated to increase the slitting yield. A computerized maintenance system (CMS) is used to maximize the line availability. In this case time- or event-triggered maintenance stops secure maximum uptime by avoiding longer unplanned line stops. In order to get anytime access to the performance of each production line, a mobile solution (MOS) is available that brings the key performance indicators to smart phones or tablets. Thus the production or top management is always informed about the uptime, capacity, and yield of each production line.

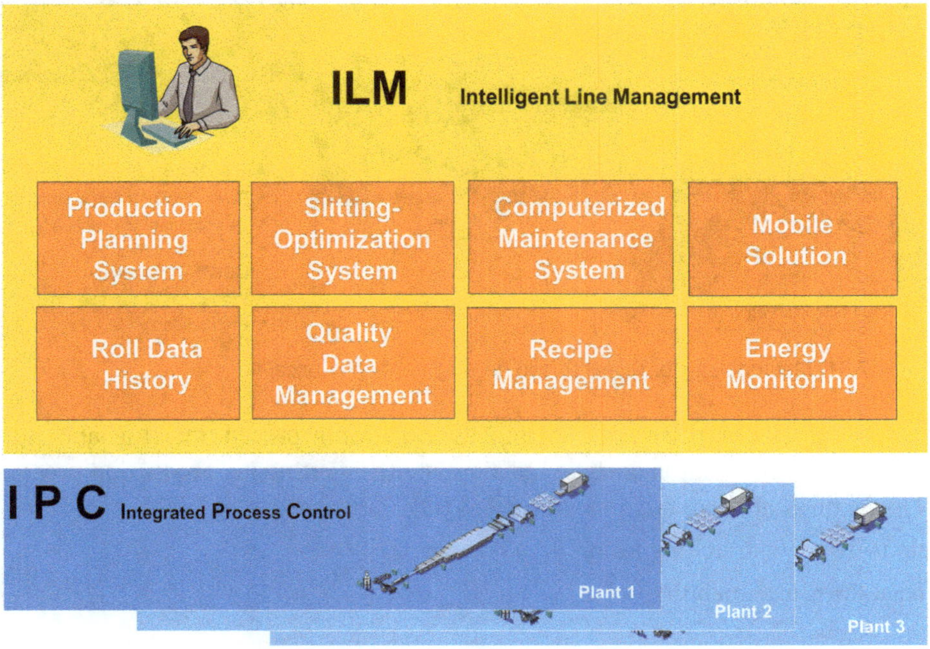

Figure 7.30 Intelligent line management (ILM) system

■ 7.4 Development Environment for Biaxial Oriented Films

In view of the diversity of biaxial oriented films and the dynamic shift of markets, continuing research and development is required, not only for the film producers but also for raw material suppliers and machine manufacturers. Most film producers of biaxial oriented films do not have their own infrastructure or pilot line with testing facilities, so the question is, how can tests be performed that are necessary for the development of new film types? In some cases existing production lines are used to test recipe variations for a modified process setting. The disadvantage in this case is a loss of valuable production time and a high raw material consumption for such tests. This situation can be improved with a smaller, more flexible pilot line. The extrusion and orientation process is performed on a much smaller scale, and new developments are done with a minimum amount of raw material and without interference to the running production. In order to meet the demands of the oriented film industry, Brückner Maschinenbau GmbH & Co. KG, Germany, operates a technology center that is available for the use of customers on a rental basis. A three-step method of research and development action is performed in order to derive result data and to obtain information required for the design layout of production lines for newly developed films (Fig. 7.31).

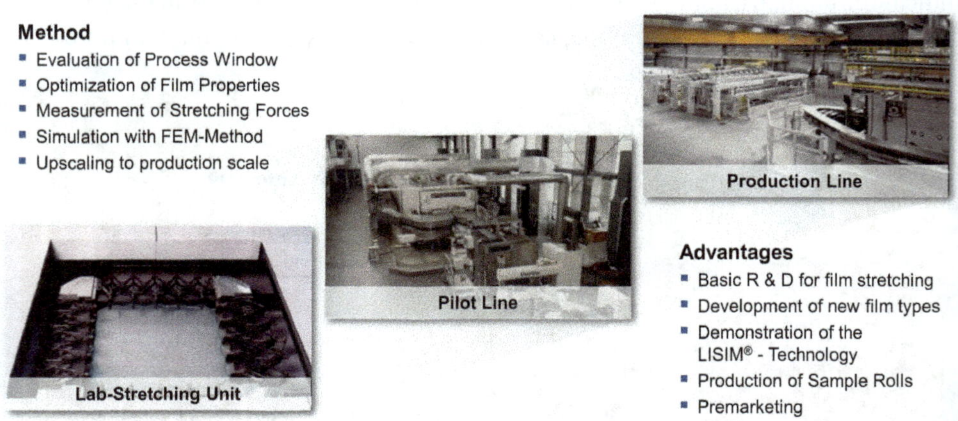

Figure 7.31 Methodology for research and development and upscaling

- The first step comprises a batch-process stretching procedure on a laboratory stretching frame that is designed to simulate the continuous production process. The features of lab stretching equipment (Fig. 7.32) are:
- up to 10 × 10 stretching ratio,
- back-drawing capability (< 1),

- MD retardation for optical films
  (MDX < 1 while TDX > 1),
- three flexible heating modules, and
- 400 °C high temperature.

The cast film for this batch stretching process is produced on a laboratory extruder with a multilayer die. The oriented samples from the lab stretcher are characterized by representative film properties and can be analyzed in the laboratory for chemical, physical, electrical, shrink, and barrier properties. Process data such as temperature, stretching ratio, and speed can be transferred to the continuous process of a film stretching line.

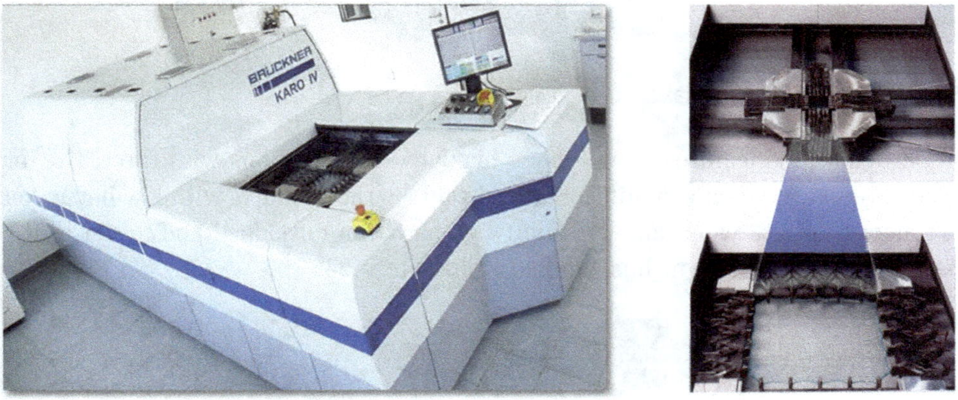

Figure 7.32 Labstretching equipment Karo IV

The pilot line is designed to be multifunctional and very flexible so that all relevant structures and film types can be produced on a pilot scale. Any important information like production stability, thickness tolerances, and product performance can be derived from such continuous tests. The pilot line is designed to operate MD, TD, sequential, and simultaneous stretching processes (Fig. 7.33). In combination with a flexible extrusion system for nearly all extrudable polymers in combination with multilayer structures by coextrusion it is possible to realize a lot of structures [5]. Furthermore, an in-line coater is available in order to apply thin-film coatings for primer, antiblock, release, protection, barrier, or optical enhancements and other functions. The film orienting is realized by a multigap stretching MDO followed by a TDO or by using the simultaneous LISIM technology. This technology allows adjustment of the stretching ratios, stretching curves, relaxation curves, and process temperatures in the most flexible way in order to meet the required film properties of newly developed films.

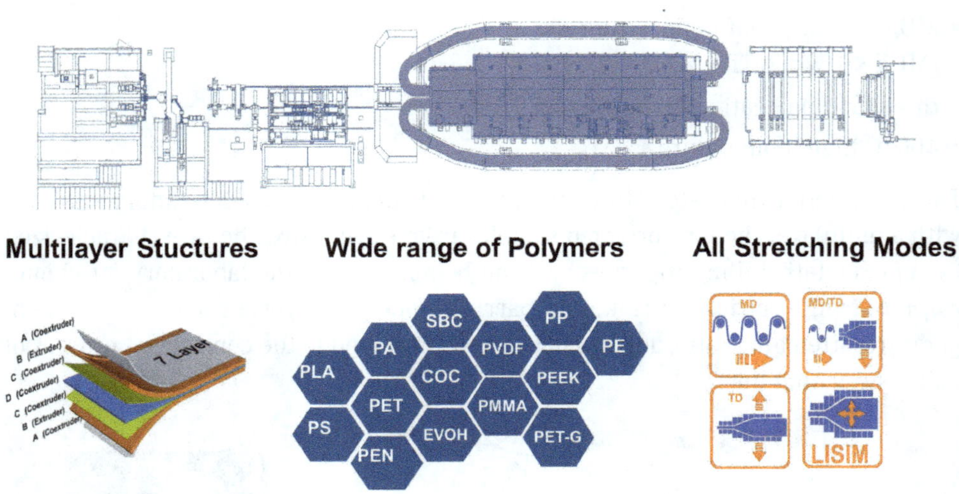

Figure 7.33 Pilot line for biaxial oriented films

In particular, the features of multilayer biaxially oriented films with three, five, or even seven layers offer a significant added value to the products with low impact on production cost. A wide range of film types has been developed and tested in the past on this pilot line, which is summarized in Fig. 7.34.

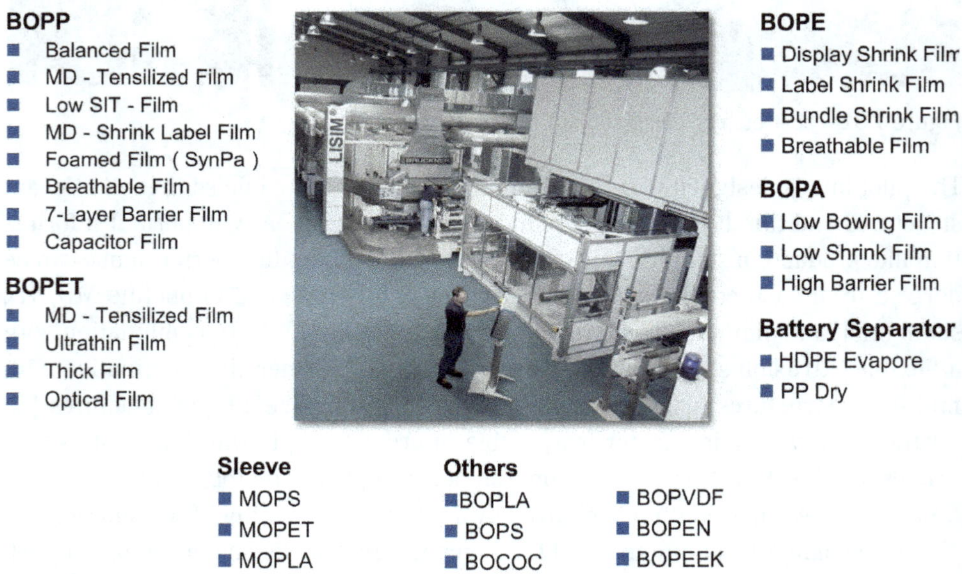

Figure 7.34 Experiences on pilot line scale

Based on these experiences this environment offers the best conditions for future development of oriented films. Some of these have been transferred to production scale using the basic stretching data from the pilot line for a dedicated line layout of the production line.

■ 7.5 Market for Biaxial Oriented Films

With the enhancement of properties in combination with very economic production, biaxial oriented films were widely propagated in packaging as well as in technical applications. The breakdown of raw materials used for oriented films is shown in Fig. 7.35 according to the worldwide installed line capacity [6].

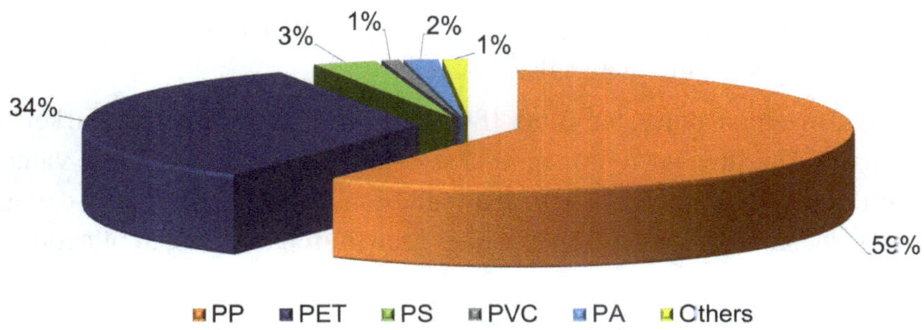

Raw material	PP	PET	PS	PVC	PA	Others	Total
x1000 tpa	11.520	6.600	603	170	322	230	19.450

Figure 7.35 Worldwide production capacity for biaxial oriented films (tpa = metric tons per annum)

BOPP represents the dominant fraction of about 60% of all oriented films with an installed worldwide capacity of nine million metric tons per year. The biggest portion of BOPP is used for packaging applications; just a small fraction is used for capacitor and other technical applications. The reason for this is that BOPP is a very suitable material because it combines good overall properties in combination with an attractive cost situation and a good yield due to the low density of 0.905 g/cm^3. The applications in packaging are very diverse and include single-layer and multilayer film structures. Single-layer structures are used for instance as flower wrapping directly but are more often laminated with other films or processed, such as for adhesive tapes. Typical applications for laminates are noodle packaging, where BOPP and cast PP are laminated together in order to combine the positive properties of both film types (mechanical strength and puncture resistance). Multilayer BOPP films are produced by coextrusion, where in most cases for each individual layer one extruder is used in order to allow maximum flexibility for covering a wide range of products. The most common three-layer coextruded BOPP films contain in the core layer a PP homopolymer, whereas in the skin layers PP copolymers with low melting temperatures are used so that the sealing procedure that is used for most packaging applications can be applied in the temperature range where the skin layer seals without deforming the core layer. This five-layer coextrusion technology offers additional enhancements, like improved optics and opacity but also cost advantages by employing expensive additives in thinner intermediate layers instead of in the core layer. Besides the transparent applications, white opaque film types are also used, which are applied in packaging for confectionary and labels.

The market for BOPP has been growing with a stable rate of more than 6% per year for many years (Fig. 7.36) [6].

The strongest growth rates have been observed in the far eastern area, especially in China, where 44% of all BOPP production takes place. This trend is caused by the ongoing urbanization effect, where a growing portion of the population achieves a higher living standard, causing a corresponding influence on consumer behavior and more use of packaging materials.

With biaxial oriented polyester films (BOPET), over time different market trends have occurred. The turndown of magnetic storage media such as audio-, video-, and computer tapes and floppy disks, which are all based on BOPET film as a carrier, has been counterbalanced by a disproportionate growth in packaging applications.

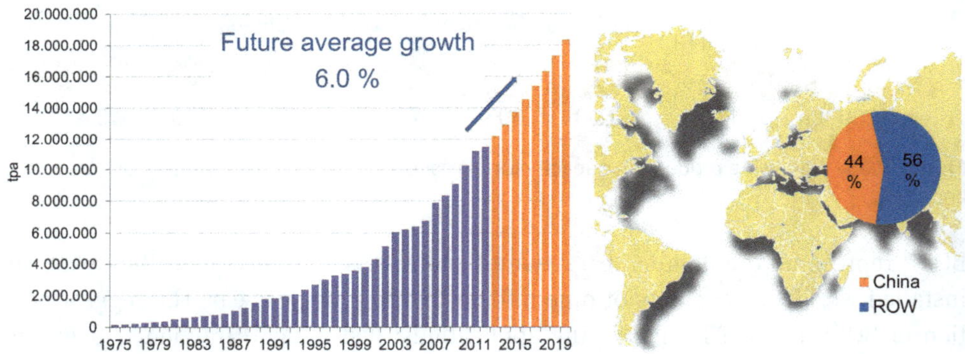

Figure 7.36 Market trend for BOPP

Besides the use in the packaging industry there is also a large field of technical applications, like capacitor films, electrical insulation, thermal-transfer film, optical film for flat-panel displays, solar back sheets, and substrates for organic electronics. Both market segments together, the technical and the packaging, show a stable growth rate of 6.8% per year, which is also likely in the near future. As for BOPP the core areas move more and more into the Asian region, especially to China, where already 42% of the worldwide BOPET capacity is installed (Fig. 7.37) [6].

Other markets for biaxial oriented films are substantially smaller. So is the installed capacity for BOPA (biaxially oriented polyamide) at 267,000 metric tons per year, where the most common applications are in the packaging area. Due to its excellent puncture resistance in combination with being a good oxygen and aroma barrier, BOPA is preferred for flexible packaging of meat, sausages, cheese, fish, and liquids. The thickness range is typically 12–25 micron. BOPA film is produced by sequential stretching as well as simultaneous and double-bubble processes.

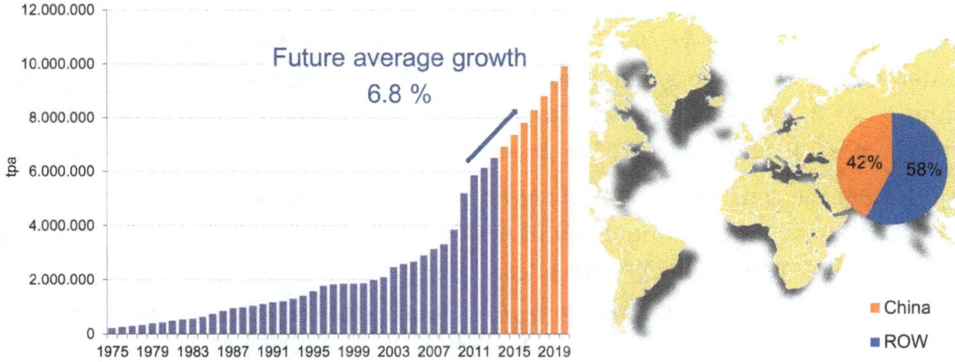

Figure 7.37 Market trend for BOPET

BOPS (biaxially oriented polystyrene) film is used in two segments. The thinner range of 30–150 micron is used as window film for envelopes and separating film in photo albums, whereas the thicker range (150–800 micron) is mostly applied as thermoforming sheet for highly transparent packaging containers. A specialty is BOPLA (biaxially oriented polylactide), which is a representative of plastics from renewable sources. The optical and mechanical properties are excellent after the orientation process, but widespread use is limited by the water-vapor barrier, the thermal stability, and the higher price for the raw material. For applications where a certain transmission of water vapor is required, like for bread and vegetables, the properties of BOPLA can be an advantage.

Another segment for oriented films is BOPE (biaxially oriented polyethylene), which is mostly used as shrinkage film, where for each application a specific property profile with shrinkage values, shrink forces, mechanical properties, and barriers are specifically adapted. BOPE shrink films are mostly produced with a double-bubble process.

The outlook for biaxial oriented films and the development of markets in the individual fields of application is very promising. The entire packaging sector is characterized by strong growth in regions with high population growth rates and continuing urbanization, which cause changes within the distribution chains for foodstuffs and other consumer goods, leading to increased consumption of packaging materials.

Another trend can be observed in the ever stronger discussion about CO_2 emissions into the atmosphere and the political guidelines on the matter, resulting in specific and binding measures to reduce CO_2 emissions. In the future these factors will have an increased influence on the entire packaging industry and therefore also on biaxial stretched films. In some countries, for example France, there will be a legal obligation to print information about the CO_2 footprint on consumer packaging. Because the CO_2 footprint of packaging films is mainly dependent on the applied raw material (resin carries 85% of the CO_2 balance in BOPP), it is logical to reduce the CO_2 footprint by down-gauging films. In principle down-gauging favors biaxial oriented film, which allows for a maximum packaging effect with the minimum raw material use. Further potential to improve the CO_2 footprint can be found in the substitution of individual layers within packaging laminates (for example aluminum foil) by coextruded or metallized high-barrier films based on biaxial oriented films.

Apart from the constant growth in conventional packaging applications for biaxial stretched films, there are a number of newer technical applications represented in future markets with strong growth rates.

Some changes in the markets result from technological developments or even innovation leaps. Within a few years products that did not exist before, incorporating biaxial stretched films, can be on the market. A typical example is optical films for flat screens, which created a substantial market not only for thick BOPET films in the range of 188 µm to 400 µm but also for other films such as COC (cyclo olefine), PC (polycarbonate), PMMA (polymethylmethacrylate) , TAC (triacetate), and others. Due to the global substitution of cathode-ray tube screens by flat screens (LCD, plasma, OLED) this trend will continue in the years to come and will result in respective demand for the aforementioned film types. Another progressive development is expected to happen in the field of flexible electronic devices: here large potentials have been identified in electronic applications that can be produced in roll-to-roll processes. These include flexible photovoltaic panels, e-paper, flexible displays, flexible printed circuits, and flat-surface illumination devices.

Another trend that came up in the recent past is the requirement for ecologically friendly mobility: hybrid and electrical power are playing an important role in the automotive industry. Forecasts show a continued substitution of combustion engines by electrical motors in the next 20 years.

Globally the development in this field is driven by the fact that battery technology is key to the ratio of cost versus range. Lithium-ion batteries show the best potential for a large market share because the required performance data can be obtained with existing technology. All lithium-ion batteries include a separator film, in which the major part of these films consists of biaxial stretched membranes. Therefore a strong growth for this kind of specialized membrane can be expected. Representing 20% of the material cost, the separator to date carries a significant share of the overall cost, and because the battery technology as a whole must become substantially cheaper (the midterm goal is to cut overall costs by half), highly productive and efficient processes for the production of battery separator films are required. Further potential for lithium-ion batteries is seen in existing and further growing markets (notebooks, portable phones) as well as in new applications such as stationary energy storage, which gain momentum through renewable energy sources such as wind and solar energy.

References

1. Jabarin, S.A. "Orientation and Properties of Polypropylene," presented at *The Society of Plastic Engineers Annual Technical Conference*, May (1992)
2. Briston, J.H., Katan, L.L., *Plastic Films* (1989) 3rd ed., Longman Scientific & Technical, Harlow
3. Breil, J., "Added Value Speciality Films Produced with Sequential and Simultaneous Stretching Lines," *Special Plastics Film Conference, 18th Annual World Congress*, Zürich, Switzerland, October 29–30 (2002)
4. Koerber, A., Lund, R., Langowski, H.-C., "Geometrical Bowing and Molecular Orientation Angle in Biaxially Stretched Poly(ethylene terephthalate) Films," *J. Appl. Polym. Sci.* (2013) 127, pp. 2928–2937
5. Breil, J., "Oriented Film Technology," *Multilayer Oriented Films,* Wagner, John R. (Ed.) (2010) Elsevier, Amsterdam
6. Brueckner Sales Data Base

References

8 Biaxially Oriented Tentering Film

Toshitaka Kanai

8.1 Introduction

There are two processes used to produce biaxially oriented films: the tentering process and the double bubble tubular film process.

The tentering process is divided into a sequential stretching method (a step-by-step stretching method) and a simultaneous stretching one. These two methods are employed independently in accordance with the characteristics of resins, but the sequential stretching method is globally the most commonly used. A sheet produced by a sheeting machine is stretched in the longitudinal direction and then stretched in the transverse direction. The tentering process has the advantage of being suitable for mass production and obtaining good uniformity, but the machine costs are expensive.

The double bubble tubular film process is composed of the tubular film process, followed by the preheating, and then the multiaxial stretching process in the second bubble. This is multiaxially inflated with much higher force resulting from pressure inside the bubble and higher take-up force because the film is below the melting temperature. Then the heat set process is accomplished by the third bubble operating at higher temperature than the stretching temperature. This process has the advantage of using low-cost machines, and various products using only one machine can be produced. Furthermore, because this process is simultaneous stretching, it can produce a good balanced film in MD (machine direction) and TD (transverse direction), and a high stretching stress during the process, which are appropriate for shrinkage film and high strength film. However, it has a low production speed, a narrow process window, and less thickness uniformity when compared to the tentering process.

8.2 Tentering Process

It is indicated in Table 8.1 that the sequentially biaxial stretching method is applied for polypropylene (PP), polyethylene terephthalate (PET), polystyrene (PS), and polyamide (PA), and simultaneous stretching is applied for PA [1].

Table 8.1 Biaxially Oriented Film Processes and Their Features

Features		Tubular film process	Tentering process: Step by step	Tentering process: Simultaneous
Productivity	1. Producible film thickness range	△ (Medium)	◎ (Thin-thick)	○ (Thin-medium)
	2. Flexibility on the change of process conditions	△	○	△
	3. Total productivity in conjunction with high speed and width broadening	△	◎	○
	4. Small lot production	◎	△	○
Quality	5. Thickness accuracy	△	◎	◎
	6. Isotropy in physical properties	○	△	◎
	7. Range of polymers that can be used	○	○	◎
Corresponding film		PP, PE Polyvinyl chloride (PVC) Polystyrene (PS)	PET, PP Polystyrene (PS) Polyvinyl chloride (PVC) Some grades of nylon (PA)	☆ Nylon 6 ☆ Nylon 66 ☆ Poly hexamethylene adipamide ☆ Polyvinyl alcohol ☆ PMMA Copolymer ☆ Nitrile film *PET Special polyester *PPS Polyarylate *Aromatic polyamide *Polyimide

◎, very good; ○, good; △, poor
☆, Step by step stretching technology is under study.
*, Application of step by step stretching technology is under consideration.

The sequentially biaxial stretching process basically consists of MD stretching and then TD stretching. Generally, biaxially oriented PP (BOPP) film is produced by MD and then TD stretching. For biaxially oriented PET (BOPET) film, the following method is used [1].

MD→TD single-stretching method

MD→TD→MD multistretching method

MD→TD→MD→TD multistretching method

Because the MD→TD stretching process can lead to low-cost film production, biaxially oriented film is mainly produced by this process. The multistretching process is adapted to the particular stretching machine.

8.3 Biaxially Oriented Tentering Machine

The biaxially oriented tentering film process is shown in Fig. 8.1 and is composed of an extruder, a die, melt casting, longitudinal stretching, transverse stretching, heat treatment, and a winding process [2].

The extruder plays an important role in the high output rate and uniform kneading and mixing, and the flat die produces a uniform thickness distribution in the width. The tandem type is more popular than the conventional single extruder as an extruder for a tenter system machine in order to get a high output rate.

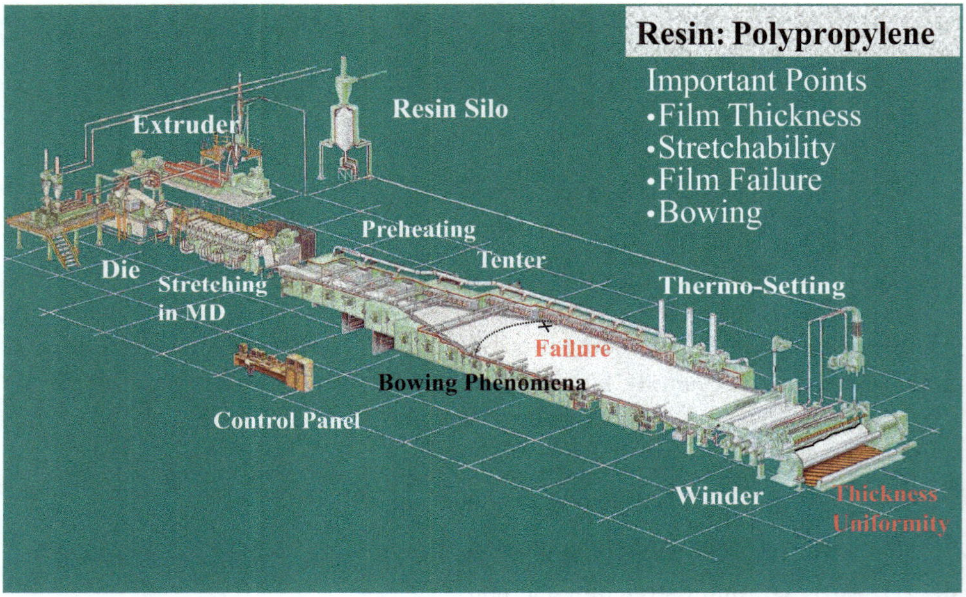

Figure 8.1 Biaxially oriented tentering film process

The tandem extruder system is divided functionally into two sections shown in Fig. 8.2. The primary extruder, focused on the melting function, is a high-speed, small extruder with a high melting efficiency. The secondary extruder, aimed at homogenizing, maintaining lower melt temperature, and metering, is a low-speed extruder with a larger screw diameter. By controlling the resin pressure at the inlet of the secondary extruder automatically, the machine can be used as if both extruders were one.

Recently, a twin screw extruder has also become popular for this purpose.

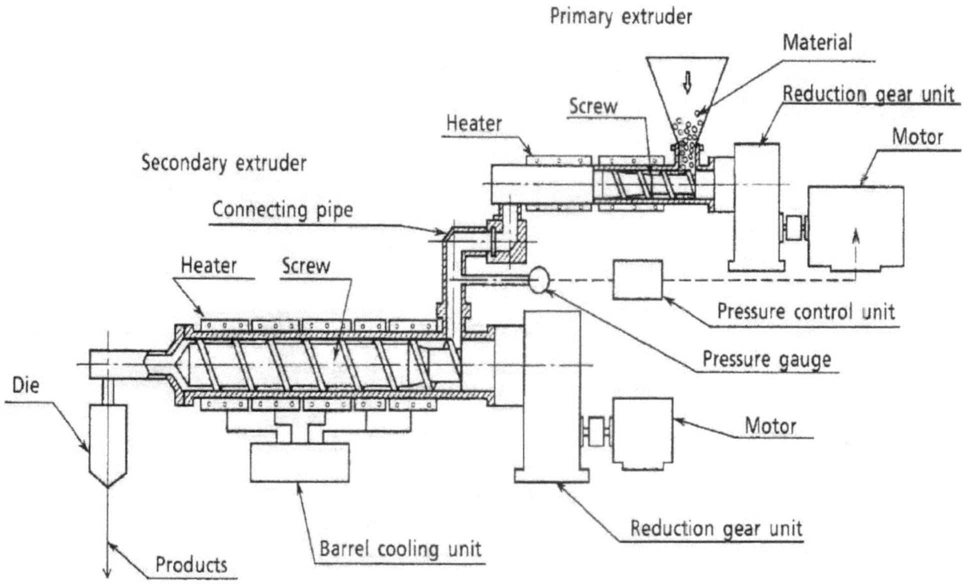

Figure 8.2 Tandem extruder

The die must have an excellent adjusting function for the film thickness and be easy to disassemble, clean, regrind, and reassemble. The coat-hanger die with an adjusting system of both heat bolt or lip heater shown in Figs. 8.3 and 8.4 is generally used as a die. The thickness adjusting methods are divided into two types [1]. One is a die lip gap adjustment and the other is a viscosity adjusting method involving a lip heater.

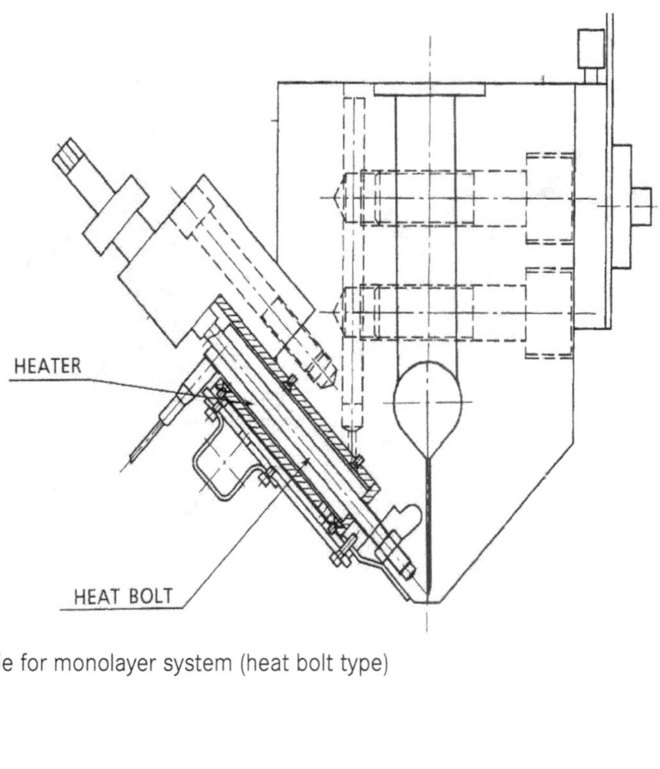

Figure 8.3 Die for monolayer system (heat bolt type)

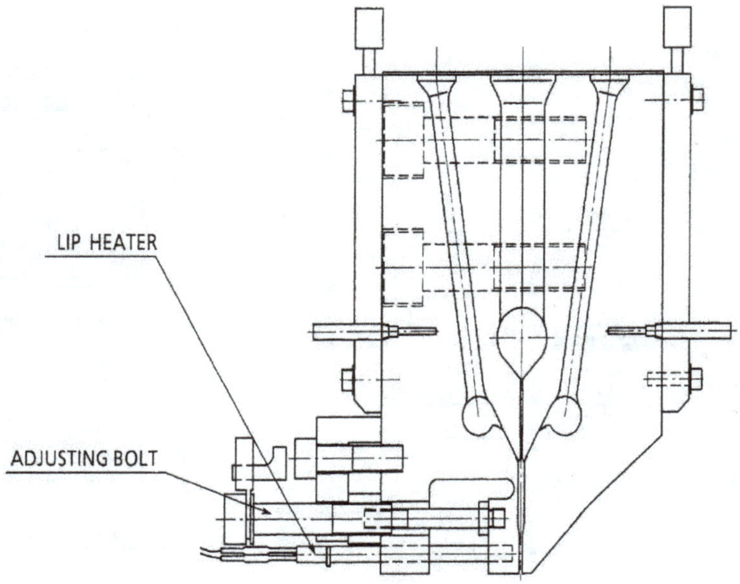

Figure 8.4 Die for three layer system (robot and lip heater type)

It is important for the melt casting process to reduce air gap and crystallinity by using a high cooling speed. An air gap, the distance between the die and the chill roll, should be short to reduce the necking phenomenon that leads to width reduction compared with die width. For this reason, an air knife shown in Fig. 8.5 is used for OPP (oriented polypropylene). Because PET has low melt strength, electrostatic pinning is commonly used.

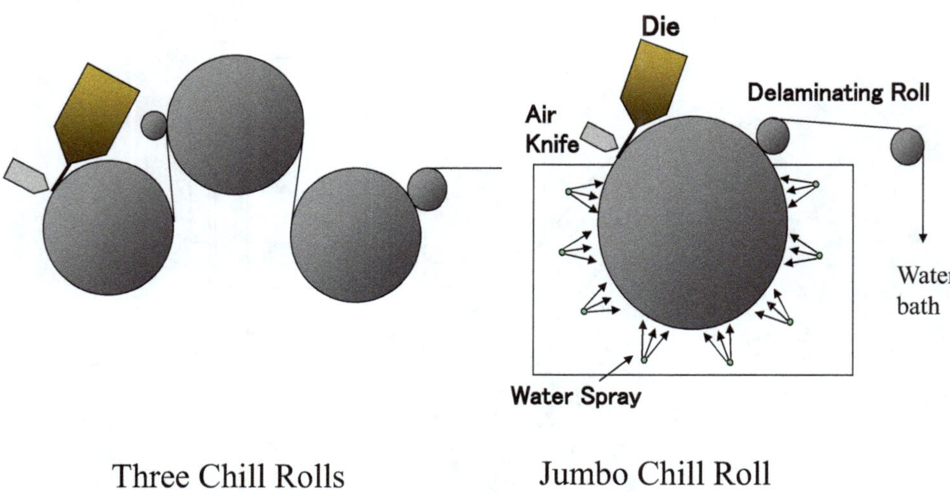

Figure 8.5 Schematic view of casting machine with cooling system

In order to reduce crystallinity a high cooling rate is needed. One jumbo chill roll with a cooling water shower rather than three chill rolls has become more widely used to accomplish this as shown in Fig. 8.5. The jumbo roll diameter sometimes reaches 2500 mm. It is important to minimize fluctuation at the stretching point, reduce crystallinity using a high cooling speed, obtain uniform temperature distribution in the film, and use proper tacticity of PP when a high stretching speed is required.

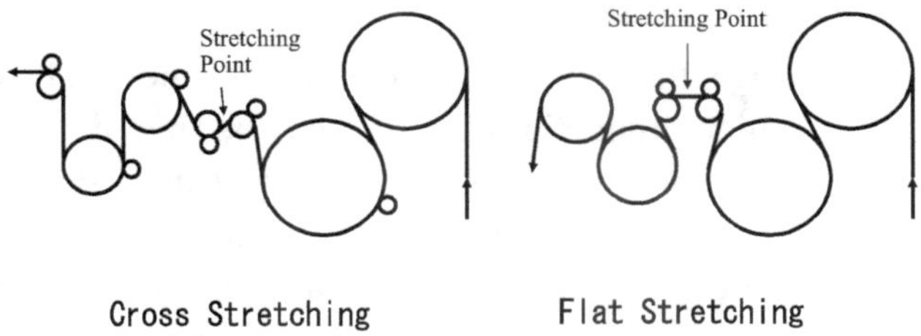

Figure 8.6 Stretching method in longitudinal stretching

The sheet is heated by several heated rolls in the preheating zone, and the longitudinal stretching is carried out by using rolls with different peripheral speeds. There are two methods for the roll stretching process, cross stretching and flat stretching, as shown in Fig. 8.6 [1]. The merit of the cross-stretching method is a smaller stretching gap. This method is used for PP, and so on, which can develop large necking. This method is preferable because the smaller stretching gap and reduced neck-in lead to more stable stretching and better gauge control. On the other hand, because the flat stretching cannot set the stretching gap smaller than the diameter of the stretching roll, the stretching gap is larger than the stretching gap of the cross stretching. Flat die stretching is used for OPS (oriented polystyrene), OPET oriented polyethylene terephthalate, and others. Process auxiliary heating by electric heaters at the gap space prevents the film from sticking to the roll. It can increase stretching stability because the gap between the last heater and the high-speed side stretching roll is small.

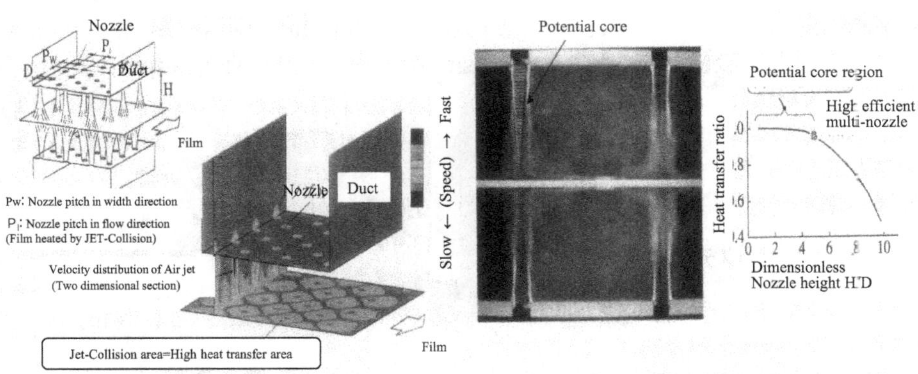

Figure 8.7 Comparison of multinozzle type with slit nozzle type in stretching method in the tenter oven

After that, both edges of film are clipped in the tenter oven and stretched in the transverse direction at a lower temperature than the melting point. The tentering oven length increases with increased speed of the biaxially oriented film process. In order to obtain the high heat transfer coefficient in the oven, the multinozzle type is used rather than slit nozzle type as shown in Fig. 8.7 [3].

The stretched film is then heat set at a higher temperature than the stretching temperature and then is wound by a winder.

More than two decades ago, the BOPP machine was 5 m wide and had a 200 m/min line speed. Recently, a high-production line that has become more widely used is 10 m wide and has a 530 m/min take-up speed; one line can produce about 7 ton/hr and more than 30,000 ton/year of biaxially oriented film. In this case, as good thickness uniformity and continuous production without break are required, low tacticity PP that can reduce crystallinity before stretching is generally used. Furthermore, polypropylene having MFR (melt fracture rate) 2 and 3 and a wide molecular weight distribution is preferable for good uniformity and high-speed production.

8.4 Theoretical and Experimental Analyses and Polymer Design for Biaxially Oriented Film

The biaxially oriented film produced by systems like that depicted in Fig. 8.1 is required to have uniform thickness, uniform orientation distribution, high transparency, and also no breakage during high-speed production, which has become more desirable because of the reduced cost of the film. To cope with these requirements, evaluation technology for resin in terms of stretchability, thickness uniformity, and film quality has been necessary, while only using a small amount of the sample. From this point of view, a theoretical analysis that can predict the stress development, processability, and thickness uniformity shown in Fig. 8.8 is very important [4].

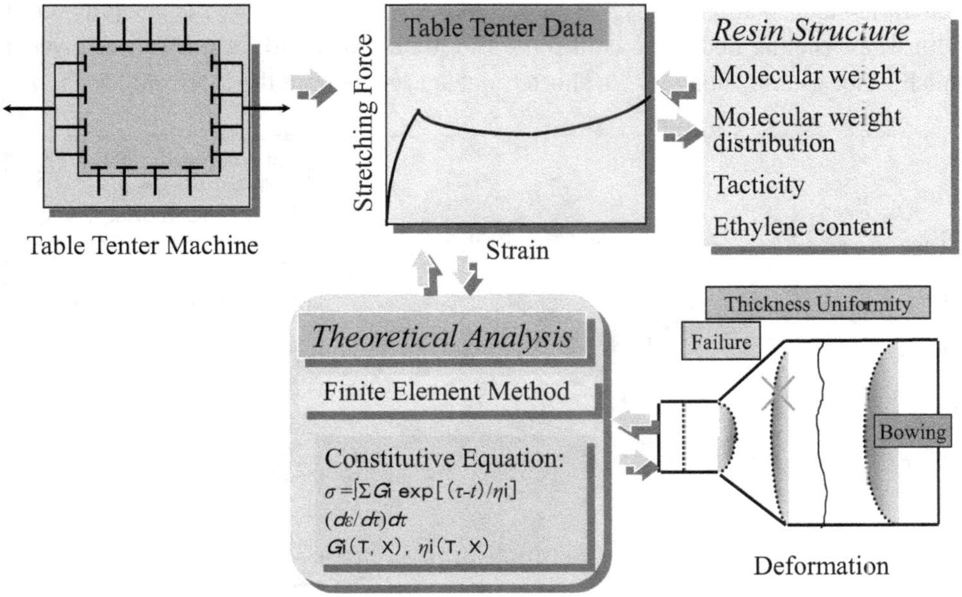

Figure 8.8 Theoretical analysis of stress development and deformation during biaxially oriented film process

8.4.1 Cooling Process Analysis

The model of the cross section of the primary film in contact with chill roll or cooling water must be considered. The heat-conduction equations of the primary film and the boundary condition between film surface and cooling water are shown as follows.

$$\frac{\partial T}{\partial t} = \frac{k}{C_p \cdot \rho} \cdot \frac{\partial^2 T}{\partial x^2} \tag{8.1}$$

$$k\frac{\partial T}{\partial x} = h\left(T_W - T_\infty\right) \tag{8.2}$$

where T_W is film surface temperature, k is film thermal conductivity, C_p is heat capacity, ρ is film density, T is film temperature, x is film position, h is heat transfer coefficient, and T_∞ is temperature of a fluid or air. The crystallization of resin was incorporated into the analysis by using crystallization latent heat. The primary film cooling program was developed and used as a theoretical analysis.

A high-speed production line requires a high-speed cooling process that reduces the crystallinity of film. For this reason, a jumbo roll with water shower is used rather than three rolls. Figures 8.9 and 8.10 are examples of sheet temperature distributions as a function of cooling time and position across the sheet. It was found that the plateau region that is the result of a heat-balanced region between crystallization heat generation and cooling causes a major difference when comparing the one jumbo roll with water cooling and three chill rolls. The jumbo rolls with water shower gives a much faster cooling along with a shorter plateau region than the three chill rolls.

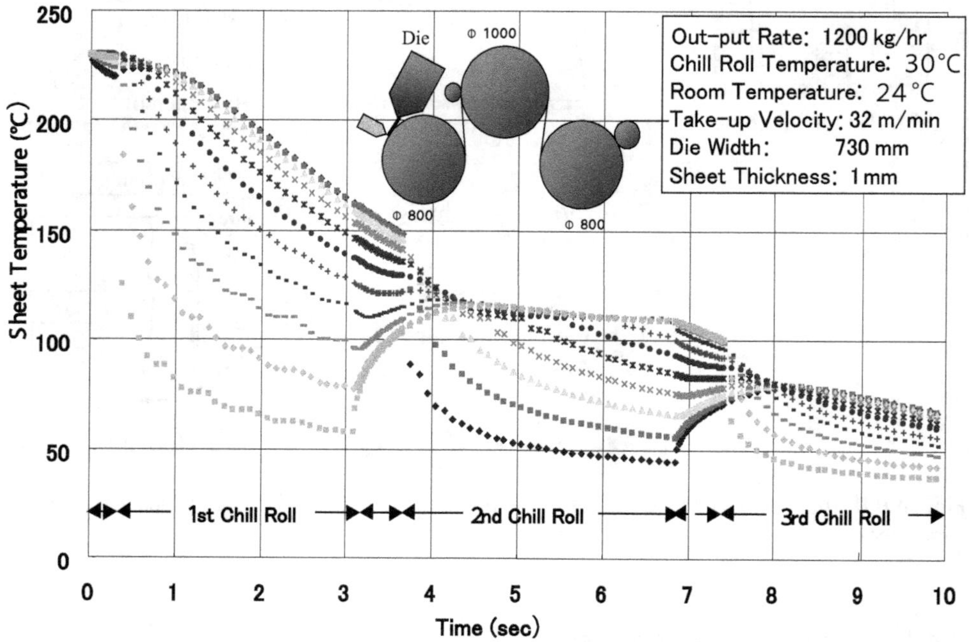

Figure 8.9 Temperature distribution of PP sheet during three chill rolls cooling

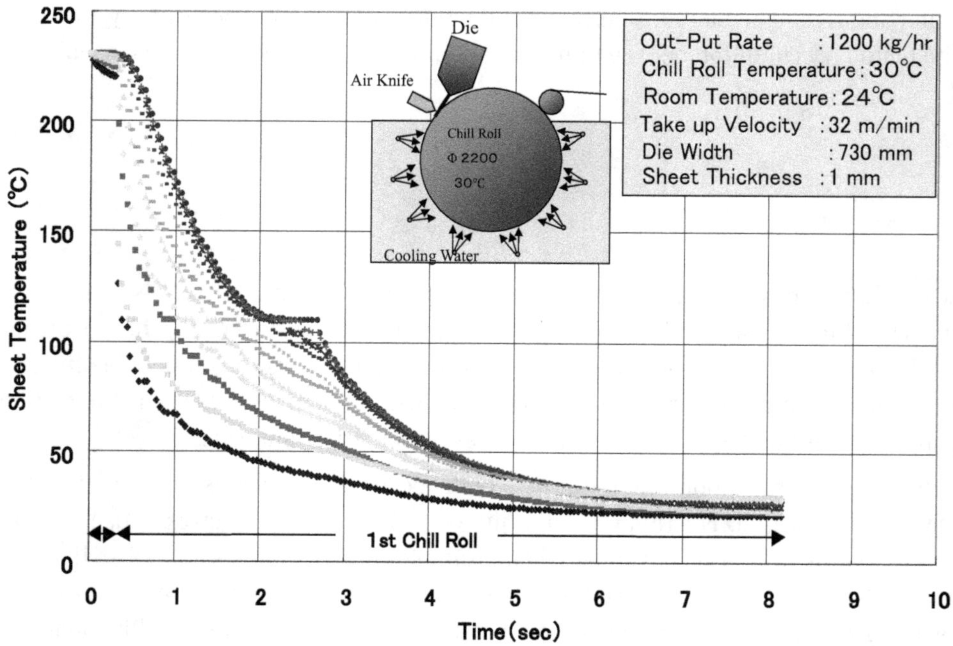

Figure 8.10 Temperature distribution of PP sheet during chill roll with water cooling

The cooling system is very important for the superstructure of the film. For example, the spherulite size produced by an ordinary chill roll is much larger than by a chill roll with water cooling (Fig. 8.11). The cooling temperature is also one of the factors that control the spherulites, as shown in Fig. 8.12.

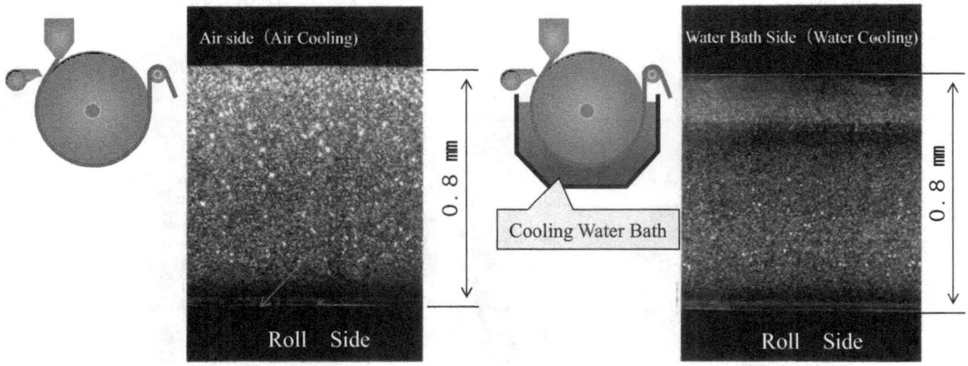

Figure 8.11 Cross-sectional view of PP film by polarized optical microscope

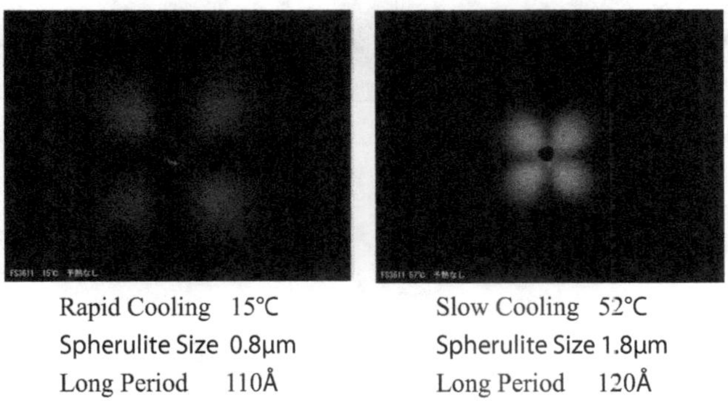

Figure 8.12 Relationship between cooling water temperature and spherulite size of rancom PP copolymer (ethylene content 4 wt%) taken by light scattering

8.4.2 Stretching Process Analysis

1. Longitudinal Stretching and Its Stability

The longitudinal stretching, namely the roll stretching, is shown in Fig. 8.13. The longitudinal stretching is also important for the quality of film and stretchability. This process also has the necking phenomenon observed as found in T-die casting. In the case of high-speed stretching, the stretching line should be stable, and the stretchability is dependent on the resin quality and temperature control. The stretching point should be uniform in the transverse direction and stable, but it is sometimes different along the transverse direction and unstable under the high-speed condition shown in Fig. 8.14. This instability phenomenon is similar to the draw resonance of the T-die casting. Furthermore, the thickness distribution is not uniform in the transverse direction especially at both edges. The stretching condition should be controlled by changing the stretching temperature, which is adjusted by the roll temperature, roll-to-roll distance, and resin design. Under the optimum conditions the stretching line is straight along the transverse direction and stable and the necking is short.

2. Stretching in the Transverse Direction

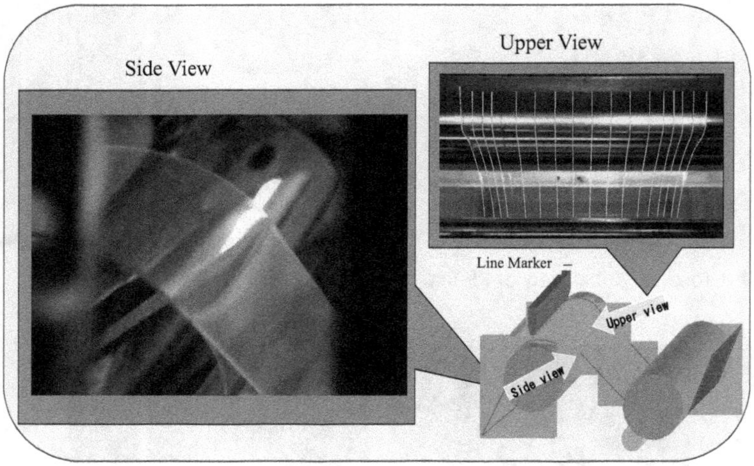

Figure 8.13 Longitudinal stretching

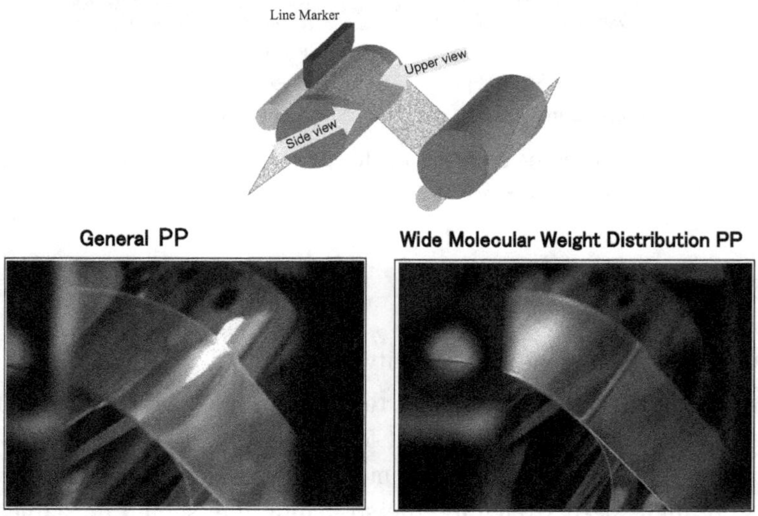

Figure 8.14 High-speed longitudinal stretching of BOPP

The schematic view of transverse stretching is shown in Fig. 8.15. The stretching in the transverse direction for PP is experimentally obtained as shown in Fig. 8.16, and the necking stretching phenomenon is clearly observed. The stretching position is changed, and after the stretching ratio reaches more than 5, uniform stretching seems to have taken place. The upper figure in Fig. 8.16 shows a typical stretched pattern in the TD stretching oven, and the lower figure shows how each increment of the film is stretched. This necking phenomenon is closely related to the stress-strain curve. After the stretching stress reaches the yield value, the stress decreases gradually and the necking phenomenon is observed. As the stress increases over the stretching ratio 5, the necking phenomenon decreases and is not observed over stretching ratio 7.5. For this reason, the stretching ratio for PP is set at more than 8, at which no more necking phenomenon is observed.

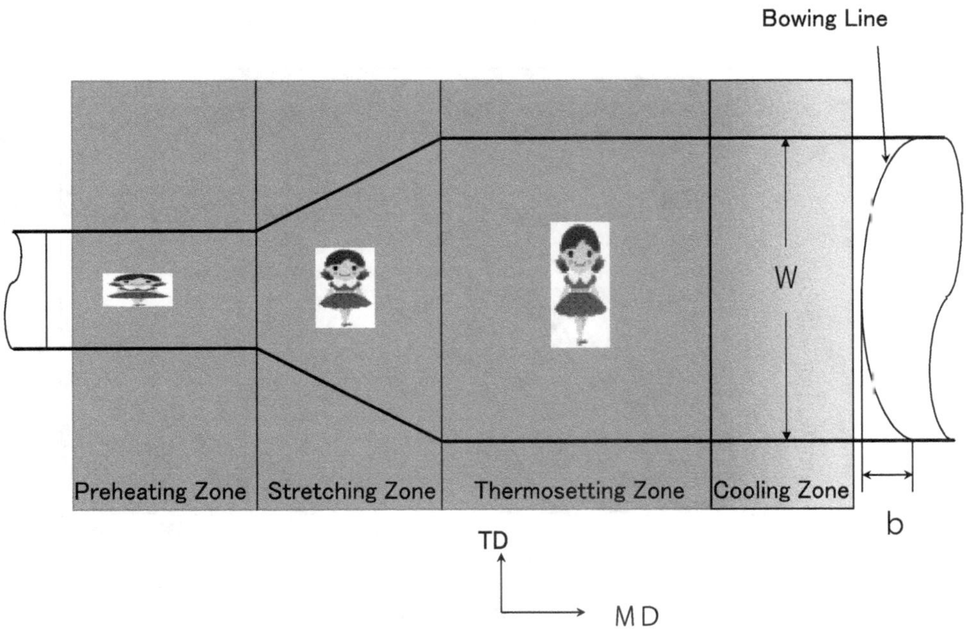

Figure 8.15 Schematic view of biaxially stretching tenter

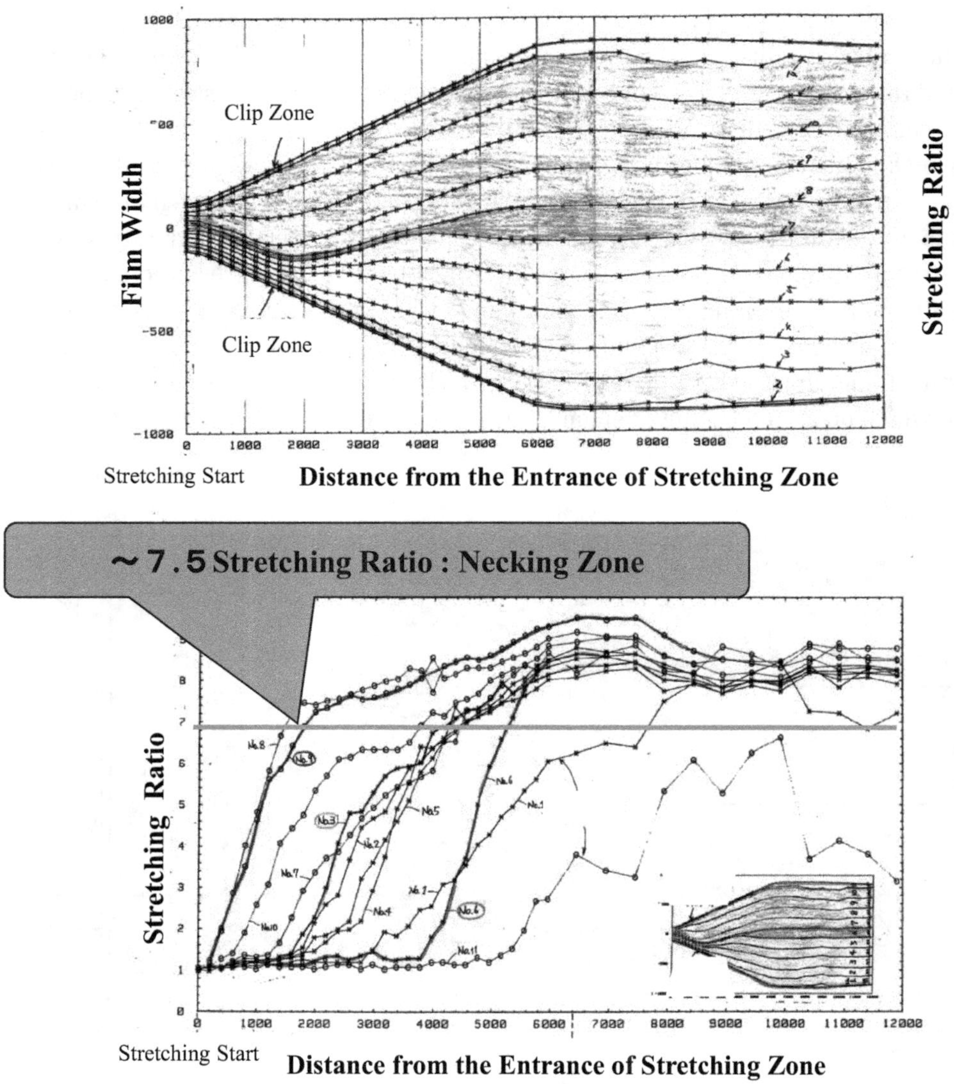

Figure 8.16 Necking phenomenon in the tentering zone

The temperature range of stretchability is very much dependent on the composition distribution of PP. Stretchability and uniform thickness are dependent on the molecular weight distribution and tacticity of PP, which influence stretching force buildup at the higher stretching ratio and initial stretching force, respectively. It is possible to predict the processability by theoretical analysis and the laboratory tenter data obtained with a small amount of resin that has the same resin characteristics.

The viscoelastic model is assumed as a mechanical model, and the factors of viscosity parameter η_i and elastic moduli G_i are obtained by the curve of stretching force and stretching ratio when the resin is produced by the small-size tentering machine.

Figure 8.8 shows the theoretical analysis of stress development and deformation during the biaxially oriented film process.

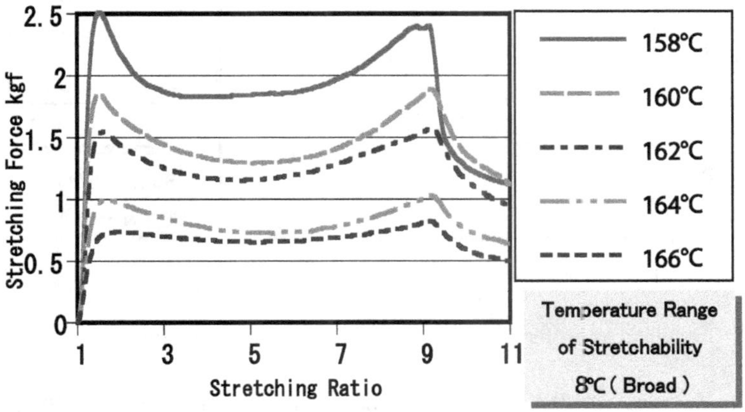

Figure 8.17 (a) Stretching force versus stretching ratio for BOPP grade

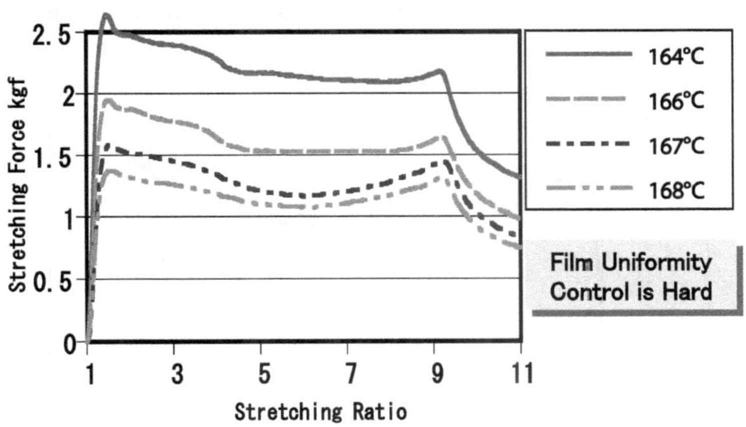

Figure 8.17 (b) Stretching force-strain curve of high tacticity BOPP resin

Figures 8.17 (a) and (b) show stress-strain curves of various polypropylenes. The low tacticity or ethylene copolymerization leads to low elastic modulus. The stretching stress decreases with increasing stretching temperature, decreasing stretching rate, and decreasing stretching ratio as process conditions, but also with decreasing tacticity and molecular weight and increasing ethylene content as resin characteristics. Figure 8.18 shows that the single-site catalyzed polypropylene has a strain hardening curve, because this PP has low crystallization speed when it is produced during the sheeting process. At the same time, a narrow composition distribution leads to a narrow stretching process window. So composition distribution should also be controlled by blending or by using multireactors.

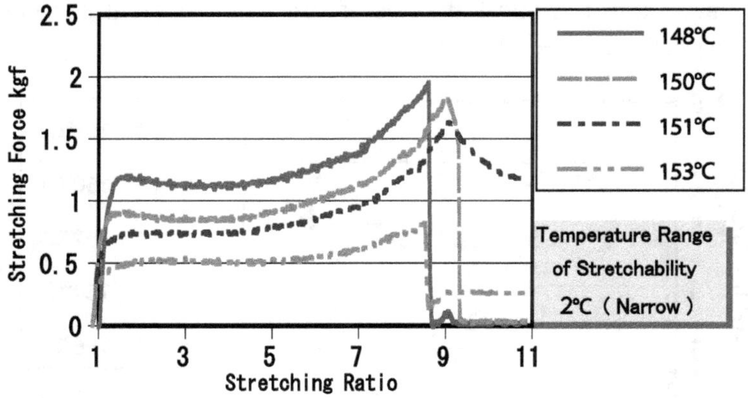

Figure 8.18 Stretching force versus stretching ratio for metallocene PP

The wide molecular weight distribution has a strain hardening pattern at a high strain ratio as shown in Fig. 8.19 and gives better thickness uniformity as shown in Fig. 8.20. If thickness unevenness is assumed to be at maximum 10% of the initial sheet before stretching, thickness unevenness of biaxially oriented film is predicted by changing relaxation time.

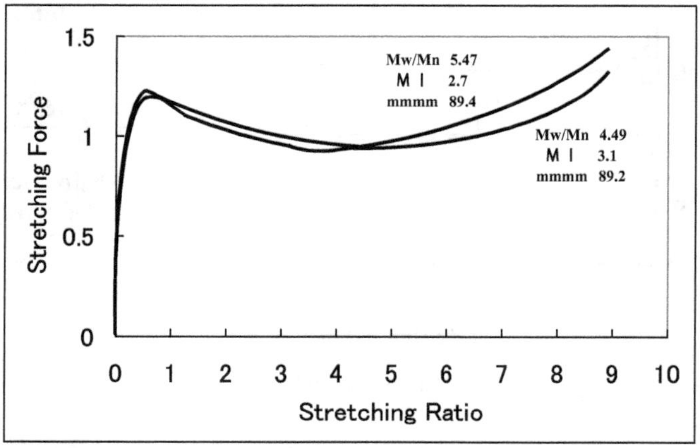

Figure 8.19 Stretching force-strain curve of different molecular weight distributions

Figure 8.20 portrays the long relaxation time portion (1→2→3) for several samples, which is closely related to the high molecular weight portion, and long chain branching causes strain hardening. This strain hardening in the high stretching strain is caused by the molecular chain entanglement of high molecular weight polymers and gives a good thickness uniformity. As this concept is the same as the blow molding, the high molecular weight portion influences the molecular chain entanglement.

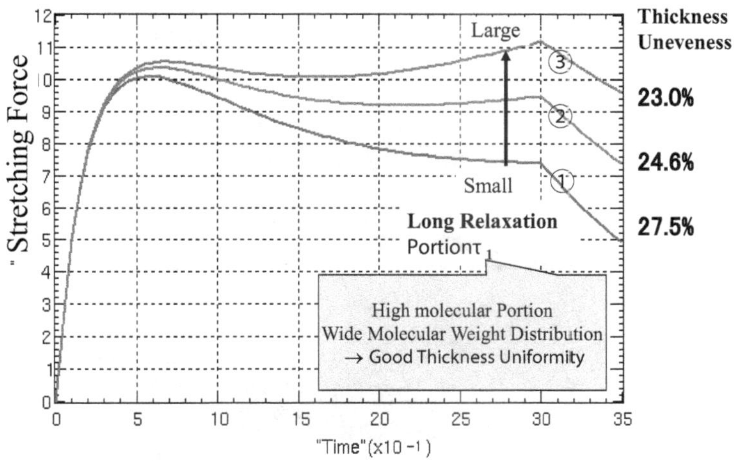

Figure 8.20 Relationship between stress-strain curve and thickness uniformity

The polymer design that can reduce crystallinity in a prestretched sheet and have strain hardening during the stretching process has the potential of giving good thickness uniformity, an ultrathin film, and high-speed stretching [5].

The results of the deformation pattern and stress during the tentering process are shown in Figs. 8.21 (a) and (b). These results suggest that bowing phenomena, stretchability, stress buildup, and thickness uniformity are predictable. In the tentering oven, the bowing phenomenon is drastically changed.

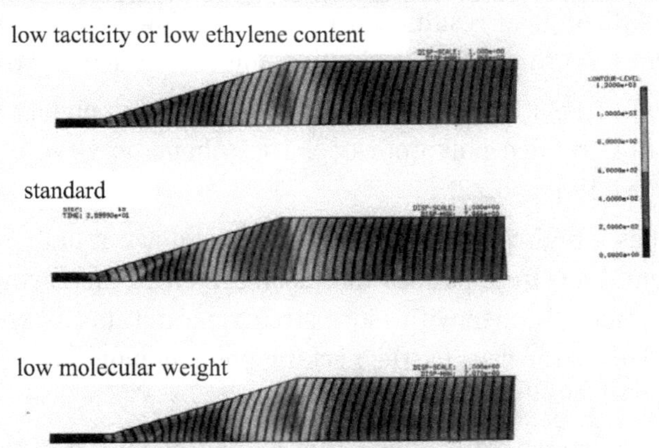

Figure 8.21 (a) Stress distribution of various polymers

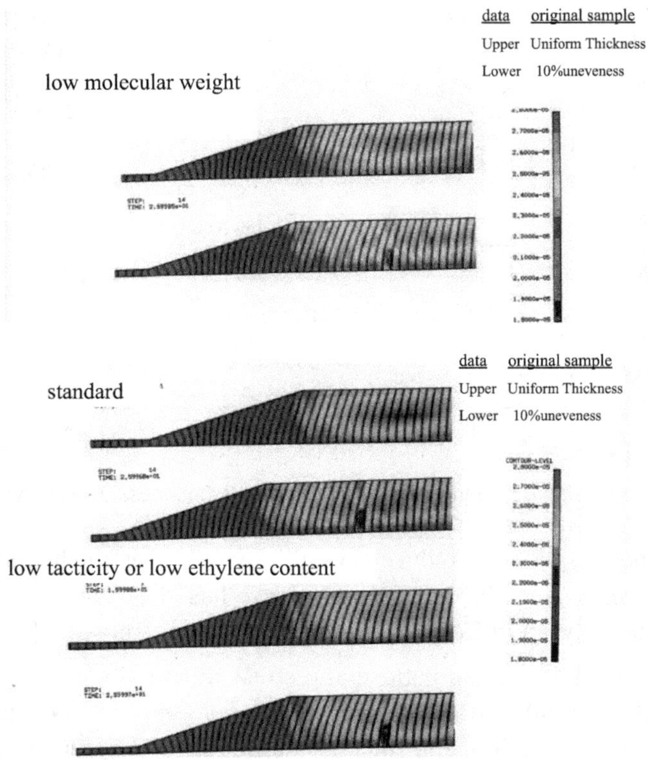

Figure 8.21 (b) Stress distributions of various polymers

For example, the theoretical results can predict that low tacticity or low ethylene content PP has uniform thickness distribution and low stretching stress.

From this kind of information, processability and optimum polymer design are predicted for biaxially oriented polypropylene. The schematic view of film thickness uniformity is shown in Fig. 8.22.

The stress that has a high yield value and then goes down rapidly is shown in the figure on the right. The stress pattern that decreases with increasing strain means that the part becomes thinner with more stretching and its thickness uniformity becomes worse. This is the case for the necking phenomenon that occurs at the low stretching ratio.

On the contrary, the stress, which has strain hardening shown in the left figure in Fig. 8.22, increases with increasing strain. This pattern means that the thinner the part becomes the less stretching and the better thickness uniformity.

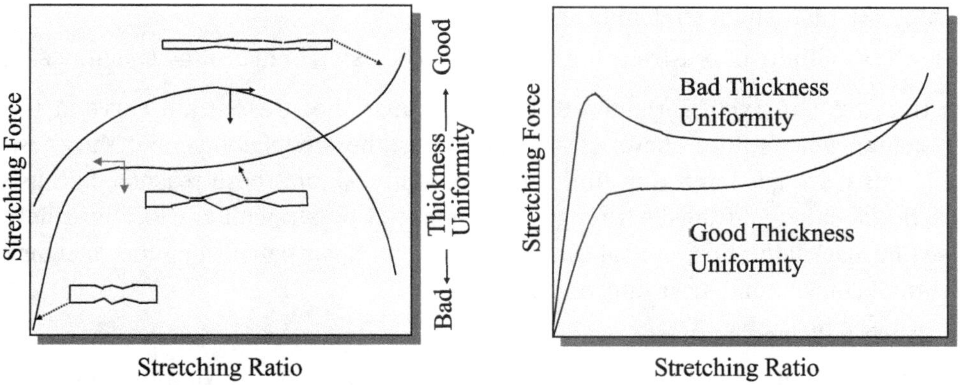

Figure 8.22 Stretching force-strain curve and thickness uniformity

From this information, the strain hardening pattern is preferable for thickness uniformity and stretchability. Furthermore, low stress sensitivity as a function of temperature may have a wider stretchable window and give better stretchability.

In order to reduce the stretching stress, raise the stretching temperature and decrease the stretching speed as process conditions; polymers designed with low tacticity or copolymer with ethylene and low molecular weight are effective.

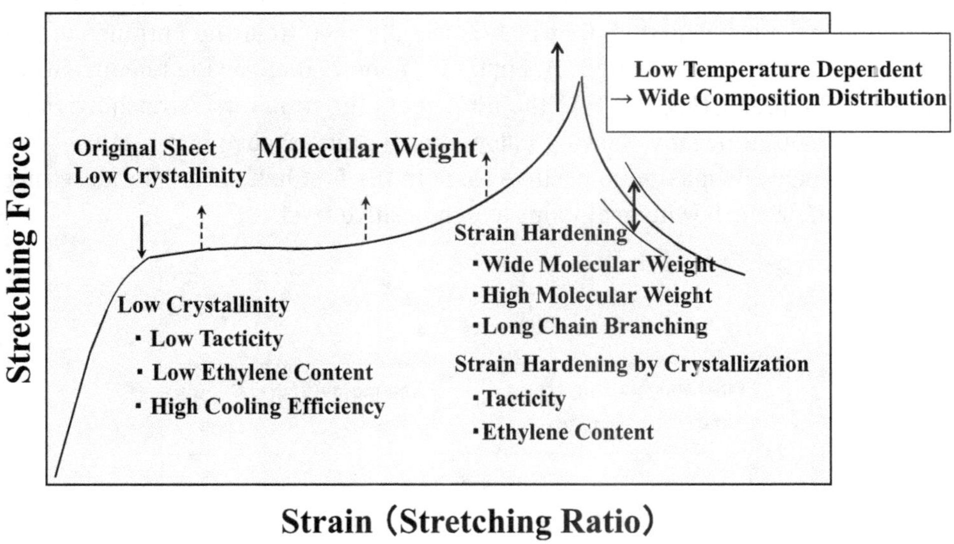

Figure 8.23 Ideal stretching force versus strain curve

The above results can be graphically depicted as seen in Fig. 8.23, and the good biaxial stretching conditions for PP are listed as follows [5]:

- wide molecular weight distribution
- low tacticity or low ethylene content
- wide composition distribution

- high MFR for a high stretching speed
- low crystallinity of prestretching sheet produced using high cooling conditions.

In the case of biaxially oriented PET film, because the stress-strain curve at the stretching temperature shows strain hardening, the stretchability and thickness uniformity are good and thin film can be produced. From these reasons, because elastic modulus and tensile strength are improved by stretching, OPET is widely used for packaging film, optical film, back sheet for solar panel, computer memory magnetic tapes, condenser film, and so on.

3. Bowing Phenomena

Bowing is a kind of uneven stretching in which a straight line drawn transversely on the film entering the tenter bends in a bow shape as the film goes through the TD stretching zone and thermosetting zone [6–8]. As shown in Fig. 8.15, the distortion of bowing is expressed as the bow height b as a percentage of the film width W. When the film center lags behind the film edges, the bow height b is considered positive. The bowing distortion b/W measured at different positions within the tentering oven consists of four zones: preheating, TD stretching, thermosetting, and cooling, as indicated in Fig. 8.15. Shown in Fig. 8.24 are the changes in distortion b/W of experimentally observed bowing as the film goes through the TD stretching zone and thermosetting zone [9]. In Fig. 8.24 the position in the tenter is expressed as the dimensionless length L/W, which is the distance from the entrance of the tenter L divided by the film length W. Figure 8.24 shows that bowing remains insignificant in the preheating zone. As the film enters the transverse stretching zone, the film develops a negative bowing followed by a quick change into positive bowing. Bowing takes its maximum positive value in the first half of the thermosetting zone. Thereafter, the bowing maintains a high positive level.

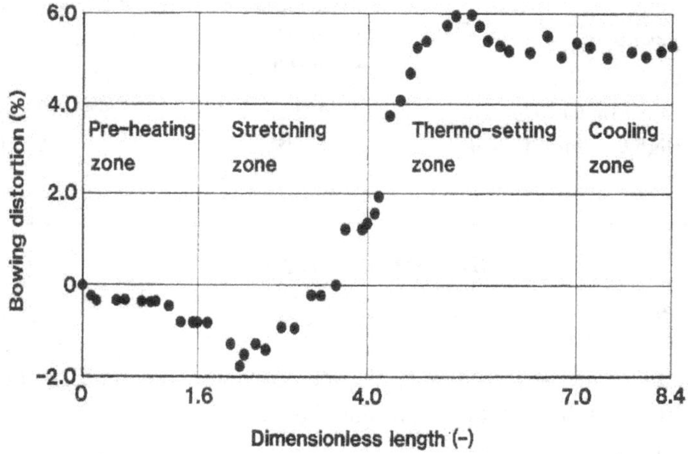

Figure 8.24 Experimentally observed deformation behavior in tenter expressed as bowing distortion versus dimensionless length

Figure 8.25 shows that higher thermosetting temperatures tend to increase the distortion of bowing, while changing the temperature of cooling does not affect bowing as Fig. 8.26 shows.

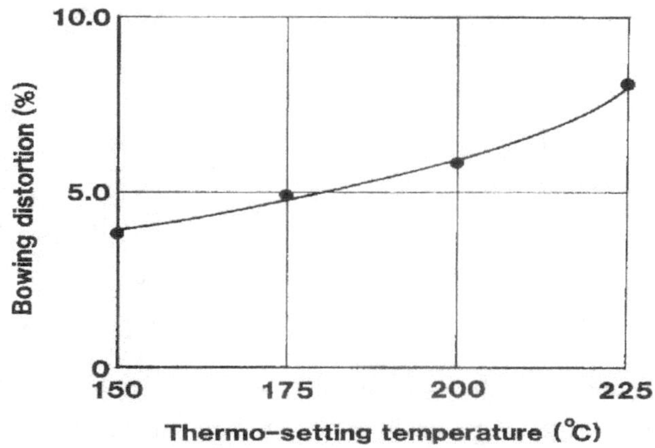

Figure 8.25 Bowing distortion at tenter exit versus thermoset temperature

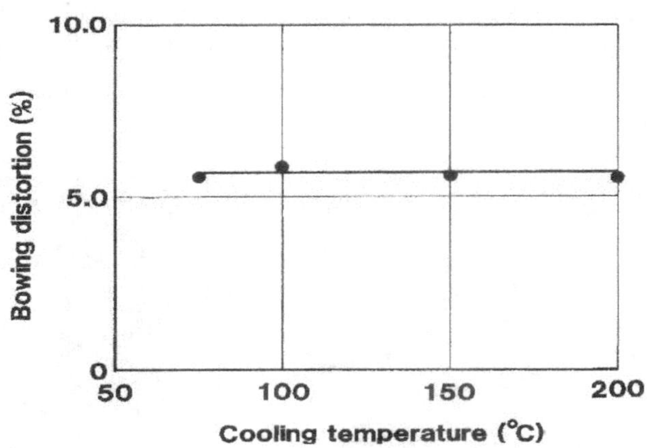

Figure 8.26 Bowing distortion at tenter exit versus cooling temperature

It was found that allowing the film to shrink transversely in the thermosetting zone tends to increase bowing as shown in Fig. 8.27. Negative relaxing, that is, restretching, in the transverse zone tends to suppress bowing.

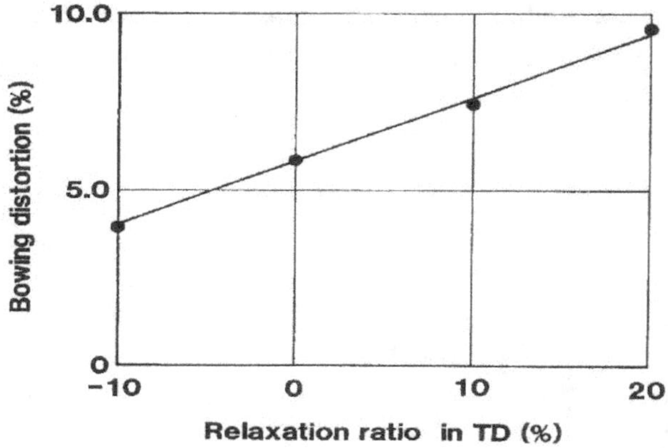

Figure 8.27 Bowing distortion at tenter exit versus relaxation ratio in TD within thermoset zone

The bowing phenomenon is undesirable and needs to be reduced. A new tentering machine reported to reduce the bowing phenomenon shown in Fig. 8.28 was developed by Bruckner Co. A linear motor type simultaneous biaxially oriented film machine (LISIM) can produce balanced film in MD and TD and also scar-free film and low-temperature heat sealable film because there is no roll used in the longitudinal stretching (Fig. 8.28). LISIM can produce low shrinkage film under both relaxations in MD and TD, but because the machine cost is very expensive, the application is limited at present.

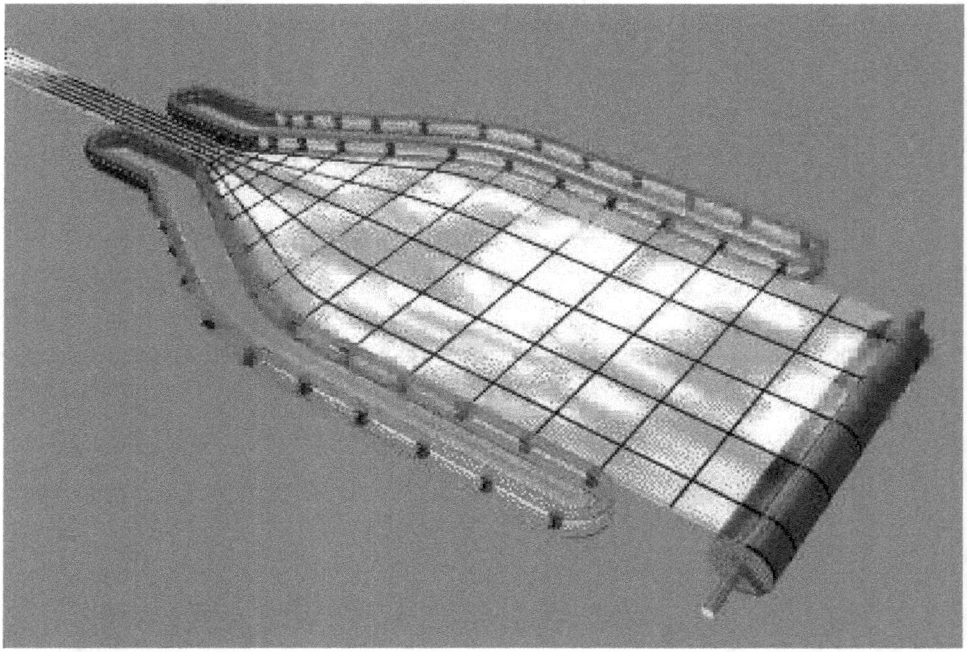

Figure 8.28 Schematic view of LISIM

In the case of sequential biaxial stretching, the stretching speed in MD reaches a maximum at 5000%/sec, but the stretching speed in TD is about 400%/sec, so it is possible to produce a thin stretched film. For example, BOPET film can be commercially produced at less than 1 μm and BOPP film can be produced at less than 3 μm, which is especially useful for condenser film.

8.5 Visualization of Stretching Process

The in situ observation of structure development during the stretching process such as birefringence, light scattering, small angle X-ray scattering (SAXS), and stress-strain buildup is studied by using the equipment shown in Fig. 8.29. This in situ stretching evaluation method can observe the structure development during the uniaxial stretching at any temperature. From this experiment, stretchability, thickness uniformity, spherulite size, and molecular orientation shown in Fig. 8.30 can be obtained at the same time. The comparison of high tacticity PP with a low tacticity one is shown in Fig. 8.31. This figure suggests that high tacticity PP has higher yield value of the stretching stress and larger spherulite size at the initial strain and less uniform birefringence pattern than low tacticity PP. The spherulite is broken up at the yield point and then made to disappear by the further stretching. Even though there is a very small amount of sample, it is possible to know the potential of the orientation, stretchability, thickness uniformity, and spherulite deformation during the stretching process.

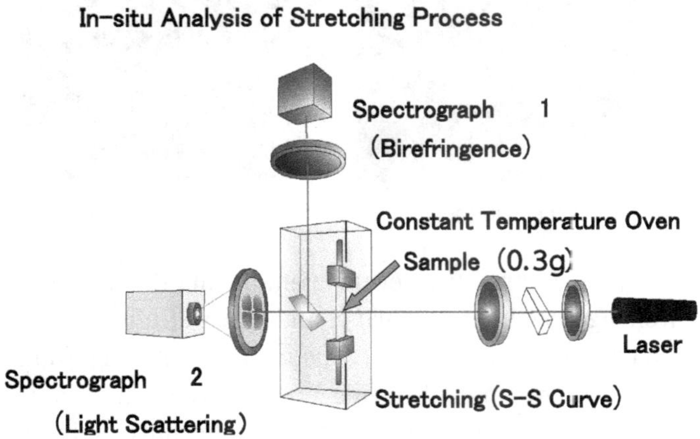

Figure 8.29 Basic research of stretching film

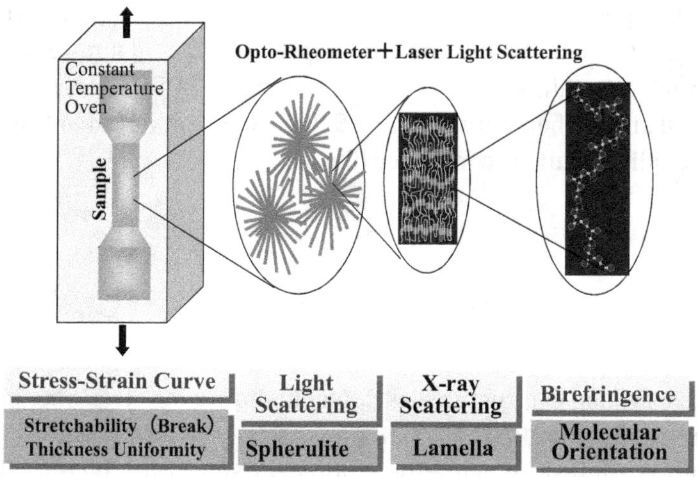

Figure 8.30 In situ stretching evaluation method for structure development

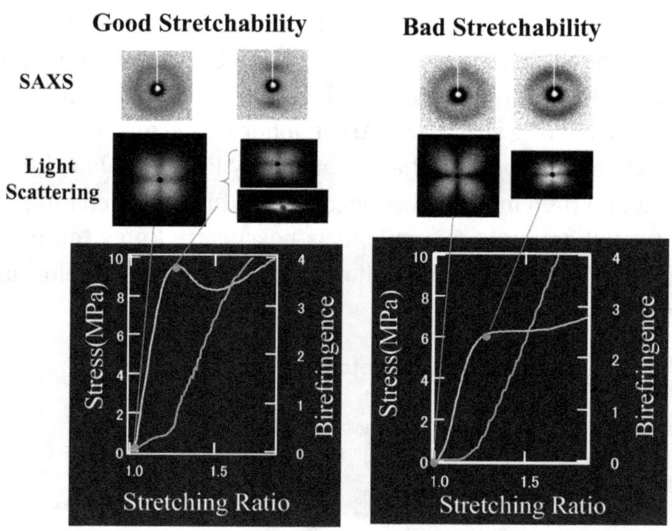

Figure 8.31 Simultaneous measurement of stress, birefringence, and light scattering

The in situ observation of structure development during the biaxial stretching process such as birefringence, light scattering, and stress-strain buildup is studied by using the equipment shown in Fig. 8.32. This in situ stretching evaluation method can observe structure development during biaxial stretching at any temperature. From this experiment, stretchability, thickness uniformity, spherulite size, and three-dimensional molecular orientation shown in Fig. 8.33 and Fig. 8.34 can be obtained at the same time.

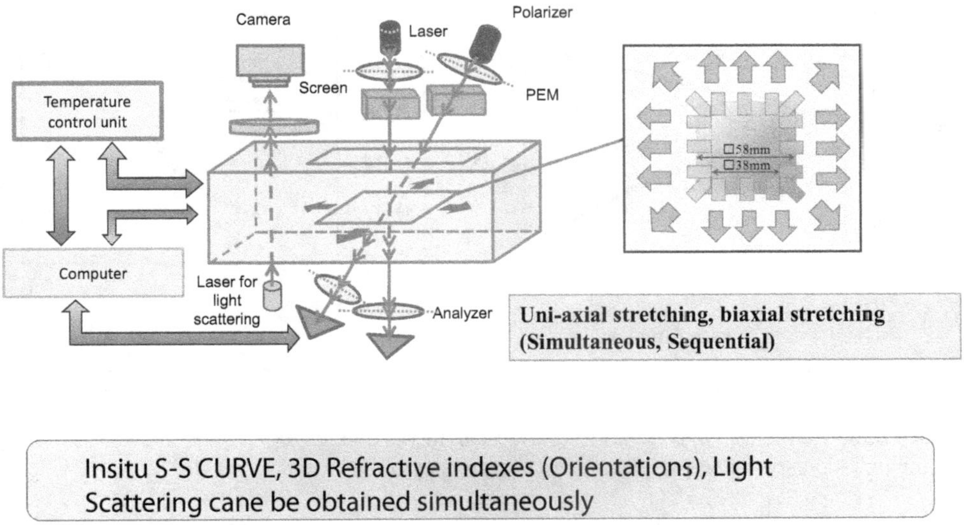

Figure 8.32 Biaxial stretching machine developed by our group; Kanazawa University and Eto Co., Ltd.

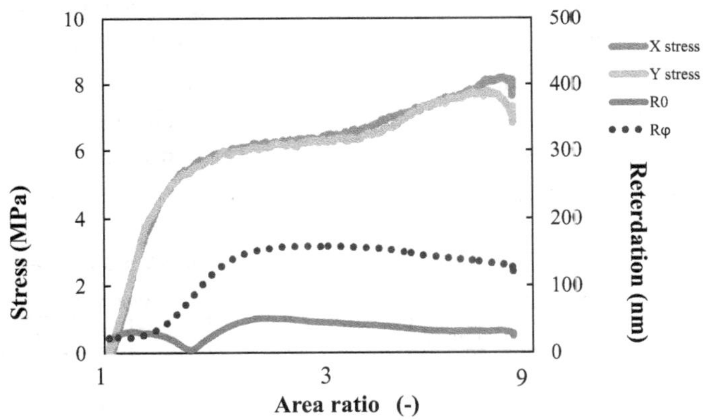

Figure 8.33 Stress-area stretching ratio curves and retardation-area stretching ratio curve of simultaneous biaxial stretching for PA6

The higher order structure and stretching behaviors of PA6 using this equipment is presented in Figs. 8.33 and 8.34. In the simultaneous biaxial stretching of these materials, the molecular orientation was equal both in the machine direction and in the transverse direction of the stretched plane.

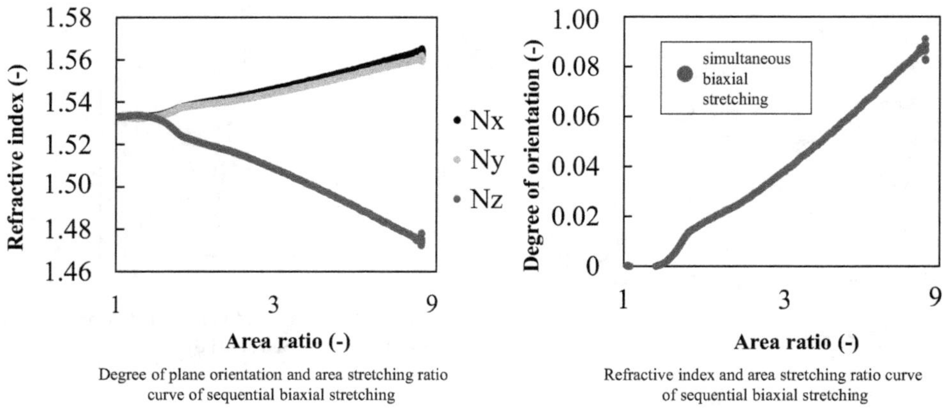

Figure 8.34 Refractive index-area stretching ratio curves and degree of orientation-area stretching ratio curve of simultaneous biaxial stretching for PA6

Recent research reported that stretching force parameters obtained by a table tenter can predict the stretchability and can be used to develop new materials to satisfy the demands of higher production speeds and thinner films [11]. Fm/Fy (Fm: maximum stretching force at maximum stretching ratio; Fy: yield stretching force at initial stretching stage) predicts that the stretchability of PP resin will improve by widening the MWD and reducing the crystallinity (Fig. 8.35). Reducing the crystallinity is effective in widening the process window of the stretching temperature in the stretching process. Reducing the crystallinity of PP resin and widening the MWD were effective in improving the film thickness uniformity (Fig. 8.36).

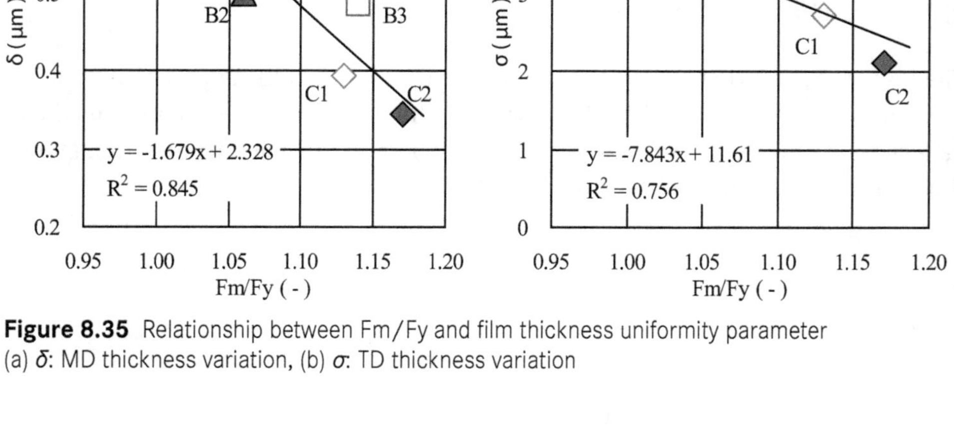

Figure 8.35 Relationship between Fm/Fy and film thickness uniformity parameter (a) δ: MD thickness variation, (b) σ: TD thickness variation

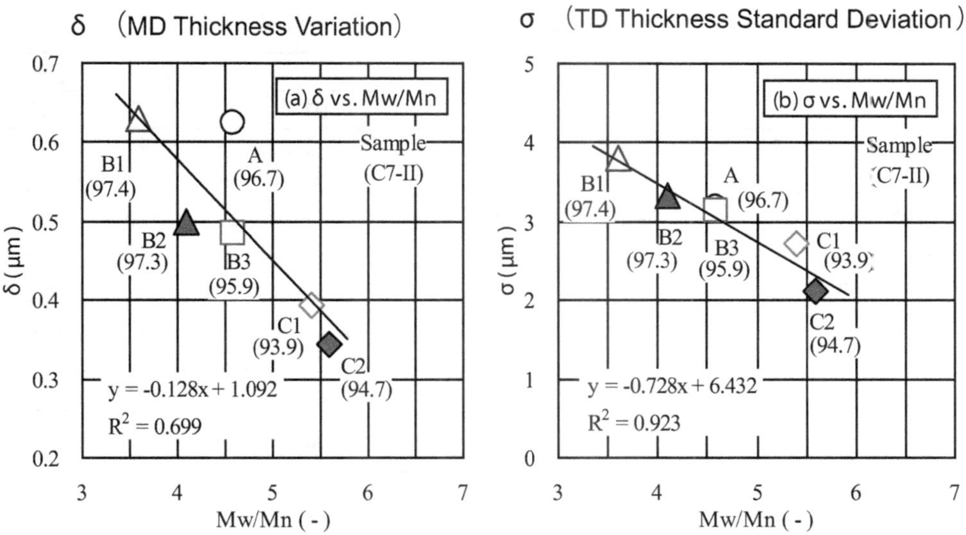

Figure 8.36 Relationship between molecular weight distribution Mw/Mn and film thickness uniformity parameter (a) *δ*: MD thickness variation, (b) *σ*: TD thickness variation

It is thought that bad film thickness uniformity is the cause of film breakage during the stretching process. Therefore, it is important to improve the film thickness uniformity to produce BOPP films at higher rate and to make thinner films.

The film thickness uniformity obtained with a sequential and biaxially oriented stretching machine had a good correlating relationship with the stretching force parameter obtained by a table tenter. So the analysis of the stretching force obtained by a table tenter with a small amount of resin was an effective technique in predicting the stretchability of BOPP film.

8.6 Film Physical Properties of Biaxially Oriented Film

The comparison of the film physical properties of biaxially oriented film with those of other process films is provided in Table 8.2. This table shows that the biaxially stretched PP film has properties superior to other films.

This table shows that the biaxially oriented film has high elastic modulus, high tensile strength, low elongation, low haze, and a high barrier property.

Table 8.2 Mechanical Properties of Stretched Polypropylene Films [10]

	unit	test	OPP		CPP			IPP
			F300	F309	F607	F651	F652	F650
Film thickness	µm	–	25	25	25	25	25	25
Haze	%	–	0.5	0.4	3.0	3.0	2.5	4.5
Gloss	%	–	135	145	100	100	120	90
Tensile strength at yield point	MPa	MD	–	–	21.6	18.6	17.7	17.7
		TD	–	–	21.6	18.6	17.7	17.2
Tensile strength at break	MPa	MD	118	147	48.0	53.9	58.8	58.8
		TD	255	294	44.1	48.0	53.9	53.0
Elongation at break	%	MD	60	45	550	600	600	580
		TD	15	10	600	650	650	600
Young's modulus	MPa	MD	2648	2942	1079	785	776	716
		TD	5001	5884	1177	794	726	706
Friction coefficient			–		0.20	0.25	0.25	0.25
Blocking force			0	0	0	1.0	1.5	2.0
Impact strength	N/m	23°C	38.2	44.1	27.5	24.5	34.3	48.0
Heat seal strength	N/15 mm	130°C	–	–	–	0.9	2.1	1.0
		140°C	–	–	–	3.5	8.3	6.4
		150°C	–	–	–	11.8	10.8	11.8

PET is an easily stretched resin. Because the biaxially oriented film of PET has a high elastic modulus and a high tensile strength, it is widely used as packaging film, optical film, magnetic film for computer memory, capacitor film, and so on. Table 8.3 shows the influence of the stretching condition on the physical properties [12]. PET film is widely used under the sequential stretching conditions of stretching ratio in MD 4 and then in TD 4, and stretching improves PET properties.

Table 8.3 Mechanical Properties of Stretched Polyethylene Terephthalate Films [12]

Method of stretching	Stretching ratio	Tensile strength MPa		Elongation at break %		Young's modulus MPa		F5 value MFa	
	MD × TD	MD	TD	MD	TD	MD	TD	MD	TD
Unstretched, no heat set	1.0 × 1.0	58	67	410	435	2000	2350	37	39
Unstretched, heat set	1.0 × 1.0	60	50	640	580	1800	1700	38	37
Uniaxially stretched under free width, no heat set	1.5 × 1’	72	34	380	430	1500	1500	47	37
	2.0 × 1’	119	35	350	480	1600	1500	52	36
	2.5 × 1’	193	42	270	450	3000	1600	87	30
	3.0 × 1’	235	45	140	600	3200	1700	91	28
	3.5 × 1’	246	31	71	830	4700	1500	113	33
	4.0 × 1’	275	30	62	910	4900	1000	156	30
Uniaxially stretched under constant width, no heat set	1.5 × 1	60	58	530	610	1800	1600	39	36
	2.0 × 1	111	48	370	480	1800	1600	45	35
	2.5 × 1	200	56	140	480	4100	1700	89	40
	3.0 × 1	222	58	96	490	4800	1900	120	42
	3.5 × 1	245	59	73	470	6000	2100	142	43
	4.0 × 1	256	57	50	500	6600	2400	169	69
Two-way successively biaxially stretched, heat set	4.0 × 1.5	247	81	46	380	7000	2300	213	81
	4.0 × 2.0	230	116	43	130	6000	3100	168	103
	4.0 × 2.5	220	158	34	140	5200	3000	150	105
	4.0 × 3.0	205	183	44	88	4900	3800	132	123
	4.0 × 3.5	196	192	47	62	4300	4100	133	125
	4.0 × 4.0	165	184	39	32	3600	4700	122	160
Simultaneously biaxially stretched, heat set	1.5 × 1.5	52	50	420	410	1900	1800	37	37
	2.0 × 2.0	105	108	330	340	2200	2300	64	63
	2.5 × 2.5	184	182	110	100	3600	3300	105	104
	3.0 × 3.0	187	203	66	70	3800	3800	118	118
	3.5 × 3.5	222	220	61	62	4000	3900	130	120
	4.0 × 4.0	240	241	41	36	3900	4000	136	129

Various resins other than PET and PP are also widely used and their stretched films are applied to various uses that require their characteristics.

8.7 Surface Roughness Control of Stretched Film

The surface roughness control of films without adding any additives is important for industrial films. Many studies have been conducted regarding the surface structure of BOPP, especially its crater-like structure [13, 14]. A new hypothesis of the crater-like film surface roughness formation mechanism is proposed by observing the transformation of craters from sheet to BOPP film and by investigating the relationship between the stress-strain curve and surface roughness change.

Formation behavior of the crater was observed with SEM (scattered electron microscopy) using the PP films at various stretching ratios using a table tenter (Fig. 8.37).

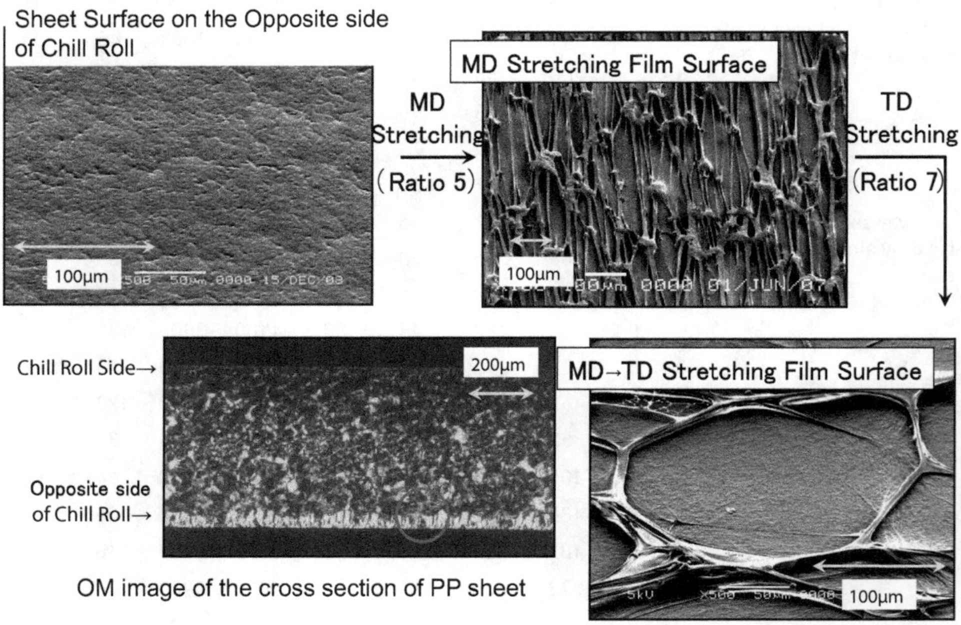

Figure 8.37 SEM image of the surface of BOPP film stretched from A80C 500 μm thickness sheet at different stretching ratios at 156°C

In an attempt to clarify the formation mechanism of crater-like surface roughness, forming conditions shown in Fig. 8.37 were used. The chill roll temperature had a great effect on the shape of the crater, and the crater size of the BOPP film stretched at 80°C was larger than the PP sheet cast at 30°C. When chill roll temperature rose, the crystal grain of the PP sheet (transcrystal structure) at the surface layer of the opposite side of the chill roll grew larger. This means that the shape of the crater is closely related to the crystal grain shape, namely the transcrystal structure.

Furthermore, the crater grew from the hollow that appeared at the beginning stage of the stretching process. When the stretching temperature was low, the craters were rarely formed at the end of the stretching process because the apertures that appeared at the beginning of the stretching process changed into shallow dents. Meanwhile, craters were formed at the end of the stretching process at high temperature because the apertures were deep enough to change into the craters. Finally, it was assumed that there must be another crater formation mechanism related to the collapse of crystalline grain on the surface layer of the sheet other than the crystal dislocation system from β crystal to α crystal.

References

1. Tobita, K., Miki, T., Takeuchi, N., Chapter 6.1 in *Film Processing*, Kanai, T., Campbell, G. A. (Eds.) (1999) Hanser Publications, Cincinnati
2. Bruckner Pamphlet of BOPP machine
3. Yoshida, A., *Seikei-Kakou, Japan Soc. Polym. Process.* (2006) 18, p. 194
4. Kanai, T., *17th Polym. Proc. Society Annual Meeting Abstracts*, (2001) 17, CD-ROM
5. Kanai, T., *Seikei-Kakou, Japan Soc. Polym. Process.* (2006) 18, p. 53
6. Sakamoto, K., *Japan Soc. Polym. Process. (Seikei-Kakou)* (1991) 3, p. 496
7. Sakamoto, K., *Kobunshi Ronbunshu* (1991) 48, p. 671
8. Yamada, T., Nonomura, C., *J. Appl. Polym. Sci.* (1993) 48, p. 1399
9. Yamada, T., Chapter 6.4 in *Film Processing*, Kanai, T., Campbell, G. A. (Eds.) (1999) Hanser Publications, Cincinnati
10. Moore, Jr., E. P. (Ed.), *Polypropylene Handbook* (1996) Hanser Publications, Cincinnati
11. Tamura, S., Kuramoto, I., Kanai, T., *Polym. Eng. Sci.* (2012) 52, pp. 1383–1393
12. Tsunashima, K., Toyoda, K., Yoshii, T., Chapter 6.3 in *Film Processing*, Kanai, T., Campbell, G. A. (Ed.) (1999) Hanser Publications, Cincinnati
13. Tamura, S., Ohta, K., Kanai, T., *J. Appl. Polym. Sci.* (2012) 124, pp. 2725–2735
14. Tamura, S., Takino, K., Yamada, T., Kanai, T., *J. Appl. Polym. Sci.* (2012) 126, p. 50

9 Structure Development in Uniaxial and Biaxial Film Stretching

T. Kikutani and W. Takarada

9.1 Introduction

Fibers, films, bottles, and various plastic products of three-dimensional shape are produced from polymeric materials through polymer processing technologies such as fiber spinning, film casting and drawing, tubular film blowing, blow molding, thermoforming, injection molding, and so on. The most important characteristic of polymeric materials is that the properties of polymer products vary significantly depending on the higher-order structure developed in the products during polymer processing. Especially for fibers and films, control of higher-order structure plays an important role in the improvement of properties of the final products.

From the viewpoint of the control of higher-order structure formation in film processing, the most important factors are considered to be (i) improvement of mechanical properties through the orientation of molecular chains in the film plane, (ii) control of planar orientation considering a balance between properties in the machine and transverse directions, and (iii) enhancement of thermal resistance and dimensional stability at elevated temperatures through crystallization. When the films are to be applied in optical devices, three-dimensional control of optical retardation is another important issue.

Poly(ethylene terephthalate) (PET) is known as one of the most important crystalline polymers for the production of base film because of its mechanical properties, good thermal resistance, and low cost. In this chapter, results of fundamental studies on structure development behavior of PET film during the three typical elongation modes, that is, uniaxial, planar, and biaxial elongation (Fig. 9.1), will be

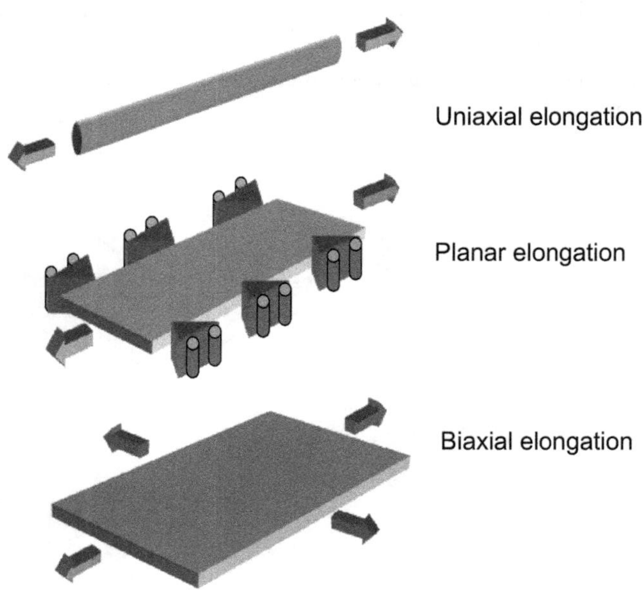

Figure 9.1 Three typical elongation modes

reported. Particular attention was paid to the development of oriented structure. It also should be noted that crystallization behavior can be accelerated through the development of molecular orientation, while the state of molecular orientation can be affected through the spontaneous structure formation caused by the crystallization [1]. Although there have been various publications on the uniaxial and biaxial stretching behavior of PET films [2–12], research on in situ measurement of stretching behavior and three-dimensional analyses of orientation development is limited [13, 14].

9.2 Equipment for In Situ Measurement of Optical Retardation

The stretching test of PET films was carried out using a batch-type biaxial film drawing machine equipped with an optical system for the in situ measurement of optical retardation. A schematic illustration of the experimental setup is shown in Fig. 9.2.

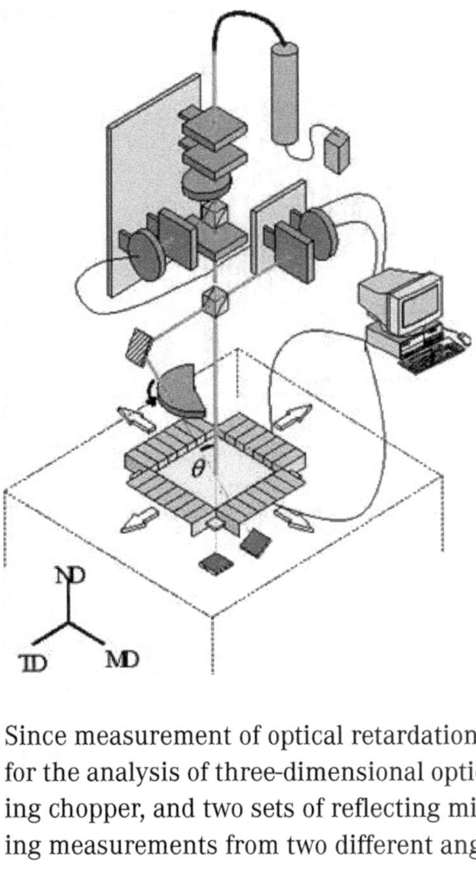

Figure 9.2 Batch-type biaxial film stretching device equipped with an optical retardation measurement system. A beam splitter, a rotating chopper, and two sets of reflecting mirrors are installed to realize the alternating measurements from two directions

Since measurement of optical retardation from two different directions is necessary for the analysis of three-dimensional optical anisotropy [15], a beam splitter, a rotating chopper, and two sets of reflecting mirrors were installed to realize the alternating measurements from two different angles to the film plane.

The optical system for the measurement of optical retardation is composed of a He-Ne laser light source, polarizing filters, quarter-wave plates, a rotating polarizing filter, a beam splitter, a beam expander, and photodetectors as shown in Fig. 9.3. A typical example of the measurement results is shown in Fig. 9.4.

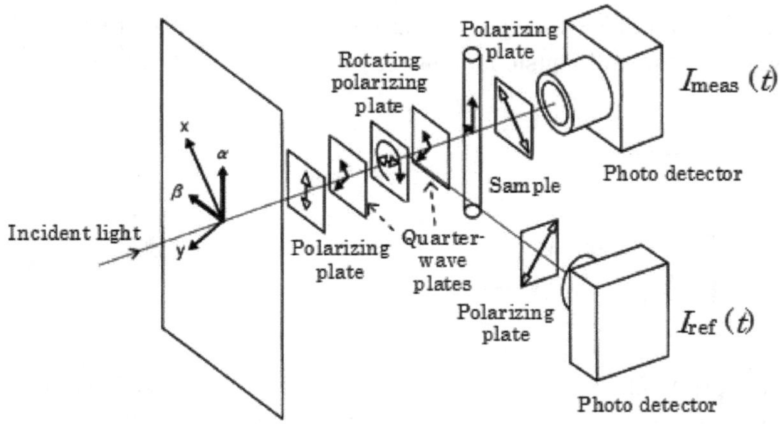

Figure 9.3 Optical system with rotating polarizing plate for the measurement of optical retardation using a laser light source

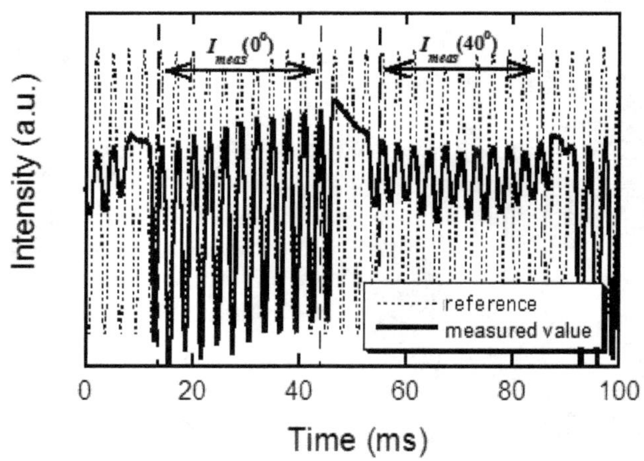

Figure 9.4 Typical example of raw data from optical retardation measurement system during alternating measurements with two different incident angles to film surface

In this system, intensities of measured light and reference light show periodic change because of rotation of the polarizing filter. Birefringence Δn can be estimated from the phase shift between the measured and reference lights using Equation 9.1.

$$\Delta n = \frac{(2\pi k + \delta)\lambda}{2\pi L} \tag{9.1}$$

where γ is the wavelength of incident light and L is the optical path length of light in the film. The order of interference k needs to be estimated separately. The principle of optical measurement has been reported elsewhere [16]. In this measurement, the direction of incident light was changed every 30 ms using a rotating chopper as shown in Fig. 9.2.

9.3 In Situ Measurement during Batch-Type Film Stretching Experiments

9.3.1 Variation of In-Plane Birefringence during Uniaxial Stretching and Relaxation Processes

Variations of true stress and birefringence with time during uniaxial stretching and relaxation processes of PET film under the stretching conditions of strain rate 0.1 s^{-1} and stretching temperature 90°C are shown in Fig. 9.5.

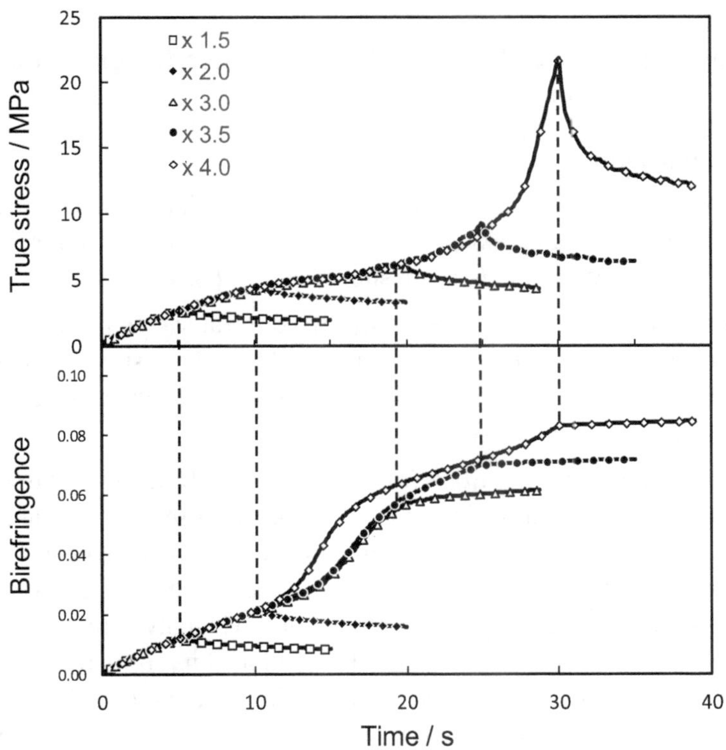

Figure 9.5 Variations of true stress and birefringence with time during uniaxial stretching and relaxation processes of PET film under the stretching conditions of strain rate 0.1 s^{-1} and stretching temperature 90°C. Stretching was halted at various draw ratios indicated in the figure

Stretching was stopped at various draw ratios indicated in the figure, and the subsequent relaxation process was also analyzed. In this experiment, birefringence was measured only from the direction normal to the film surface.

In the initial stage of stretching (0 to 10 s, draw ratio 1.0 to 2.0), true stress and birefringence showed similar behavior. In this region, a linear relationship between true stress and birefringence was confirmed. In other words, the stress-optical rule was found to be applicable. In the middle stage (10 to 20 s, draw ratio 2.0 to 3.0), birefringence started to increase steeply, whereas the true stress showed a tendency of saturation. In the late stage of stretching (after 20 s, draw ratio > 3.0), birefringence started to saturate whereas true stress started to increase steeply; this can be called strain-hardening.

When the stretching was stopped in the initial stage, true stress and birefringence relaxed in a similar manner following the stress-optical rule. On the other hand, when the stretching was stopped in the middle or late stage, stress relaxed as expected whereas birefringence increased spontaneously. In the late stage where the strain-hardening was observed, steep stress reduction started to appear at the beginning of stress relaxation.

9.3.2 Three-Dimensional Analysis of Birefringence Development in Film Stretching

9.3.2.1 Uniaxial Elongation

Results of in situ three-dimensional analyses for the variation of refractive indices and birefringences during uniaxial stretching of PET film at the strain rate of 0.1 s^{-1} and stretching temperature of 90°C are shown in Fig. 9.6.

Representations of the machine direction (MD), the transverse direction (TD), and the normal direction (ND) are also shown in the figure. True stress in MD was also measured simultaneously. This result corresponds to the drawing stage of the uniaxial elongation shown previously in Fig. 9.5.

At the initial stage of drawing, the refractive index in MD (n_{MD}) increased whereas refractive indices in TD and ND (n_{TD} and n_{ND}) decreased in a similar manner with increase of the draw ratio. Accordingly, birefringence between MD and TD ($\Delta n_{MD\text{-}TD}$) and birefringence between MD and ND ($\Delta n_{MD\text{-}ND}$) increased similarly whereas birefringence between TD and ND ($\Delta n_{TD\text{-}ND}$) remained at virtually zero. Since the mean refractive index was assumed to be constant, variation of n_{MD} from the starting value was two times larger than those for n_{TD} and n_{ND}. This is the typical behavior of uniaxial stretching with the MD axis-symmetry. When there started to be a steep increase of $\Delta n_{MD\text{-}TD}$ and $\Delta n_{MD\text{-}ND}$, however, balance between the two birefringences, that is, the balance between n_{TD} and n_{ND}, was lost for a certain period. The balance was recovered after the start of strain-hardening.

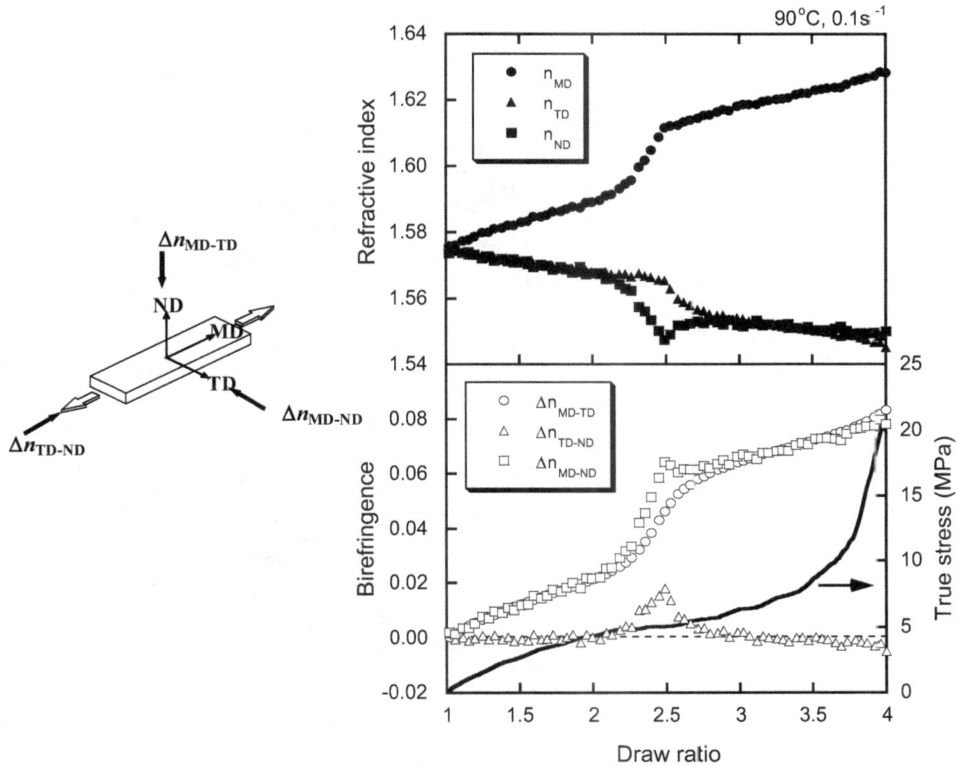

Figure 9.6 Results of in situ three-dimensional analyses for the variation of refractive indices and birefringences during uniaxial elongation of PET film at the strain rate of 0.1 s^{-1} and stretching temperature of 90°C. True stress in stretching direction (MD) is also shown

9.3.2.2 Planar Elongation

Result of in situ measurement for the planar film stretching is shown in Fig. 9.7.

In this case, the PET film was stretched in MD, whereas the width of film in TD was kept constant. With increase in the draw ratio, n_{MD} increased and n_{ND} decreased as in the case of uniaxial stretching, whereas n_{TD} showed only a slight reduction. Accordingly, both $\Delta n_{MD\text{-}TD}$ and $\Delta n_{MD\text{-}ND}$ increased with the draw ratio where $\Delta n_{MD\text{-}ND}$ was always larger than $\Delta n_{MD\text{-}TD}$. $\Delta n_{TD\text{-}ND}$ also showed a positive value. True stresses in MD and TD increased with the draw ratio where the MD stress was always significantly larger than the TD stress. Both stresses showed a tendency toward saturation in the middle stage followed by the subsequent steep increase in the late stage. The stress behavior in MD was similar to the one observed in the uniaxial stretching shown in Fig. 9.6, but the strain-hardening behavior was more significant.

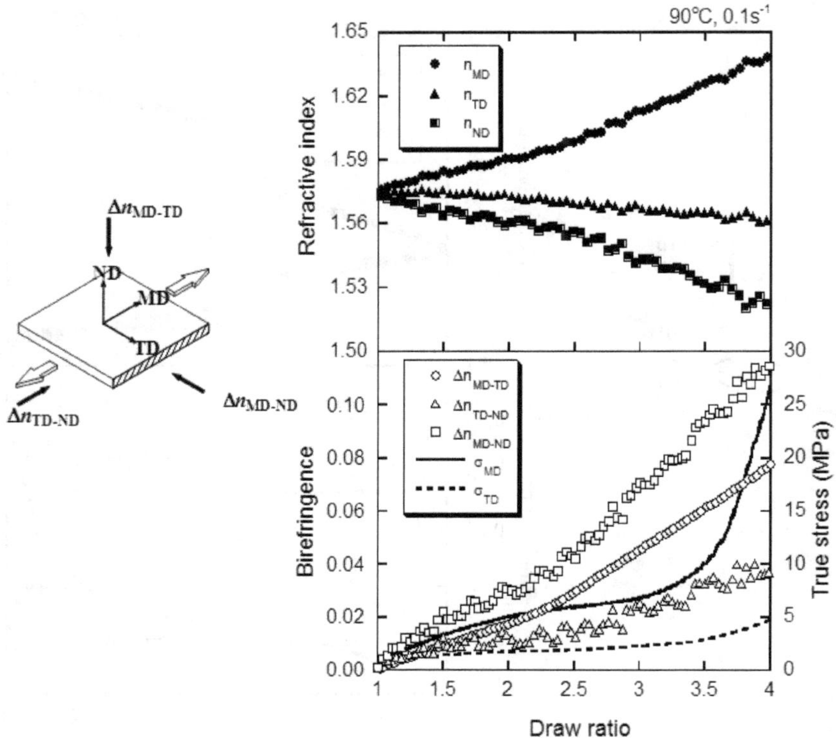

Figure 9.7 Results of in situ three-dimensional analyses for the variation of refractive indices and birefringences during planar elongation of PET film at the strain rate of 0.1 s^{-1} and stretching temperature of 90°C. True stress in stretching direction (MD) and transverse direction (TD) are also shown

9.3.2.3 Simultaneous Biaxial Elongation

Results of in situ measurement for the simultaneous equi-biaxial stretching of PET film are shown in Fig. 9.8.

In the initial stage, n_{MD} and n_{TD} increased with an increase in the draw ratio in a similar manner, whereas n_{ND} decreased with the stretching. The variation of n_{ND} from the starting value was two times larger than those of n_{MD} and n_{TD}. Accordingly, $\Delta n_{MD\text{-}ND}$ and $\Delta n_{TD\text{-}ND}$ increased with the draw ratio whereas $\Delta n_{MD\text{-}TD}$ stayed at nearly zero value. This is the typical behavior of simultaneous equi-biaxial stretching. In the middle stage of stretching, however, there was a hesitation of increase of $\Delta n_{MD\text{-}ND}$ and $\Delta n_{TD\text{-}ND}$. At the same time, the balance of n_{MD} and n_{TD} was lost, and $\Delta n_{MD\text{-}TD}$ deviated from zero. With further increase of the draw ratio, the balance between MD and TD was recovered along with the start of strain-hardening. The behavior of losing and recovering of balance between MD and TD is similar to the one observed in the uniaxial stretching shown in Fig. 9.6, where losing and recovering of balance between TD and ND occurred.

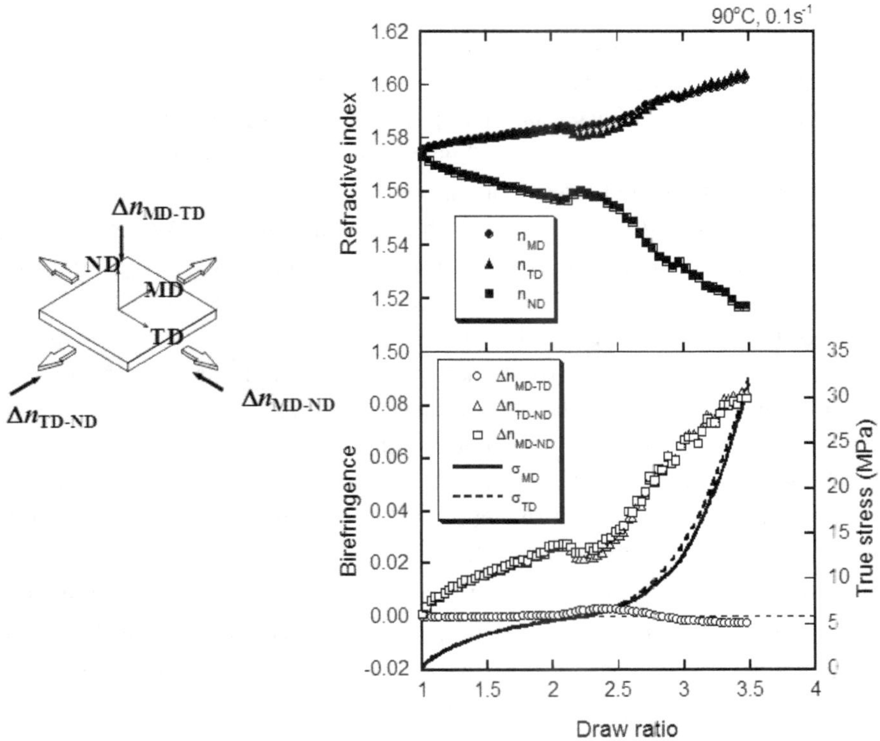

Figure 9.8 Results of in situ three-dimensional analyses for the variation of refractive indices and birefringences during simultaneous equi-biaxial elongation of PET film at the strain rate of 0.1 s^{-1} and stretching temperature of 90 °C. True stresses in two stretching directions (MD and TD) are also shown

9.3.3 Stress versus Birefringence Behavior

The relation between stress and birefringence obtained from the three types of stretching modes are summarized in Figs. 9.9(a), (b), and (c).

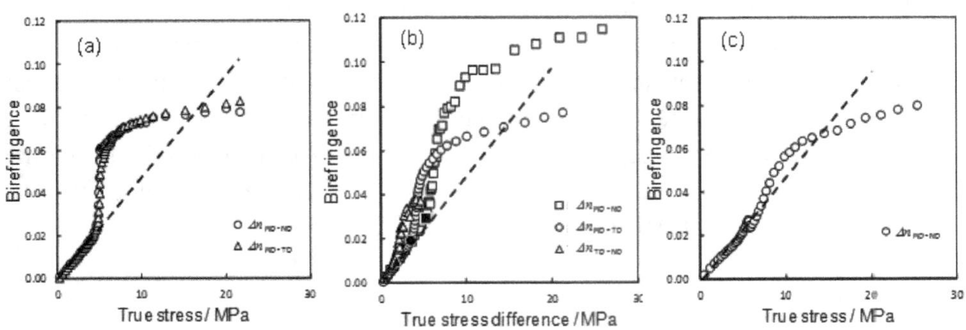

Figure 9.9 True stress versus birefringence plots for (a) uniaxial elongation, (b) planar elongation, and (c) simultaneous equi-biaxial elongation of PET film. The dotted lines in the figure correspond to the stress-optical coefficient of 0.49 GPa^{-1}

In the uniaxial elongation, the stress-optical rule was applicable up to the birefringence of about 0.02. The estimated stress-optical coefficient of 0.49 GPa^{-1} agreed well with the reported value [16]. Subsequently, there was an upward deviation of the birefringence versus stress relationship from the stress-optical rule followed by the tendency toward saturation. Similar behavior was reported by Ryu et al. [17]. For the data obtained from the planar elongation, it is noteworthy that the linear relationship of birefringence versus stress with a similar slope of 0.49 GPa^{-1} was confirmed for the relationships of $\Delta n_{MD\text{-}TD}$ versus σ_{MD}-σ_{TD}, $\Delta n_{MD\text{-}ND}$ versus σ_{MD}, and $\Delta n_{TD\text{-}ND}$ versus σ_{TD}. (Note that stress in ND is null because the film has a free surface.) In other words, three-dimensional applicability of the stress-optical rule was confirmed in this data. On the other hand, the closed symbols in the figure correspond to the points where deviation from the stress-optical rule started. It also should be noted that even though the level of birefringence is different, the deviation from linearity for the three combinations of birefringence versus stress difference started at the same draw ratio of around 1.5. In the data for the simultaneous equi-biaxial stretching, the tendency of the data was similar to the one observed for the uniaxial and planar elongations, and applicability of the stress-optical rule was also confirmed.

It should be noted that similar stress versus birefringence behavior for the uniaxial stretching shown in Fig. 9.9(a) was found in the research on online birefringence measurement during the high-speed melt spinning process of PET [18]. In both cases, the critical birefringence for the start of considerable structural ordering was about 0.02, even though the levels of strain rate and elongation temperature are significantly different. In addition, even in the case of equi-biaxial stretching (Fig. 9.9(c)), the upward deviation of the stress versus birefringence relationship started at a similar critical value of about 0.02. It was found that after this critical point, spontaneous increase of birefringence occurred even in the relaxation process after the cessation of stretching as shown in Fig. 9.5. In the melt spinning of PET, the critical point where a steep birefringence increase starts to occur corresponds to the boundary where the characteristics of as-spun fibers change from UDY (undrawn yarn) to POY (partially oriented yarn) [19]. It also is noteworthy that the starting point for the appearance of imbalance between n_{TD} and n_{ND} in the uniaxial stretching and n_{MD} and n_{TD} in the equi-biaxial stretching corresponds to the critical birefringence value of 0.02. It can be speculated that the stretching behavior tends to be unstable when the spontaneous structural ordering starts. This is because a small difference between the two directions expands when the structure development behavior in one direction proceeds beyond the threshold and falls into the region of spontaneous structural ordering.

9.4 Analysis of Sequential Biaxial Elongation

For the investigation of sequential biaxial stretching, instead of conducting a batch-type experiment as described above for three different stretching modes, two types of approaches were selected. One is the theoretical prediction of the birefringence development. The other is the analysis of film samples collected from a pilot-scale film tentering machine after the sudden cessation of its operation.

9.4.1 Theoretical Prediction for Sequential Biaxial Elongation

Based on rubber elasticity theory with the assumption of a network structure, stress development was estimated considering the strain energy potential function and Kuhn statistical theory. A linear relationship between stress and birefringence was assumed. The result of the theoretical calculation of birefringence development for the sequential biaxial stretching process is shown in Fig. 9.10.

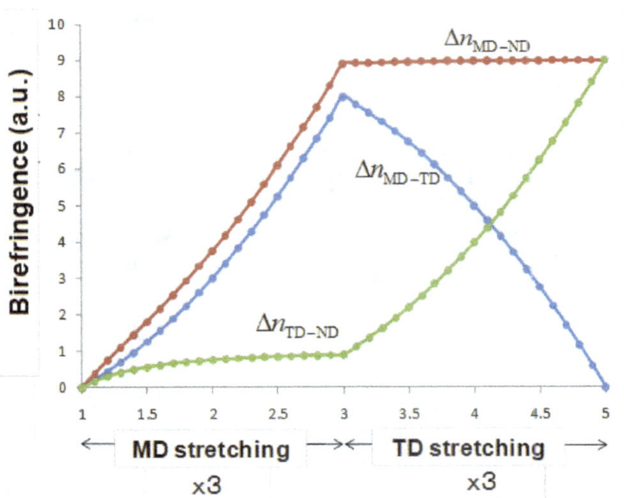

Figure 9.10 Result of theoretical prediction for the variation of three different birefringences during sequential equi-biaxial stretching

During the MD stretching with constant film width in TD, both $\Delta n_{MD\text{-}ND}$ and $\Delta n_{MD\text{-}TD}$ increased significantly where $\Delta n_{MD\text{-}ND}$ was slightly higher than the $\Delta n_{MD\text{-}TD}$. At the same time, $\Delta n_{TD\text{-}ND}$ increased slightly with the stretching. The stretching behavior up to this point corresponds to the planar stretching shown in Fig. 9.7. Experimental results and theoretical predictions showed a similar tendency of orientation development behavior. During the subsequent stretching to TD with constant width in MD, there was a slight increase of $\Delta n_{MD\text{-}ND}$, while $\Delta n_{MD\text{-}TD}$ decreased and $\Delta n_{TD\text{-}ND}$

increased rapidly. At the end of the TD stretching, $\Delta n_{MD\text{-}ND}$ and $\Delta n_{TD\text{-}ND}$ reached the same value. Accordingly $\Delta n_{MD\text{-}TD}$ reached zero. At this point, the film is supposed to be isotropic if it is observed from the direction normal to the film surface, while there is a planar orientation of molecular chains parallel to the film surface.

Variation of three-dimensional optical anisotropy during the sequential biaxial stretching process of film is schematically illustrated using the refractive index ellipsoid as shown in Fig. 9.9. Immediately after the film casting on the chill roll, no optical anisotropy is introduced in the film, and the shape of the refractive index ellipsoid is spherical. After the MD stretching with constant width, molecular chains are oriented along MD and the refractive index ellipsoid changes to the shape of a rugby ball with a slight flattening in the direction normal to the film surface. During the TD stretching in the tentering process, the refractive index ellipsoid is also stretched to TD and eventually became the shape of a flying saucer with equal size in MD and TD.

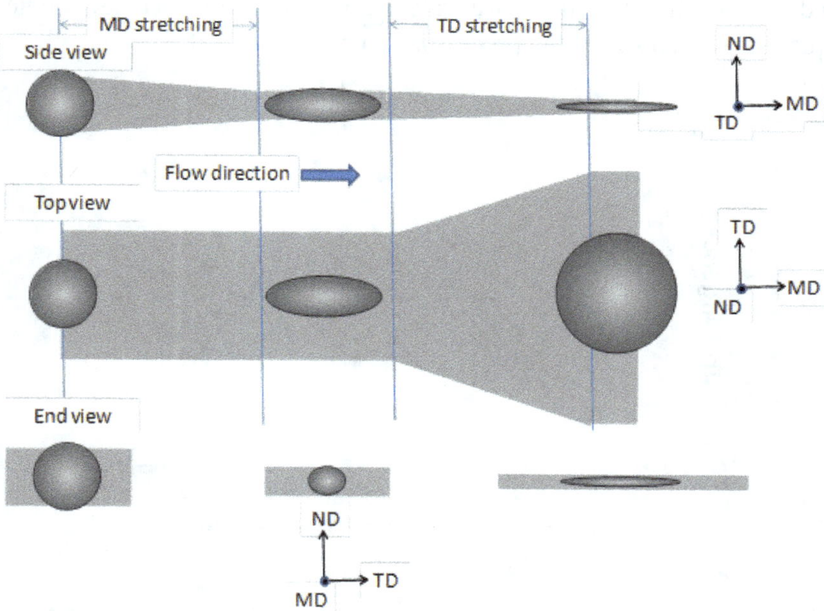

Figure 9.11 Schematic illustration for the variation of the refractive index ellipsoid during sequential equi-biaxial stretching

9.4.2 Off-Line Analysis of Film Samples from the Sequential Biaxial Stretching Process

9.4.2.1 Birefringence

Analysis of film samples collected from a pilot-scale film tentering machine after the sudden cessation of its operation was conducted. The PET film uniaxially drawn with the MD draw ratio of 3.2 was fed into the tentering machine and stretched with the TD draw ratio of 3.8. The result of birefringence measurement from the direction normal to the film surface is summarized in Fig. 9.12. In this figure, the direction and length of each arrow correspond to the direction of optical axis and birefringence value respectively. Since uniaxially drawn film was introduced into the preheating zone, the arrows in this region indicate the orientation of molecules along the machine direction. In the tentering zone, the arrows gradually became shorter along with the increase of stretching ratio in TD, and finally its direction rotated 90 degrees at around the boundary of the tentering and heat setting zones. This result indicates that the uniaxially oriented structure gradually changed to planar orientation during the tentering process, and finally the molecules showed a slight imbalance of orientation in the film plane, in that there was a preferential orientation of low degree toward TD.

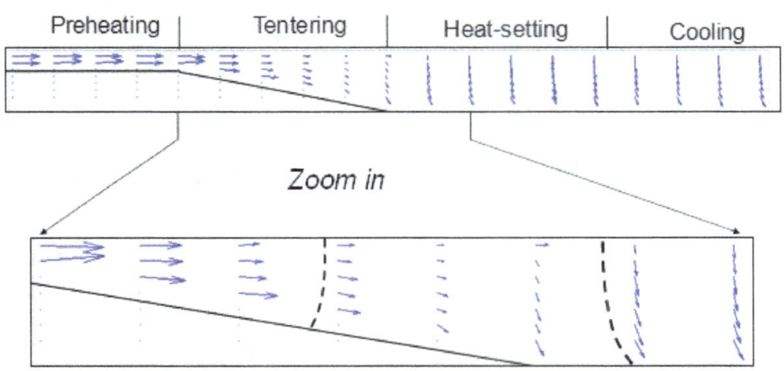

Figure 9.12 Birefringence distribution in the PET film sampled from the sequential biaxial stretching process. Length and direction of arrows indicate the absolute value of birefringence and the direction of optical axis. Broken lines in the figure represent the shape of bowing predicted from the tilting of optical axes

In the measurement of film samples from the off-center positions, a slight tilting of optical axis was also observed. The tilting angle increased gradually from the center to the side edge of the film. It is obvious that such tilting originates from the bowing phenomena. It is well known that the direction of bowing reverses from forward to backward when the film is moved from the tentering zone to the heat-setting zone [20–22]. The distribution of tilting angle of optical axis in the film exhibited a tendency as indicated by the dotted lines in the figure.

Three-dimensional analysis of the optical anisotropy was conducted along the centerline of the collected film samples of two different draw ratios, i.e., MD × TD of 2.4 × 3.0 and 2.4 × 3.8. The results are summarized in Figs. 9.13 and 9.14. In the preheating zone, n_{MD} was higher than n_{TD} and n_{ND}, while there was a certain difference between n_{TD} and n_{ND}. Accordingly, $\Delta n_{MD\text{-}ND}$ was slightly higher than the $\Delta n_{MD\text{-}TD}$ whereas $\Delta n_{TD\text{-}ND}$ showed a positive value. (A difference between n_{TD} and n_{ND} was not

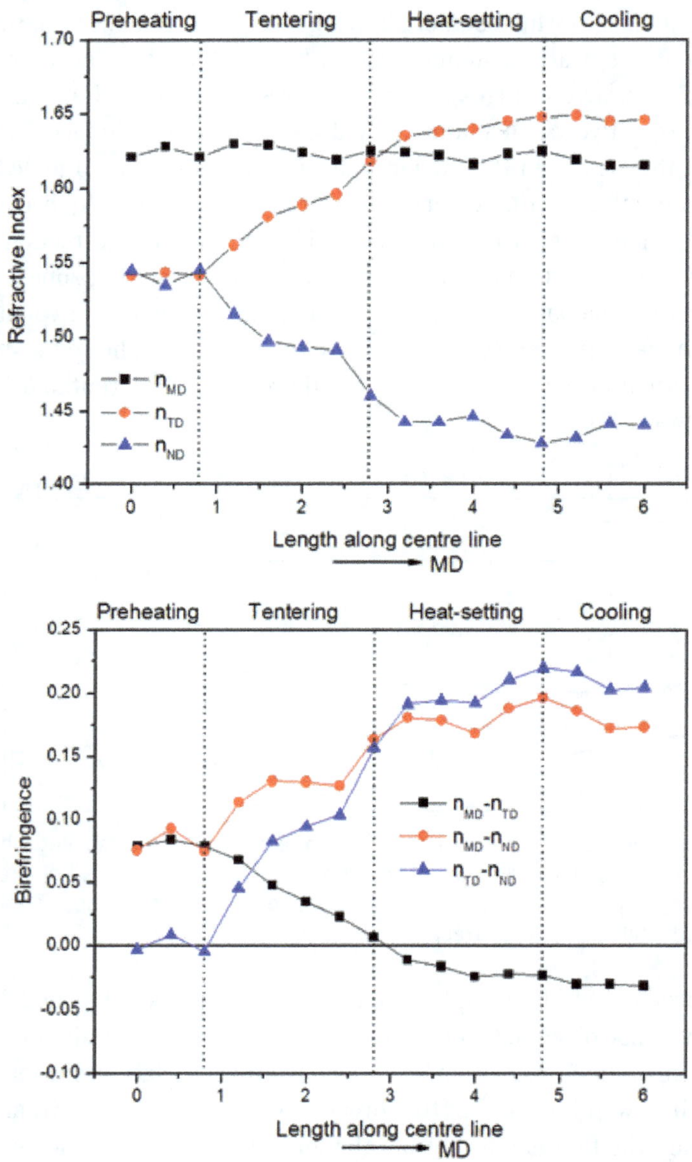

Figure 9.13 Results of three-dimensional analyses for the variation of refractive indices and birefringences along the centerline of PET film sampled from the sequential biaxial stretching process. Draw ratio MD × TD is 2.4 × 3.0

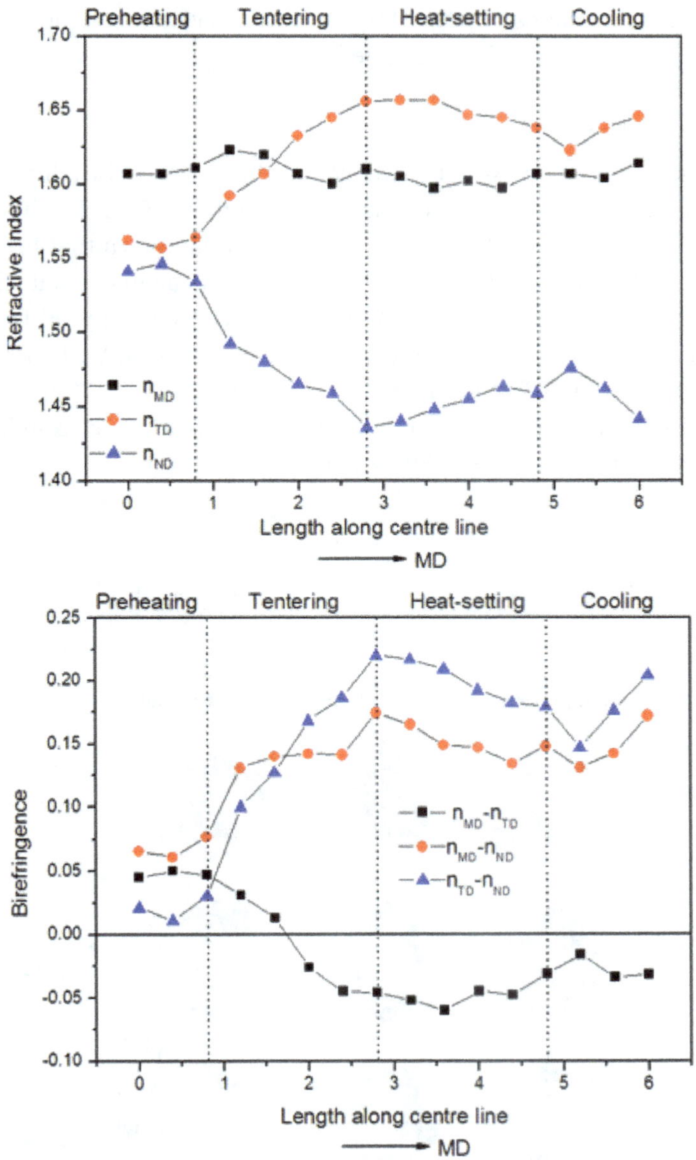

Figure 9.14 Results of three-dimensional analyses for the variation of refractive indices and birefringences along the centerline of PET film sampled from the sequential biaxial stretching process. Draw ratio MD × TD is 2.4 × 3.8

clearly detected in the case of MD × TD = 2.4 × 3.0.) In the tentering zone, there was a significant increase and decrease of n_{TD} and n_{ND} respectively, whereas n_{MD} remained almost constant. These results correspond to the steep and gradual increases of $\Delta n_{TD\text{-}ND}$ and $\Delta n_{MD\text{-}ND}$, respectively. At the end or in the middle of the tentering zone for the TD draw ratio of 3.0 or 3.8, respectively, n_{TD} became larger than n_{MD}. Accordingly, reduction of $\Delta n_{MD\text{-}TD}$ from a positive value to a negative value was also confirmed in both TD draw ratios. A larger TD draw ratio yielded the higher tendency of molecular orientation toward TD as expected.

9.4.2.2 WAXD Analysis

Wide-angle X-ray diffraction (WAXD) patterns of the collected PET film measured from the through, end, and edge directions are shown in Fig. 9.15. The film with the draw ratio of MD × TD = 3.2 × 3.8 was analyzed. It is well known that in the highly stretched and crystallized PET films, benzene rings in the main chain tend to orient parallel to the film surface [2, 23]. Such orientation can be analyzed through the detection of crystalline reflection from the (100) plane in the WAXD analysis since the benzene ring in the PET molecule is aligned parallel to the (100) plane in the unit cell of a PET triclinic crystal. The orientation of crystals can be examined utilizing the concept of an Ewald sphere (Fig. 9.16).

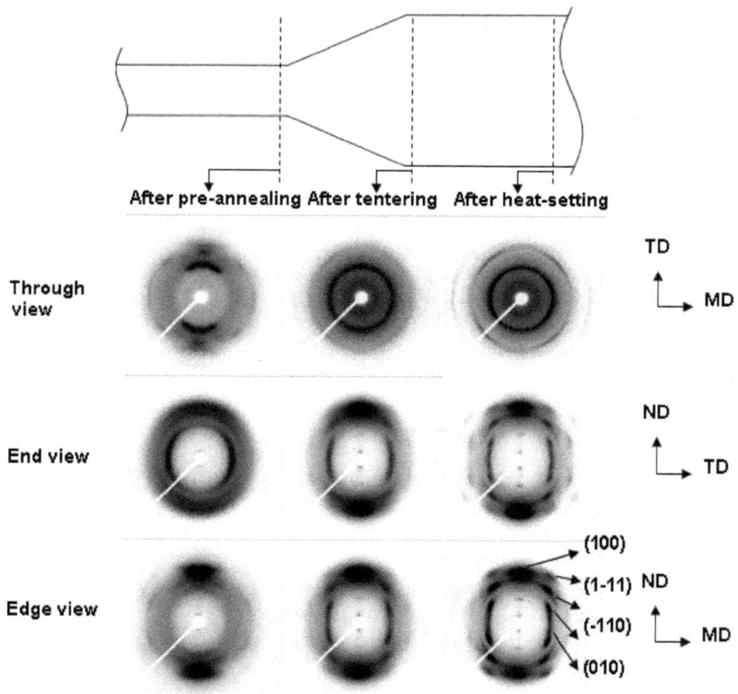

Figure 9.15 Variations of WAXD pattern along the centerline of PET film sampled from the sequential biaxial stretching process. Draw ratio MD × TD is 3.2 × 3.8

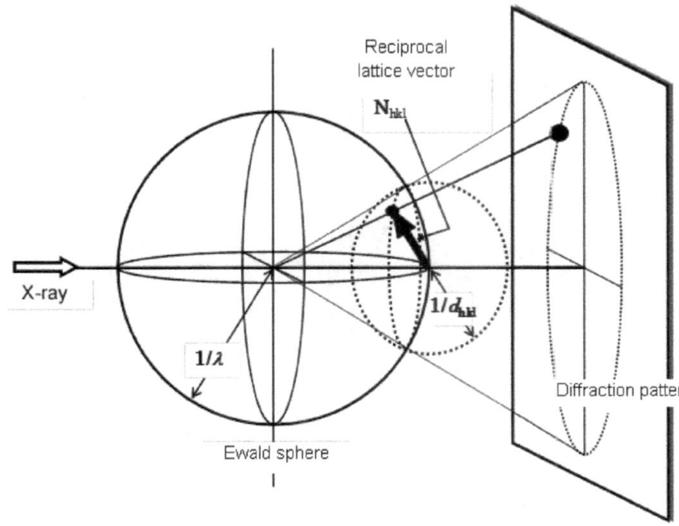

Figure 9.16 Spatial relationship between reciprocal lattice vector and the Ewald sphere for the satisfaction of Bragg's law for the appearance of diffraction

The Ewald sphere is an imaginary sphere in the reciprocal space with the radius of 1λ where λ is the wavelength of an X-ray beam [24]. Contact of the tip of the reciprocal lattice vector $\mathbf{N}_{hkl}$ with the surface of the Ewald sphere is the condition for the satisfaction of Bragg's law and for the appearance of a certain crystalline reflection. The direction of the $\mathbf{N}_{hkl}$ is normal to the crystalline plane in the real space, and the magnitude of the $\mathbf{N}_{hkl}$ is $1/d_{hkl}$ where d_{hkl} is the interplanar spacing of the (hkl) plane. For example, when the WAXD pattern of the PET film with the orientation of (100) plane parallel to the film surface is measured from the direction normal to the film surface (through view), (100) reflections are unlikely to appear because the tip of $\mathbf{N}_{100}$ locates far away from the surface of the Ewald sphere as shown in Fig. 9.17.

Observing the WAXD patterns of uniaxially drawn and preannealed PET, orientation of the *c*-axis along MD was confirmed because (hk0) reflections appeared in the direction normal to MD in the through and edge views. In the through view, (100) reflections were weak in comparison with (0-11) and (010) reflections, whereas (100) reflections were strong in the edge view. This is the indication of the orientation of the (100) plane along the film surface.

After the TD stretching, a ring of (0-11) reflection was observed clearly in the through view with a slight concentration of intensity toward MD. WAXD patterns of end and edge views were similar. These are the indications of the almost random orientation of a molecular chain in the film plane with a slight preferential orientation toward TD. On the other hand, (100) reflection appeared in the ND, whereas (0-11) and (010) reflections appeared in the MD and TD in the end and edge views. These results indicate the orientation of the (100) plane and therefore orientation of the benzene ring along the film surface.

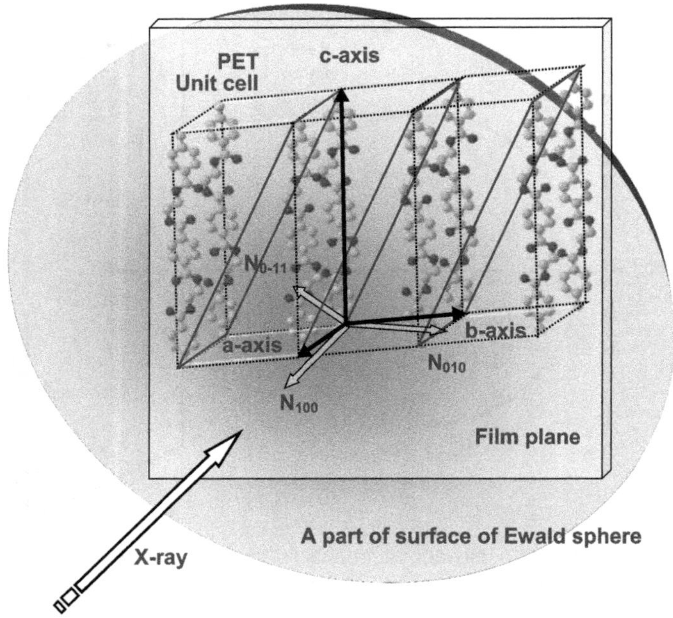

Figure 9.17 Schematic illustration of the spatial relationship between three reciprocal lattice vectors, a unit cell of PET, the film surface, and a part of the surface of an imaginary Ewald sphere for the through view WAXD measurement

Development of a highly ordered crystalline structure after the heat setting was confirmed qualitatively because the crystalline reflections became clearer and stronger. Angles between the (100) reflection and other reflections in the end view for the sample after heat setting were compared with the values predicted from the parameters of the crystalline unit cell [25] in Table 9.1. Good agreement of the experimental results with the theoretical predictions is the quantitative evidence for orientation of the (100) plane along the film surface.

Table 9.1 Comparison of the Angle between (hkl) and (100) Planes from Theoretical Calculation with the Corresponding Angles Obtained by the Analysis of a WAXD Pattern

Crystalline plane (hkl)	Angle with respect to (100) plane*	Azimuthal angle in WAXD pattern
(1-11)	38°	34°
(-110)	46°	40°
(010)	59°	60°
(0-11)	88°	81°

* Calculated using the unit cell parameters reported by Bunn [25]

9.5 Intrinsic Birefringence for Various Orientation Modes

If there is a preferential orientation of the (100) plane along the film surface, it would be useful to estimate the intrinsic birefringence of such a structure. Polarizabilities along various crystallographic axes of the triclinic unit cell of PET crystal are summarized in Table 9.2 [26]. When there is a planar orientation of molecular chain in the film plane, the *c*-axis of the unit cell is also supposed to orient in the film plane. In this case, polarizabilities for the directions parallel and perpendicular to the film surface can be defined as $\left(3\overline{P}+P_c\right)/4$ and $\left(3\overline{P}-P_c\right)/2$, respectively. On the other hand, when the (100) plane aligns parallel to the film surface, the polarizabilities for the directions parallel and perpendicular to the film surface can be defined as $\left(3\overline{P}-P_{a^*}\right)/2$ and P_{a^*}, respectively.

Table 9.2 Polarizabilities for Various Directions in a Crystallographic Unit Cell [26]

Symbol	Polarizability × 10^{25} (cm^3)	Direction
$\overline{P}$	186.96	Mean value
P_a	154.97	*a*-axis
P_b	212.46	*b*-axis
P_c	225.83	*c*-axis
P_{a^*}	124.07	*a**-axis (perpendicular to *b*- and *c*-axes)
P_{b^*}	208.60	*b**-axis (perpendicular to *a*- and *c*-axes)

By substituting the obtained polarizability into the Lorentz-Lorenz Equation (Equation 9.2), intrinsic birefringences for various orientation modes were estimated as shown in Table 9.3.

$$\frac{n^2-1}{n^2+2}\frac{M}{\rho}=\frac{4}{3}\pi NP \tag{9.2}$$

where n is the refractive index, M the molar mass of the repeating unit of PET, ρ the density, and N the Avogadro's number. For the equi-biaxial orientation, intrinsic birefringence for the planar orientation of the (100) plane along the film surface is significantly larger than that for the planar orientation of the *c*-axis. This is due to the considerable polarizability anisotropy of the benzene ring, in that the polarizability in the plane containing the benzene ring is much larger than that in the direction normal to the plane of the benzene ring.

It was indicated in Figs. 9.13 and 9.14 that the $\Delta n_{\text{MD-ND}}$ and $\Delta n_{\text{MD-TD}}$ values are in the range of 0.15 to 0.20 after the biaxial stretching. The calculation results shown in

Table 9.3 suggest that the development of such high birefringence cannot be achieved without the orientation of the (100) plane along the film surface.

Table 9.3 Calculated Intrinsic Birefringences for Various Orientation Modes

Uni-axial orientation	$\Delta n_{MD\text{-}TD}$ & $\Delta n_{MD\text{-}ND}$	0.257
Uniaxial orientation + Planar orientation*	$(\Delta)n_{MD\text{-}TD}$	0.071
	$(\Delta)n_{MD\text{-}ND}$	0.420
	$(\Delta)n_{TD\text{-}ND}$	0.350
Equi-biaxial orientation	$\Delta n_{MD\text{-}ND}$ & $\Delta n_{TD\text{-}ND}$	0.122
Equi-biaxial orientation + Planar orientation	$\Delta n_{MD\text{-}ND}$ & $\Delta n_{TD\text{-}ND}$	0.384

* planar orientation of (100) plane in film plane

9.6 Concluding Remarks

Structure formation behavior in the various elongation processes of PET film are explained in this chapter. One of the important characteristics of PET during stretching is the start of self-ordering when the degree of orientation exceeds a certain threshold. Another important behavior is the orientation of not only the molecular chain axis but also the benzene ring in the main chain parallel to the film surface. These behaviors cause considerable difficulty in predicting the structure development behavior and the properties of the final film products.

For research on the prediction of structure development behavior in polymer processing, information on the flow/deformation and temperature histories are indispensable. On the other hand, in the literature for film processing, especially for the film tentering process, it is quite rare to find the actual processing conditions such as the size of tentering frame, processing speed, temperature distribution, and so on. It also should be noted that in terms of processing speed and strain rate, the conditions for batch-type film elongation experiments conducted in laboratories are normally significantly lower than the actual processing conditions in the industry. This fact also leads to difficulty in the quantitative prediction of the structure and properties of films from the processing conditions.

Nevertheless, there should be a certain fundamental principle for biaxial stretching, which needs to be clarified. At present, even the basic understanding of the structure formation behavior in biaxial stretching is limited. Collaboration between industry and academia is necessary for further development in this area.

References

1. Kikutani, T., in *Handbook of Textile Fibre Structure,* Vol. 1, Eichnorn, S.J., Hearle, J.W.S., Jaffe, M., Kikutani, T. (Eds.) (2009) Woodhead Publishing, Cambridge, UK, pp. 157–180
2. Matsumoto, K., Ieki, H., Imamura, R., *Sen'i Gakkaishi* (1971) 27, pp. 516–524
3. Matsuo, M., Ooki, J., Harashina, Y., Ogita, T., Manley, R.St.J., *Macromol.* (1995) 28, pp. 4951–4960
4. Khan, M.B., Kenner, C., *Polym. Eng. Sci.* (1996) 36, pp. 1290–1299
5. Champchesnel, J.B.F., Tassin, J.F., Monnerie, L., Sergot, P., Lorentz, G., *Polymer* (1997) 38, pp. 4165–4173
6. Kim, H.H., Kang, C.K., Chang, C.G., Ihim, D.W., *Eur. Polym. J.* (1997) 33, pp. 1633–1638
7. Viny, M., Tassin, J.F., Lorentz, G., *Polymer* (1999) 40, pp. 397–406
8. Adams, A.M., Buckley, C.P., Jones, D.P., *Polymer* (2000) 41, pp. 771–786
9. Buckley, C.P., Jones, D.C., Jones, D.P., *Polymer* (2000) 371, pp. 2403–2414
10. Boyce, M.C., Socrate, S., Llana, P.G., *Polymer* (2000) 41, pp. 2183–2201
11. Horn, B.L.V., Winter, H.H., *Macromol.* (2003) 36, pp. 8513–8521
12. Kawakami, D., Ran, S., Burger, C., Avila-Orta, C., Sics, I., Chu, B., Hsiao, B.S., Kikutani, T., *Macromol.* (2006) 39, pp. 2909–2920
13. Carla, I., Martins, C., Cakmak, M., *Polymer* (2007) 48, pp. 2109–2123
14. Cakmak, M., Hassan, M., Unsal, E., Martins, C., *Rev. Sci. Instrum.* (2012) 83, 123901 (online)
15. Stein, R.S., *J. Polym. Sci.* (1957) 24, pp. 383–386
16. Oda, K., White, J.L., Clark, E.S., *Polym. Eng. Sci.* (1978) 18, pp. 53–59
17. Ryu, D.S., Inoue, T., Osaki, K., *Polymer* (1998) 39, pp. 2515–2520
18. Kikutani, T., Nakao, K., Takarada, W., Ito, H., *Polym. Eng. Sci.* (1999) 39, pp. 2349–2357
19. Shimizu, J., Okui, N., Kikutani, T., in *High-Speed Fiber Spinning,* Ziabicki, A., Kawai, H. (Eds.) (1985) John Wiley & Sons, New York, pp. 429–483
20. Yamada, T., Nonomura, C., *J. Appl. Polym. Sci.* (1993) 48, pp. 1399–1408
21. Yamada, T., Nonomura, C., *J. Appl. Polym. Sci.* (1994) 52, pp. 1393–1403
22. Yamada, T., Nonomura, C., Kase, S., in *Film Processing,* Kanai, T., Campbell, G.A. (Eds.) (1999) Carl Hanser Verlag, Munich, Germany, pp. 351–385
23. Cakmak, M., in *Film Processing,* Kanai, T., Campbell, G.A. (Eds.) (1999) Carl Hanser Verlag, Munich, Germany, pp. 281–318
24. Spruiell, J.E., Clark, E.S., in *Methods of Experimental Physics, Vol. 16, Part B, Polymers, Crystal Structure and Morphology,* Fava, R.A. (Ed.) (1980) Academic Press, New York, pp. 1–127
25. Bunn, C.W., Garner, E.V., *Proc. Roy. Soc.* (1947) A189, p. 39
26. Ikeda, M., *Sen'i Gakkaishi* (1969), 25, pp. 1–4

10 Double Bubble Tubular Film Extrusion

Toshitaka Kanai

10.1 Introduction

The double bubble tubular film process is a more economical way of producing biaxially oriented film when compared to the tenter biaxially oriented film process. This film manufacturing technique has been widely used to produce biaxially oriented films because of good shrinkability and excellent physical properties. For these reasons, double bubble tubular film (DBTF) is used as the shrink film for foods, PET bottles, electrical parts, and package film for retort foods.

The double bubble tubular film process has been studied using various resins such as polyvinylidene chloride (PVDC), poly(ethylene terephthalate) (PET), poly-(phenylene sulfide) (PPS), polyamide-6 (PA6), ethylene vinyl alcohol copolymer (EVOH), polypropylene (PP), and polyethylene (PE).

J. L. White [1] explained the technical trends and typical applications of those resins. Takashige and Kanai [2] reported on the double bubble tubular film process and the theoretical analysis of the stress development and the scale-up rule for polyamide-6, while S. Ree [3] investigated the double bubble tubular film process-processability and the structure development of polyamide-612. K. Song [4] researched the processability of the blending of polybutylene terephthalate (PBT) and PET.

The first application of double bubble tubular film extrusion to isotactic polypropylene is attributable to Goldman [5] of DuPont and is described in his November 6, 1958, patent application. Since then, many patents [6–10] have been applied for the double bubble tubular film process technology. The most popular double bubble tubular PP films are mainly used to package stationery, groceries, foods, and so on, because of the comparatively easier processability and relatively cheaper resin cost. Recently, shrinkage film for this usage is required to have a superior appearance and shrinkage strength. The double bubble tubular film of linear low density polyethylene (LLDPE) has good shrinkability and excellent physical properties. For this reason, many patents [11–15] have been applied for the double bubble tubular film of LLDPE. In general, it is understood that LLDPE has a narrow stretchability range for the double bubble tubular film compared with PP, so it was difficult to stretch LLDPE biaxially by this process. The improvement of LLDPE stretchability has recently been investigated.

Recently the influence of polymer design and process conditions on the stretchability and physical properties of polyolefins has been studied by Uehara [16–19].

Environmental problems have come into question in the packaging industry in recent years. Dechlorination and waste reduction have been addressed as serious problems in this industry. For example, to produce a standing repackage pouch, thin and strong biaxially oriented nylon film is indispensable. The demand for biaxially oriented PA film has been increasing [2, 20–25].

10.2 Double Bubble Tubular Machine

The double bubble tubular film process consists of two stages of bubbles. Figure 10.1 shows a schematic drawing of the double bubble tubular film process. In the first stage, the bubble is blown from a molten state. In this step the crystalline polymer is rapidly cooled by temperature-controlled water in order to reduce the crystallinity. The first bubble is flattened out by a set of nip rolls and reinflated into a larger second-stage bubble at a higher temperature than the softening temperature. The second bubble is stretched biaxially in the infrared heater oven, and after stretching the bubble is cooled by air. The second bubble is usually stretched biaxially at the same stretching ratio in the machine direction (MD) and the transverse direction (TD). The second bubble is also flattened out by a set of nip rolls. In the case of shrink film, it is slit along both edges, and each film is wound separately. In the case of biaxially oriented film except for shrink film, the third bubble is heat set at a higher temperature than the stretching temperature to prevent the shrinkage.

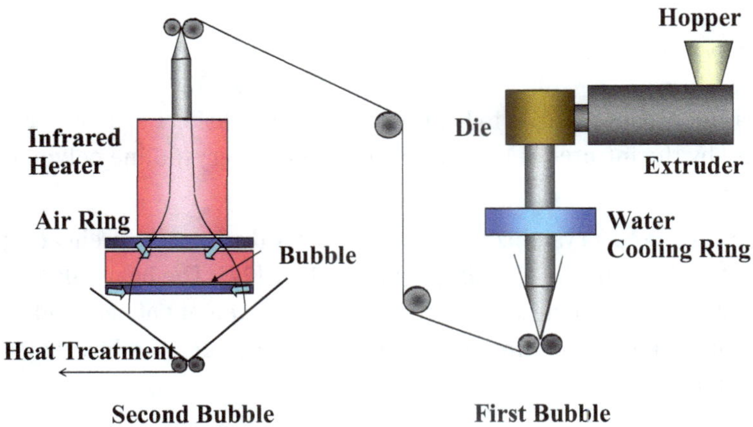

Figure 10.1 Schematic view of double bubble tubular film process

10.3 Theoretical Analysis of Double Bubble Tubular Film Process

10.3.1 Theoretical Analysis of Preheating Process and Stretching Process

The first bubble cooling analysis is almost the same as the cooling process of non-stretched sheet [26]. Figure 10.2 shows the temperature distribution of the primary LLDPE film and the plateau area cut is balanced by the crystallization latent heat and cooling. The average elapsed times of the plateau area were calculated from the analytical results for various cooling water temperature conditions. The analytical results are closely related to various experimental values such as haze, density, and spherulite size.

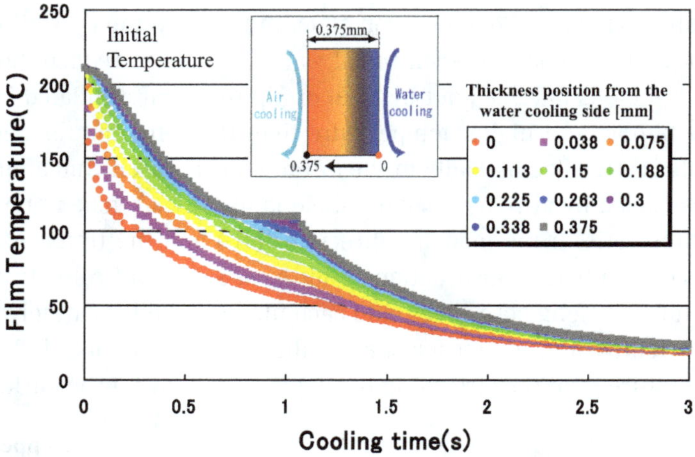

Figure 10.2 Predicted temperature distribution of primary film

The preheating process is used to heat the nonstretched sheet to a proper stretching temperature by the infrared heater. In this process, the crystalline polymer develops crystallinity.

The temperature and crystallization distributions during the preheating process were predicted theoretically by using Equation 10.1 [27]. The heat balance equation and the crystallization equation are derived by the movement of heat and the change of the crystallization in the film minute section *dz*. The derived heat balance equation is shown below.

$$\rho C_P Q dT = 2\pi R_r \sigma \varepsilon_r T_r^4 F \gamma_p \alpha_{pz} dz - 2\pi R dz \sigma \varepsilon T^4 + \rho Q \Delta H_c dx - 2\pi R dz h_a (T - T_{\text{room}}) \quad (10.1)$$

where ρ is the film density, C_p is the specific heat capacity, Q is the extrusion output, R is the film radius, H is the film thickness, v is the film velocity, R_r is the infrared heater radius, σ is the Stefan-Boltzmann constant, ε_r is infrared heater emissivity, F is the view factor, γ_p is the effective efficiency of IR radiation energy, α_{pz} is the heat absorptivity, ε is the film emissivity, ΔH_c is the enthalpy of crystallization, and h_a is the heat transfer coefficient. The crystallization equation is shown below.

$$x(t) = x(\infty)[1 - \exp\{-(\int_0^t K(T)dt)^n\}] \quad (10.2)$$

where $x(\infty)$ is the final crystallinity, K is the nonisothermal crystallization rate constant, n is the Avrami exponent, t is time, and $K(T)$ is the following expression derived by Patel.

$$K(T) = \left[\exp(A - \frac{BT}{(T - T_g - 51.6)^2} - \frac{CT_m}{T(T_{m^0} - T)}) \right]^{\frac{1}{n}} \tag{10.3}$$

where T_g is the glass transition temperature, T_m^0 is the equilibrium melting point, and A, B, and C are the polymer constants. Derived Equations 10.1 and 10.2 were transformed into the differential equations, and they were solved by using the Runge-Kutta method.

The result of the analysis of a primary film production process is shown as follows.

The analytical results of temperature and crystallization in the preheating process are shown in Fig. 10.3.

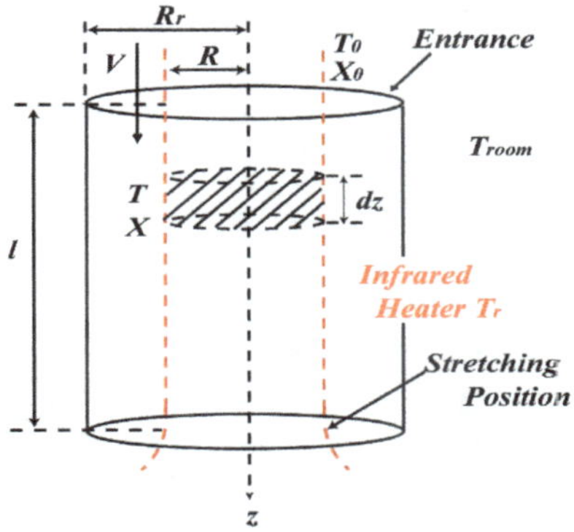

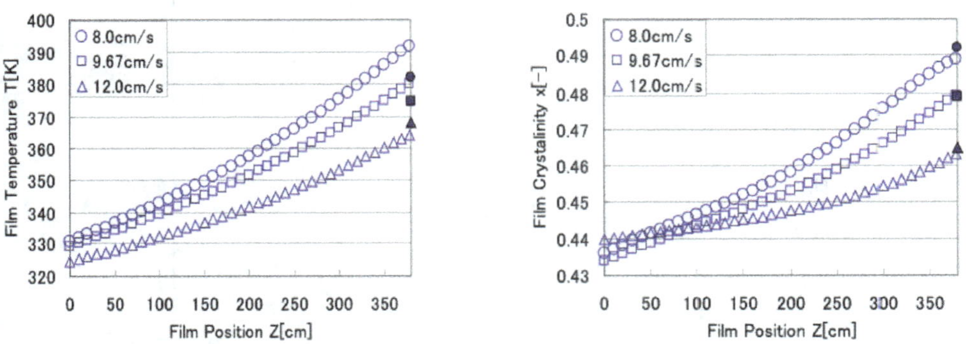

Figure 10.3 Predicted results of temperature and crystallinity during preheating process

It was found that temperature and crystallinity could be predicted in this analysis by comparing with an experimental value. As the bubble speed is increased, the change of temperature and crystallinity is slower. A change in temperature and crystallinity during the primary film production and preheating processes were predicted theoretically, and these results are effective in predicting the stretchability and the optimum process conditions. From this analysis, the appropriate preheating condition to reduce crystallinity is predicted. It is important to know the stretching ratio and at what temperature the bubble should be stretched during the stretching process for the secondary process such as shrinkability and thermoforming.

Because the tubular film and the double bubble tubular film are stretched at substantially different temperatures, the stretching stress is more than one order different. But the basic theoretical equations of the double bubble tubular film process are the same as the ones for the tubular film extrusion described in Chapter 3 of Kanai and Campbell, Film Processing [25]. Stress-strain curve data, resin properties, and process conditions are put into the above force balance equations, and then the deformation behavior can be predicted. The stretching behavior of the biaxially oriented tubular film of polystyrene is shown in Fig. 10.4 [25, 28]. In the figure, dimensionless thickness H/H_0, bubble velocity V/V_0, bubble radius R/R_0, bubble temperature T/T_0, strain rates in MD and TD d_{11}, d_{22}, total strain rate d, and stretch-

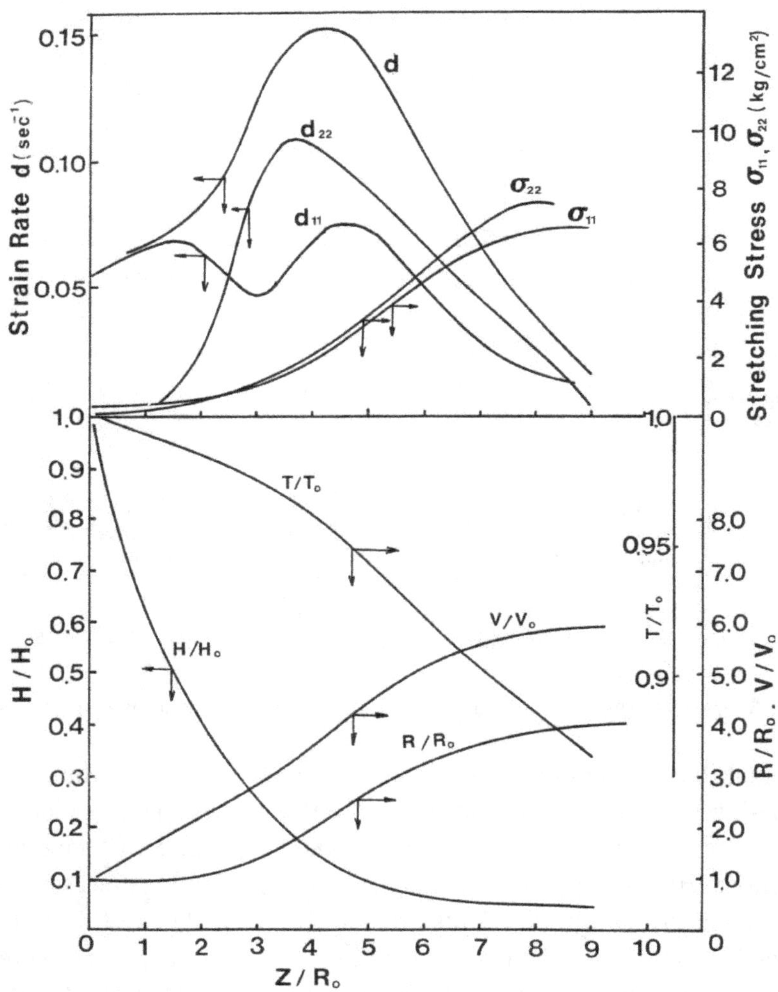

Figure 10.4 Predicted deformation behavior of double bubble tubular PS film process

ing stresses in MD and TD σ_{11}, σ_{22}, are shown. The stress and temperature histories during the stretching process influence the thermoformability as a secondary process. When the stress is low, the sheet physical properties are low. In contrast, high stress makes for low thermoformability. The polymer stress history is retained in a sheet, and the shrinkage stress of OPS remains about 75% of the maximum stretching stress during the stretching process. Stress (pressure) more than necessary to thermoform can produce a good thermoforming tray even in the corners, but, in contrast, if the stress is below the critical stress the process cannot produce a good thermoforming tray at the corner (Fig. 10.5).

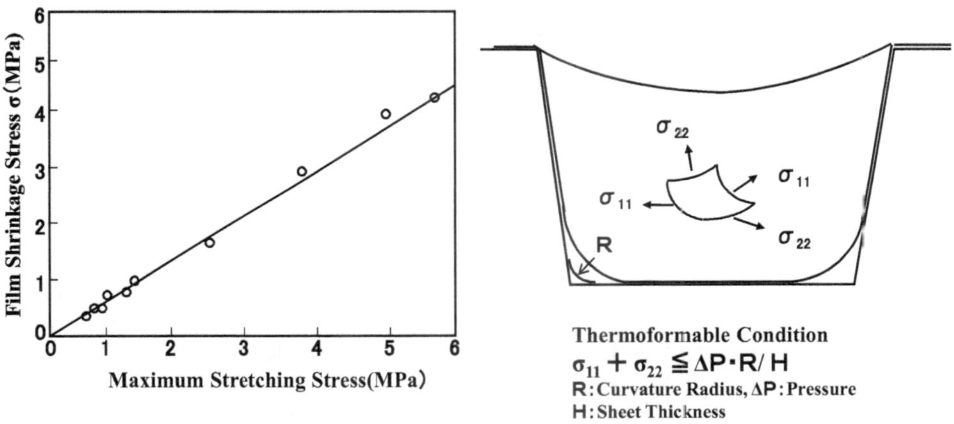

Figure 10.5 Relationship between stretching stress and thermoformability

In the case of the secondary processability of shrink film, the shrink film remembers the stretched stress and strain history at any temperature, which is very important information for the shrinkability. Recently it was found that it is very desirable to produce a wide stretchable temperature range of polyolefin films that is controllable by the composition distribution [17, 18, 20].

10.3.2 Analysis of Stretching Stress

The bubble shape in the double bubble tubular film process and also the film properties were influenced by the stretching temperatures and stretching stress. The stretching stresses during bubble stretching, which were calculated from torque, inside bubble pressure, and film thickness, are obtained by Equations 10.4 and 10.5 and the method shown in Fig. 10.6. The second bubble of the double bubble tubular film process was treated as a blown film process as reported by T. Kanai [16, 25].

The maximum stretching stresses $\sigma_{MD\max}$ and $\sigma_{TD\max}$, which are closely related to the physical properties of the film, are attained at the final stretch point where the bubble diameter is equal to the final diameter.

$$\sigma_{MD\max} = \frac{F_L}{2\pi R_L H_L} \tag{10.4}$$

$$\sigma_{TD\max} = \frac{R_L \cdot \Delta P}{H_L} \tag{10.5}$$

where H_L is the final film thickness, F_L is bubble tension, R_L is the final bubble radius, and ΔP is the inside bubble pressure.

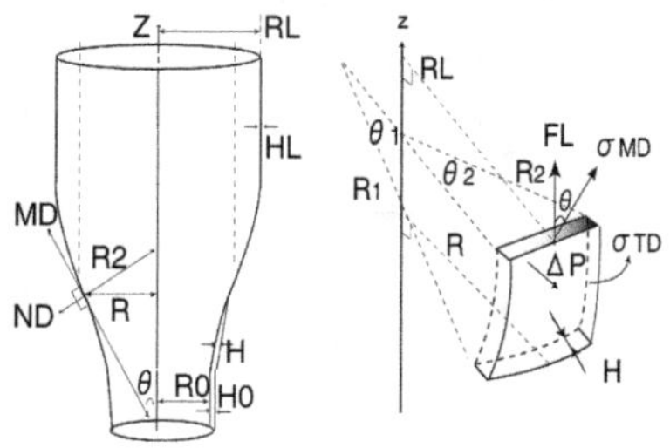

$$F_L = 2\pi R H \sigma_{MD} \cos\theta + \pi(R_L{}^2 - R^2)\cdot\Delta P \qquad (1)$$

$$\frac{H\cdot\sigma_{MD}}{R_1} + \frac{H\cdot\sigma_{TD}}{R_2} = \Delta P \qquad (2)$$

At the frost line height, $R=R_L$, $\theta=0$, $R_1=0$, $R_2=R_L$

$$\sigma_{MD\max} = \frac{F_L}{2\pi R_L H_L} \qquad \sigma_{TD\max} = \frac{R_L\cdot\Delta P}{H_L} \qquad (3)$$

Figure 10.6 Force balance equation of tubular bubble

10.4 Bubble Deformation Behavior [16]

A further experiment was done in order to obtain both the maximum and the minimum stresses of the bubble for research on bubble deformation behavior. The relationship between stretching stress and bubble shape is shown in Fig. 10.7. The photographs show that the bubble angle θ increases with an increase in the stretching stress. The grid mark on each bubble surface is visible. The heterogeneous stretching in the neck part (necking phenomenon) of the bubble is shown in Fig. 10.8.

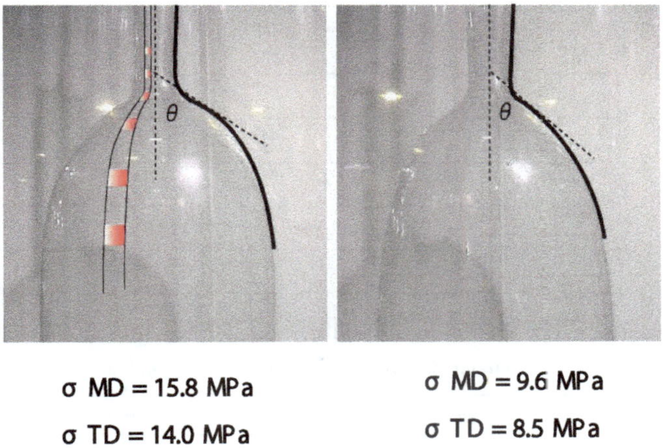

Figure 10.7 Bubble shape of the second tubular bubble

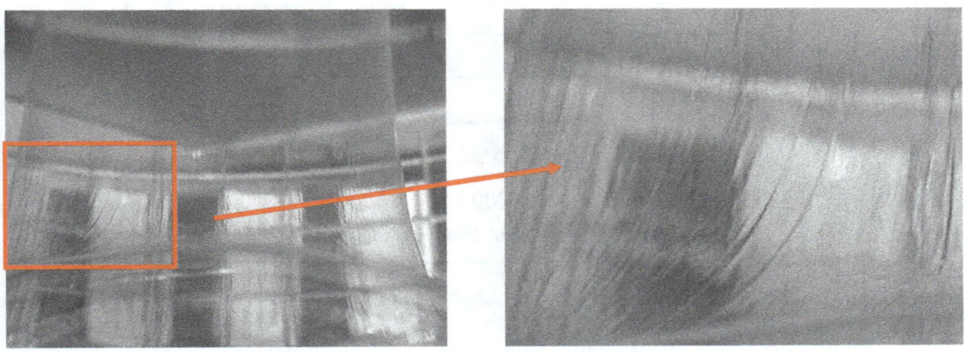

Figure 10.8 Necking phenomenon of the double bubble tubular film process

The bubble deformation behavior was quantified by measuring the marked grid on the stretch bubble. The result is shown in Fig. 10.9. This figure shows that a better uniformity of the stretch ratio is obtained under a higher stretching stress (a lower stretching temperature). It was found that this stretch ratio fluctuation occurred from the stretching starting point of the double bubble tubular film process, which was observed as heterogeneous stretching in the neck part as shown in Fig. 10.8. In addition, the comparison of strain pattern of MD and TD in Fig. 10.10 show that the bubble has been slightly stretched in the machine direction before starting to stretch in the transverse direction. This means that the double bubble tubular film process was not a perfect simultaneous biaxial stretching but a partial sequential stretching, the same as in a blown film. The bubble was stretched in a preheated oven, where a higher temperature resulted in a further increase in the prestretching in the MD.

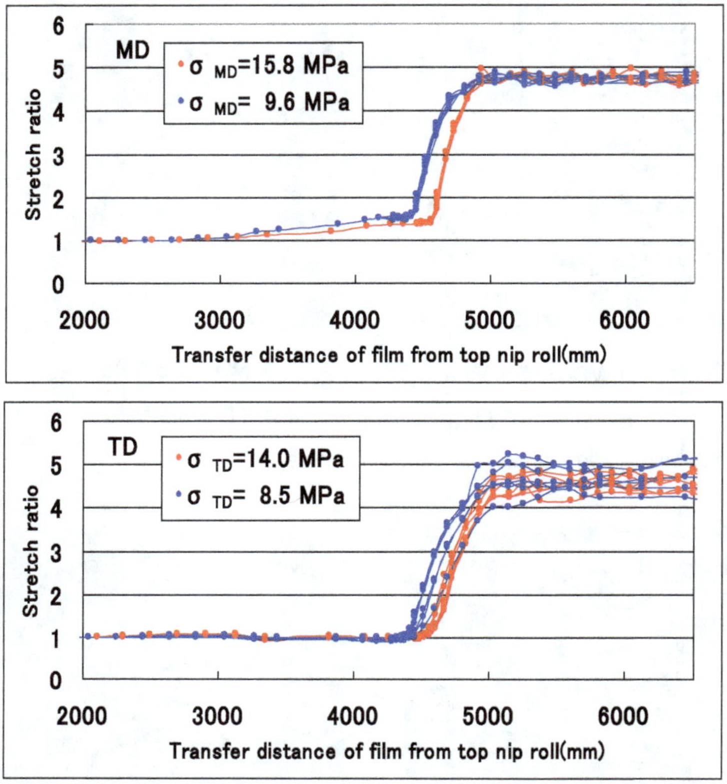

Figure 10.9 Deformation behavior of tubular bubble at different stress conditions

Figure 10.10 shows the bubble velocity and the strain rate along the bubble during the double bubble tubular film process calculated from the deformation length of the grid. These figures show that as the stretching stress increases, the maximum bubble velocity is faster in the TD and the bubble strain rate is higher. This is due to the fact that a lower stretching temperature results in an increase in the stretching stress, hence an increase in the inside bubble pressure ΔP. This increases the bubble angle θ in Fig. 10.7, finally resulting in the shortening of the stretching time.

A higher stretching stress (a lower stretching temperature) is necessary to obtain a uniform thickness in the stretched film.

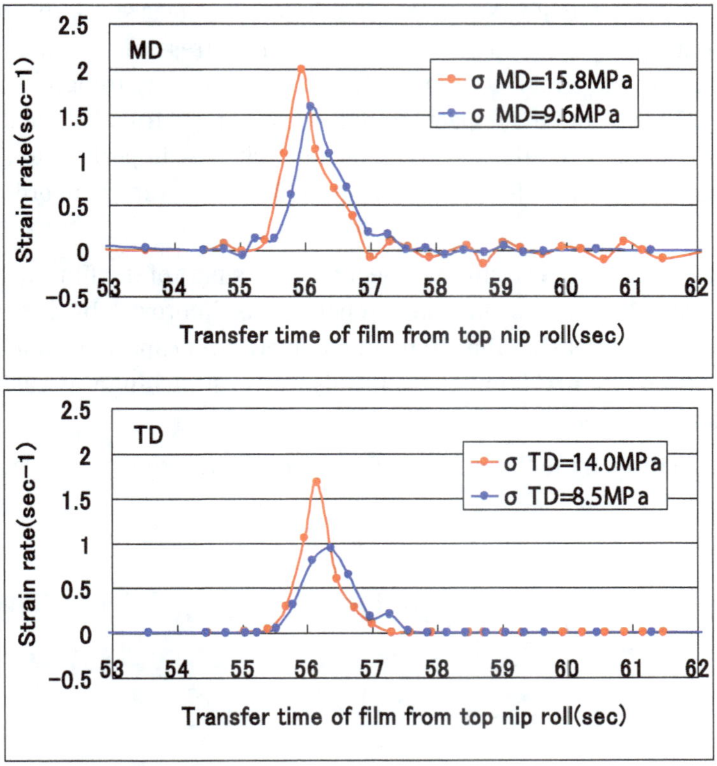

Figure 10.10 Deformation rate of tubular bubble in MD and TD at different stress conditions

10.5 Film Properties

Figure 10.11 shows the relationship between the stretching stress and the film properties tear strength and impact strength for LLDPE. The stretching stress, shrinkage, and tear strength were expressed as the average of MD and TD properties.

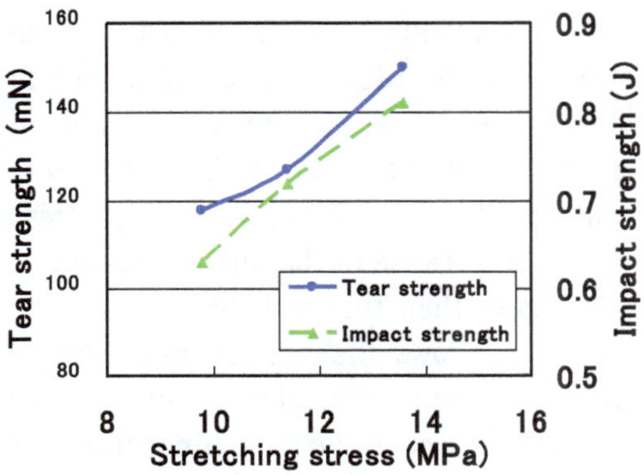

Figure 10.11 Relationship between stretching stress and film properties

An increase in the stretching stress resulted in a higher tear strength and a higher impact strength. This means that a higher stretching stress results in a more-effective orientation of the stretched film in MD and TD. The tear strength was changed from 118 to 150 mN, and the impact strength was changed from 0.63 to 0.81 J, even in the stable bubble conditions. This indicates that it is important to control the stretching stress (by controlling the mean stretching temperature) in order to obtain superior film properties.

The shrinkage reflects the strain that the amorphous part of the film was subjected to during stretching. The shrinkage property was improved by increasing the stretching stress (decreasing the stretching temperature) especially at a shrinkable temperature range from 100 to 110°C. In this range, stretching was easily affected by the stretching temperature.

Under such conditions the film haze retained good transparency.

10.6 Comparison of Double Bubble Tubular Film (DBTF) and Laboratory Tenter Stretched Film (LTSF) [17]

10.6.1 Bubble Deformation Behavior and Stretching Stress

Figure 10.12 shows the strain rate of the double bubble tubular film (DBTF) and the laboratory tenter stretched film (LTSF) for LLDPE. The strain rate of the LTSF was gradually decreased. The difference in deformation behavior is very clear from the data, and the strain rate of the LTSF was lower than that of the DBTF. Figure 10.13 shows the relationship between the stretch ratio and the stretching force for the preparation of the LTSF at various stretching temperatures. The figure shows that a lower stretching temperature resulted in a higher stretching force with an increase in the ratio between the yielding force and the final stretching force. This suggests that a lower stretching temperature results in a better film thickness uniformity. The stretching conditions and resultant stretching stresses of the LTSF process are shown in Table 10.1. The data shows that the stretching stress of the LTSF ranged from about 2.5 MPa to 15 MPa, whereas the stretching stress of the DBTF ranged from about 9 to 18 MPa as shown in Table 10.2. This means that the appropriate stretching stress range for the DBTF is narrower than for the LTSF because of the bubble instability. The maximum stretching stress of DBTF is higher than that of LTSF, because the DBTF is stretched nonisothermally. Figure 10.14 shows the relationship between stretching temperature and stretching stress. It is represented by a smooth curve showing the lower the temperature, the greater the effect on the stretching stress.

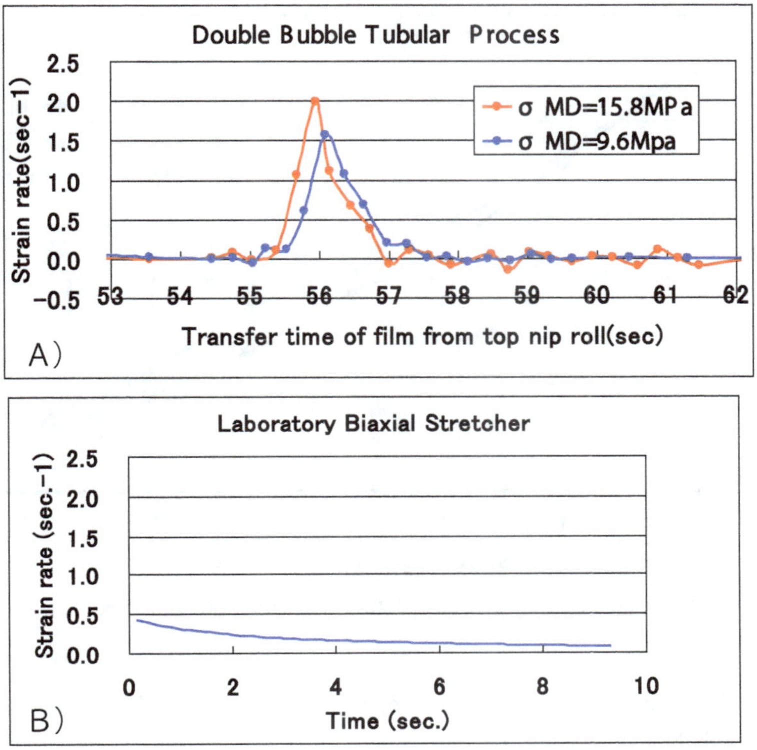

Figure 10.12 Strain rate of double bubble tubular and laboratory tenter processes

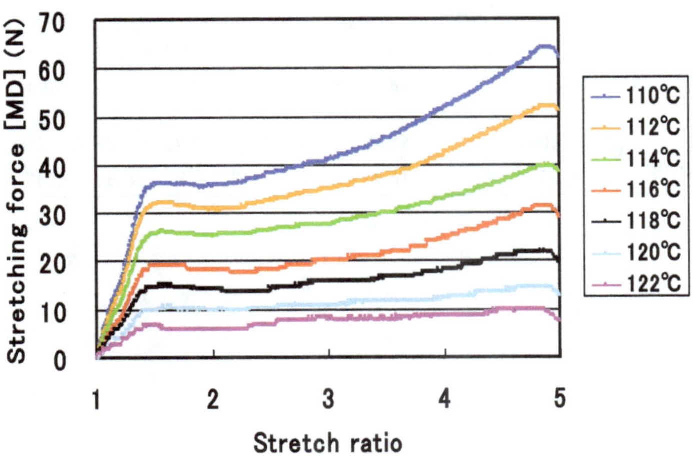

Figure 10.13 Relationship between stretching ratio and stretching force for LLDPE of LTSF at different stretching temperatures

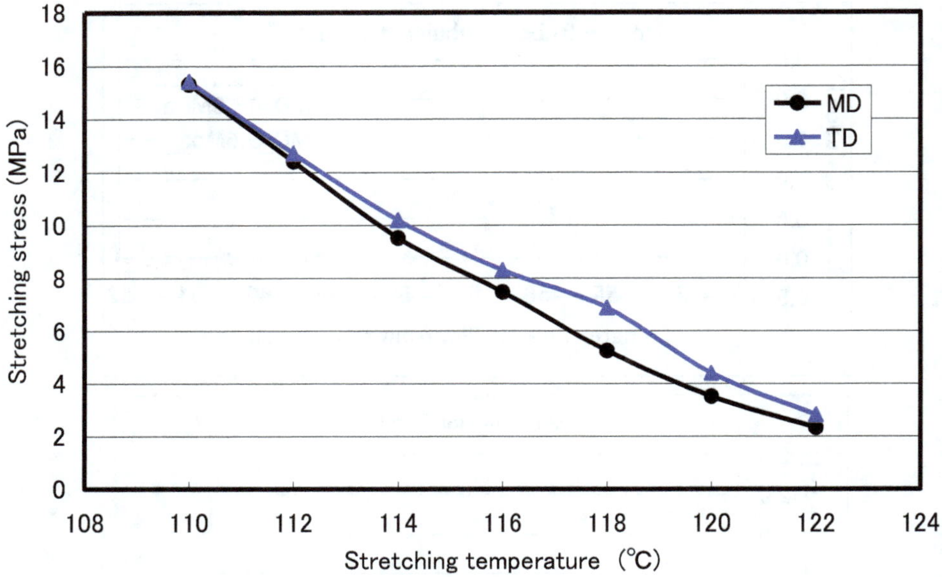

Figure 10.14 Relationship between the stretching temperature and the stretching stress of the laboratory tenter stretched film

Table 10.1 Stretching Conditions of the Laboratory Tenter Stretched LLDPE Film (thickness: 12 μm)

			1	2	3	4	5	6	7	8	9
Stretching temperature	°C	-	108 broken	110	112	114	116	118	120	122	124 melt
Maximum stretching force (stretch ratio = 5)	N	MD	-	64.1	52.1	40.0	31.4	22.1	14.7	9.8	-
		TD	-	64.5	53.3	42.8	34.9	28.9	18.4	11.8	-
Maximum stretching stress	**MPa**	**MD**	-	**15.3**	**12.4**	**9.52**	**7.47**	**5.25**	**3.50**	**2.33**	-
		TD	-	**15.4**	**12.7**	**10.2**	**8.31**	**6.88**	**4.39**	**2.80**	-

Table 10.2 Production Conditions of the Double Bubble Tubular Films

		1	2	3	4	5
Materials			LL-A LL-B LL-C	70% 15% 15%	(density 0.926) (density 0.902) (density 0.935)	
Stretch ratio MD x TD	-			5 × 5		
Thickness of stretched film	μm			15		
Ave. temp. of the preheaters	°C	289	298	307	313	319
Ave. temp. of the stretching heaters	°C	101	107	130	140	145
Film temperature after the preheater	°C	107	112	116	120	127
Stretching torque	Nm	71.0 (burst)	52.2	44.9	38.4	31.2 (unstable)
Stretching force	N	676	497	428	366	297
Bubble inside pressure	Pa	-	0.53	0.43	0.37	-
Stretching stress σ_{MD}	MPa	19.1	14.1	12.1	10.4	8.4
Stretching stress σ_{TD}	MPa	-	13.2	10.8	9.2	-

10.6.2 Comparison of LLDPE Film Properties of DBTF and LTSF

The shrinkability of the DBTF and the LTSF is shown in Fig. 10.15. A lower stretching temperature resulted in higher shrinkage for both films, but the shrinkability of the DBTF was higher than that of the LTSF. This indicates that the DBTF is oriented more effectively than the LTSF. This is because the DBTF is stretched nonisothermally, while the LTSF is stretched isothermally. In addition, the strain rate of the DBTF is higher than that of the LTSF.

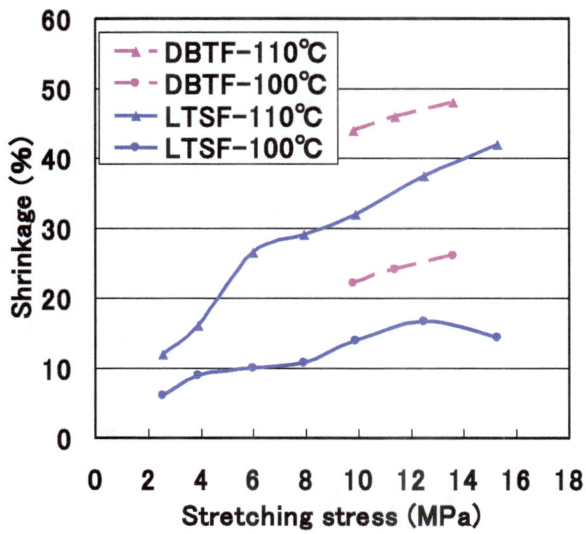

Figure 10.15 Comparison of DBTF shrinkability with LTSF

The comparison of the tear strength of the DBTF and the LTSF is shown in Fig. 10.16. A higher stretching stress had higher tear strength in both films. This physical property of the LTSF shows the same trend as that of the DBTF.

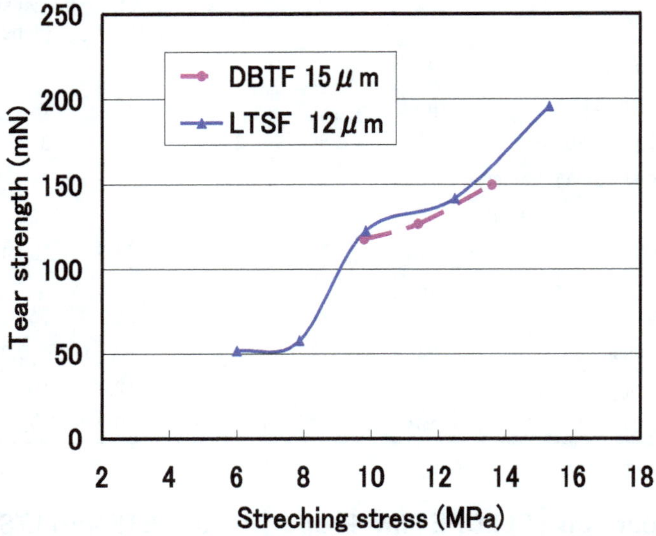

Figure 10.16 Comparison of tear strength of DBTF and LTSF

Figure 10.17 shows the relationship between stretching stress and haze. In the LTSF, haze was low and almost constant between 8 and 15 MPa, whereas it became higher when the stretching stress was lower than 8 MPa (temperature higher than 118°C). It was determined that the film surface began to melt under 8 MPa, and when it was cooled, the film surface roughened and resulted in deteriorating film transparency.

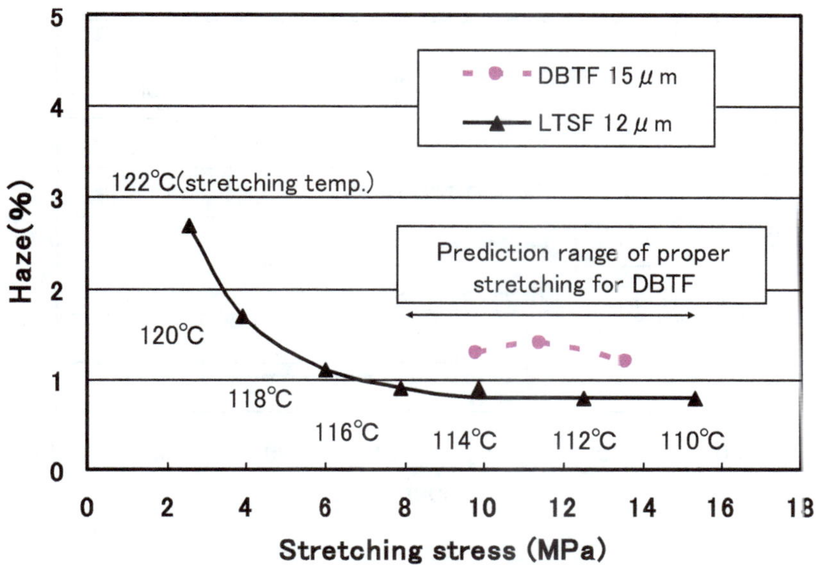

Figure 10.17 Relationship between the stretching stress and haze of the double bubble tubular film (DBTF) and the laboratory tenter stretched film (LTSF)

■ 10.7 Material Design of Polyolefin for Double Bubble Tubular Film

The most important points for the double bubble tubular film extrusion are the bubble stability and the stretchable temperature range. The bubble stability and stretchable temperature range are very much dependent on the Melt Index (MI) and composition distribution, respectively. The composition distribution of LLDPE, which is measured by temperature rising elution fractionation (TREF), is closely related to stretching temperature distribution as shown in Fig. 10.18. In order to obtain bubble stability, high melt tension is required. But too high melt tension causes the bubble to break, so proper Melt Index (MI) is very important. Because the bubble temperature distribution in TD is difficult to control by the infrared heater uniformly, a wider composition distribution of the resin is required because it leads to a wider stretchable process window. Based on this information, the resin is designed as follows.

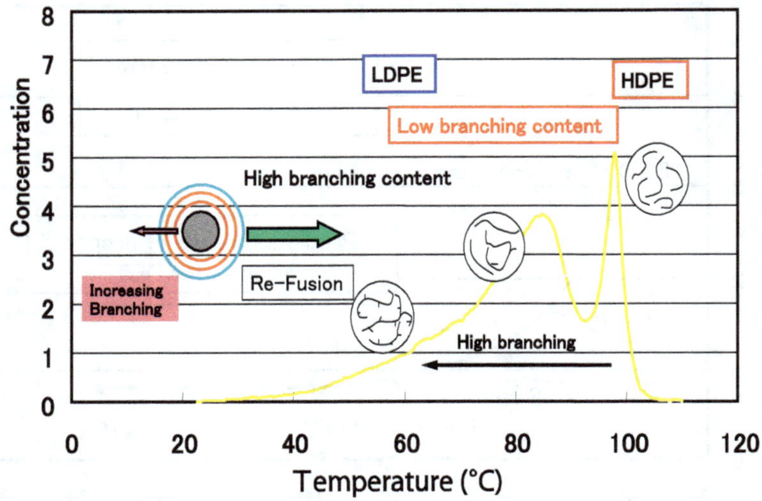

Figure 10.18 TREF data of LLDPE

10.7.1 Polyethylene [18]

In order to improve the stretchability of linear low density polyethylene (LLDPE) for double bubble tubular film (DBTF), the blending of LLDPEs has been investigated. Three LLDPEs of varying densities were blended, and several films having different densities, molecular-weight distributions, and composition distributions were produced. The TREF data of blended samples are shown in Figs. 10.19 and 10.20, which are TREF data and integral composition distribution, respectively. As a result, it is found to be effective to blend the different densities of LLDPEs to improve the stretchable temperature process window of LLDPE as double bubble tubular film grade. By blending, the composition distribution expands, and the stretchable temperature range becomes wider. The low-density material reduces the stretching stress and extends the low-temperature range for stretching. The higher density material maintains a high stretching stress. However, there are two important points to improve the stretchability of LLDPE blends. Figure 10.21 shows that the gradient (%/K) from 40 to 70% melting composition of the total in TREF needs to be around 2.5%/K, and the blended film density is achieved at about 0.915 g/cm^3 as indicated by the diameter of the data bubbles.

Predicting the stretchability of LLDPE for the double bubble tubular film using the data from laboratory tenter stretching, taking into account the haze and the stretching stress, has become more reliable.

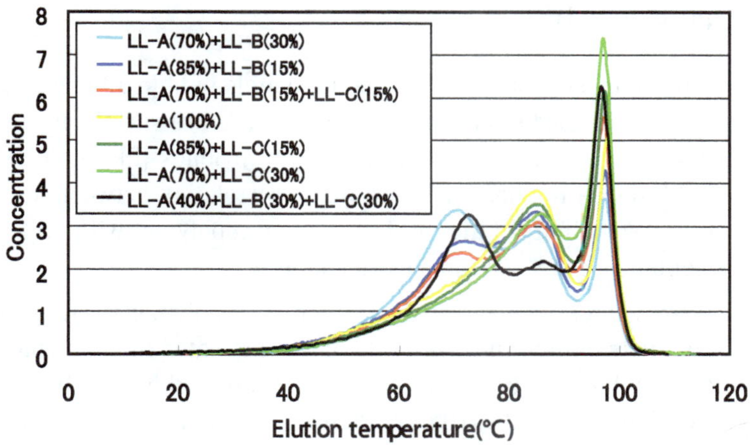

Figure 10.19 TREF data for various LLDPE blends

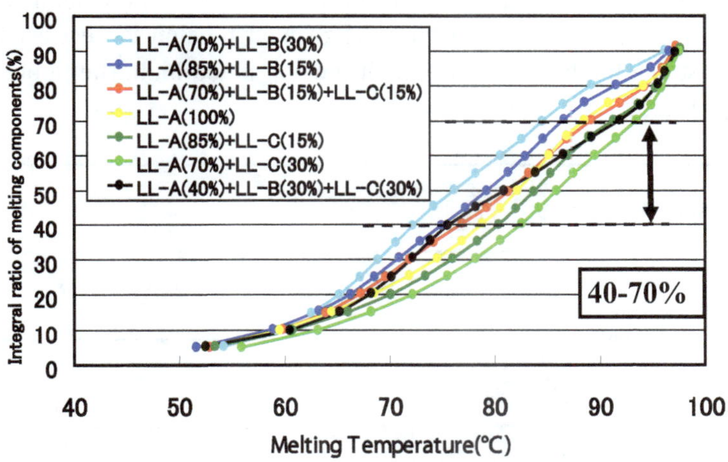

Figure 10.20 Integral components of TREF for various LLDPE blends

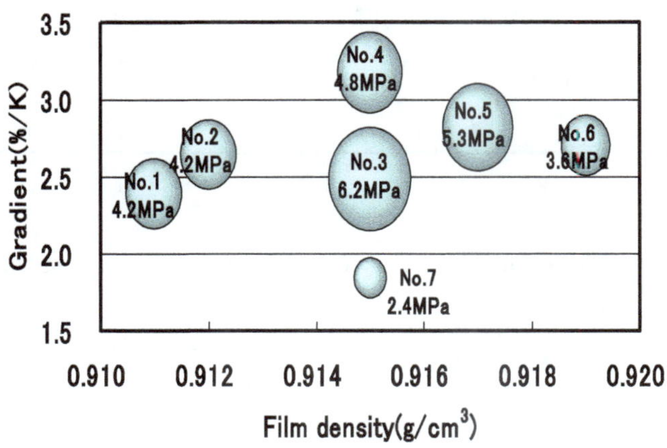

Figure 10.21 PE stretchable process window as a function of TREF gradient and density

10.7.2 Polypropylene [19]

Polypropylene is the most popular resin used for double bubble tubular film production because it exhibits good stretchability, low temperature shrinkage, good physical properties, and reasonable cost. Various polypropylenes (PP) for biaxially oriented film like PE were investigated. There was a correlation between MI, ethylene content, and the stretchability of the DBTF. A random copolymer with MI 2 g/10 min and ethylene content 4.0 wt% gave good stretchability.

Figure 10.22 shows the stretchability as a function of polymer design such as MI and composition distribution; the size of the circle indicates the stretchable range. A suitable PP for double bubble tubular film should satisfy the following conditions. The most appropriate MI is 2.2 g/10 min. PP over this value has less bubble stability, and PP below MI 2.2 shows less stretchability. The stretchability range increases with decreasing the gradient in the total melting component from 40 to 70% in TREF, which means a wider composition distribution is better for stretchability. The melt index of the film is about 2.2 g/10 min. High melt tension is necessary in order to achieve a stable bubble, but if the melt tension is too high, the bubble is easily broken (Fig. 10.23).

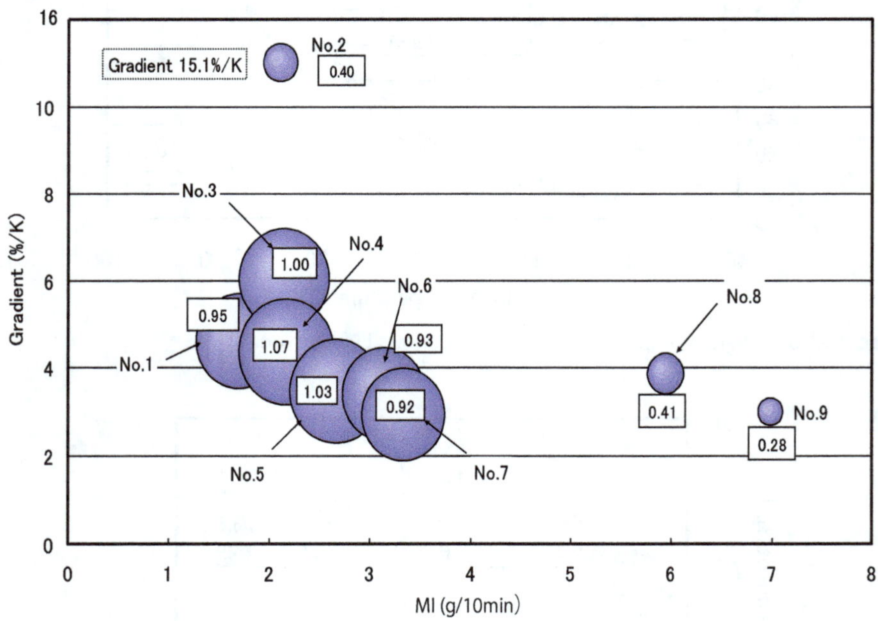

Figure 10.22 Stretchability and polymer design (MI and composition distribution)

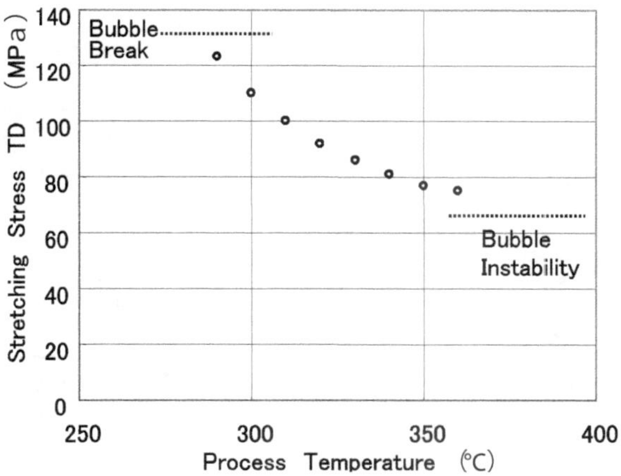

Figure 10.23 Relationship between stretching stress and infrared set temperature

10.8 Thickness Uniformity

The thickness uniformity of the film obtained by the double bubble tubular film process was investigated for PA6 [22].

An optimum stretching stress during the double bubble tubular film process exists. The bubble breaks with stress over 130 MPa, and the bubble is unstable below stress of 60 MPa. In the optimum conditions, which are a process temperature of 310 °C (temperature of heating furnace) and a stretching ratio in TD of 3.2, Figs. 10.24 and 10.25 show that stretched film uniformity was worse by about twice than the nonstretched film. The thickness uniformity of the film stretched in optimum conditions was better than that of the film stretched at a higher process temperature and at lower stretching ratio in TD. Thickness uniformity of nonstretched film significantly affects the thickness uniformity of biaxially oriented film.

The actual stretching ratio, the local stretching ratio along the film width rather than the average stretching ratio, influences tensile modulus and tensile strength. In order to obtain uniform physical properties in film, it is important to produce the film with uniform thickness. If a thick portion exists, the rate of its increasing temperature is delayed when passing through of the heating furnace. As a result, the thicker portions are less stretched, and thinner portions are more stretched. Film thickness uniformity can be improved when the stretching stress is high and the bubble stability is good. These factors are influenced by stretching temperature, stretching ratio, and air velocity from the air ring.

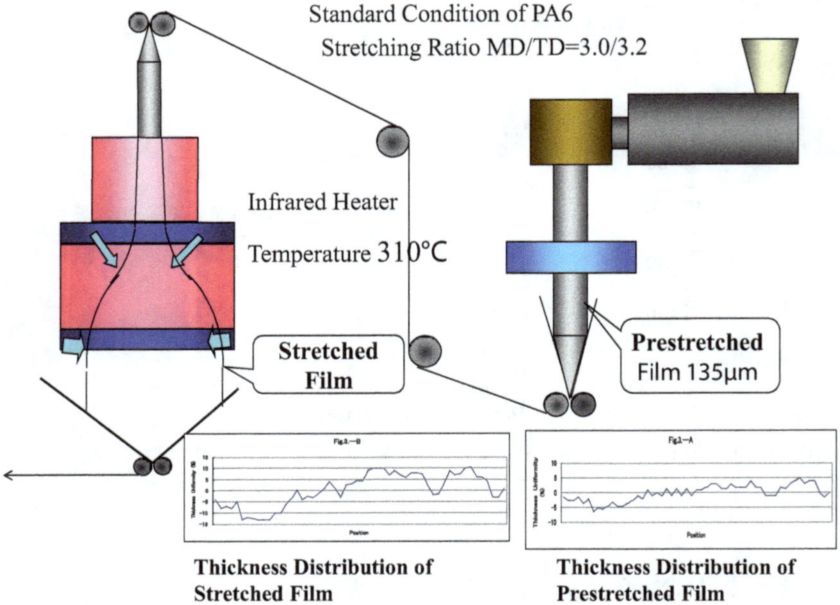

Figure 10.24 Thickness distribution of prestretched PA6 film and stretched PA6 film

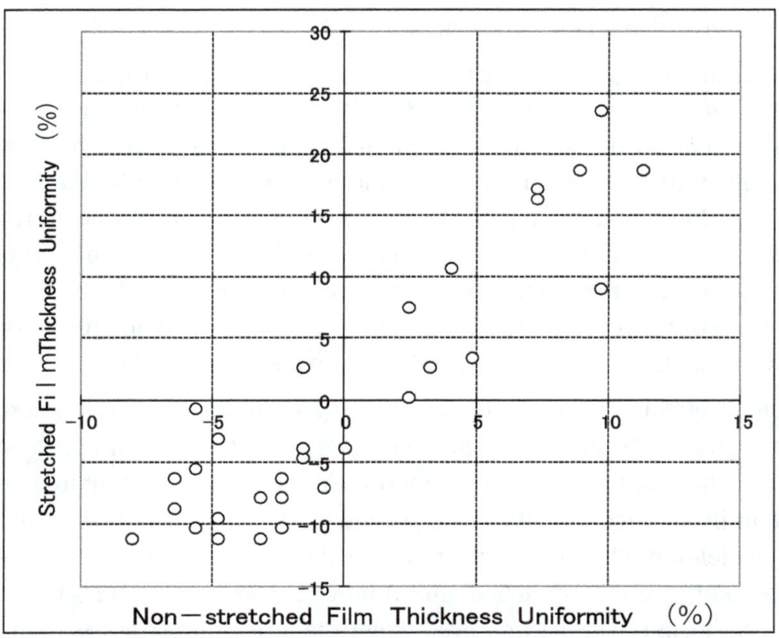

Figure 10.25 Thickness distribution of prestretched PA6 film and stretched PA6 film

10.9 High Performance Film Produced by Blend and Multilayer Stretching Process

In the marketplace, a film that can address the environmental problem and the barrier-free problem is desired. It is expected that significant progress will be made in the future in the field of applications for films made by the double bubble tubular film process. Specifically, it is foreseen that a multilayer stretching and blended stretching process will play an active role in the broadening of applications. A specific example for PA6 is discussed below.

The stretching stability and the physical properties for biaxial oriented PA6/MXD6 blending film of the double bubble tubular film process is reported. As poly (m-xylene adipamide) (MXD6) blending quantity increases, the stretching stress is lowered. Hydrogen bonding is suppressed by the steric hindrance of the MXD6 resin. The blended film can be stretched stably, and oxygen gas permeability is much lower than for PA6 film.

An MXD6-blended layer blended at 30% in the outer layers has an aligned structure in the MD direction, and an MXD6-rich layer in the center has high barrier performance (Fig. 10.26). It is found that the plate-like cylinder structure was formed even in an unstretched film from the observation by TEM shown in Fig. 10.27. The cylinder structure produced the straight line alignment of the particles. A high strength film was achieved, because it was stretched biaxially using the double bubble tubular process.

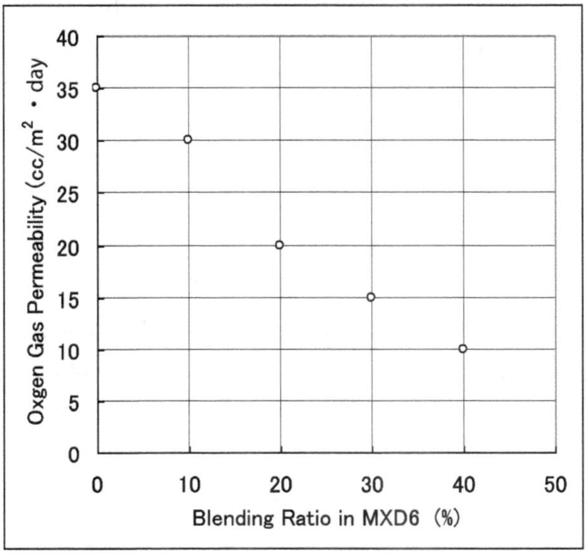

Figure 10.26 Oxygen gas permeability versus blending ratio in MXD6

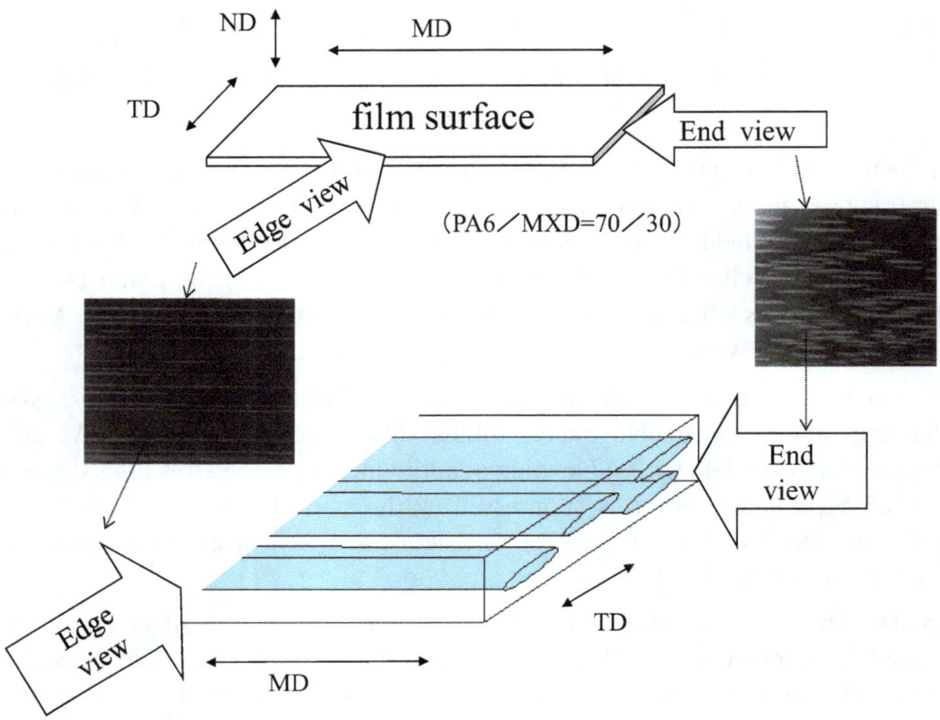

Figure 10.27 TEM observation of PA6/MXD

The multilayer double bubble tubular film composed of blending materials PA6 and MXD6 keeps the compatible performance of high strength, easy tear properties, and a high-quality barrier, and it could solve environmental and barrier-free problems.

The aim of the production technology of a film should be to solve both environmental problems and barrier-free problems.

10.10 Scale-Up Rule

The scale-up rule of the tubular film process is set up theoretically here [23, 35].

The force balance on the double bubble tubular film is developed from membrane theory. Membrane theory leads to a balance of forces on the film between positions *Z* and take-up position *L* shown in Fig. 10.28. This has the form:

$$F_L = 2\pi RH\sigma_{MD} COS\theta + \pi\left(R_L{}^2 - R^2\right)\cdot \Delta P \tag{10.6}$$

The stress σ_{MD} and σ_{TD} are related to the pressure ΔP through the expression:

$$H\cdot\sigma_{MD} / R_1 + H\cdot\sigma_{TD} / R_2 = \Delta P \tag{10.7}$$

where F_L is the bubble tension, R_L the final bubble radius, and R_1 and R_2 are appropriate radii of curvature.

$$R_1 = \left(\left(1 + dR / dZ\right)^2\right)^{3/2} / \left(d^2R / dZ^2\right) \tag{10.8}$$

$$R_2 = R / \cos\theta \tag{10.9}$$

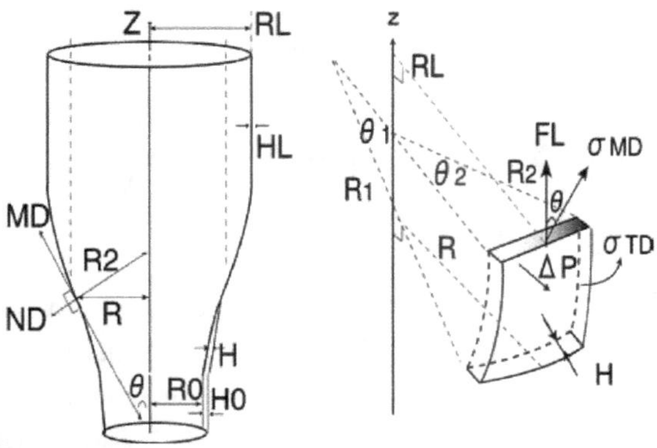

Figure 10.28 Coordinate system of tubular film bubble

The maximum stretching stress σ_{MD} and σ_{TD} are closely related to the physical properties of the film. The maximum stress at the final stretching is used to set up the scale-up rule.

At the final stretching, bubble diameter is equal to final bubble diameter.

$$\sigma_{MD} = F_L / 2\pi R_L H_L \tag{10.10}$$

$$\sigma_{TD} = \Delta P \cdot R_L / H_L \tag{10.11}$$

H_L is final film thickness.

By using the analysis of dimensionless terms, stresses σ_{MD} and σ_{TD} are presented by the following equations.

$$\sigma_{MD} = \left(A + B\left(R_L / R_0\right)^2\right) / 2\pi \cdot Q\eta_0 / R_0 R_L H_L = G_1 \cdot Q\eta_0 / R_0 R_L H_L \tag{10.12}$$

$$\sigma_{TD} = \left(B / \pi\right) \cdot Q\eta R_L / R_0{}^3 H_L = G_2 \cdot Q\eta_0 R_L / R_0{}^3 H_L \tag{10.13}$$

where Q is output rate, η_0 is viscosity, R_0 is initial bubble radius, Z_L is axial distance between beginning point and final point of bubble inflation, A and B are constants under the same conditions of R_L/R_L, V_L/V_0, and Z_L/R_0, and G_1 and G_2 are constants with the condition of constant stretching ratio of machine direction, and stretching ratio of transverse direction. These results are summed up in Fig. 10.29.

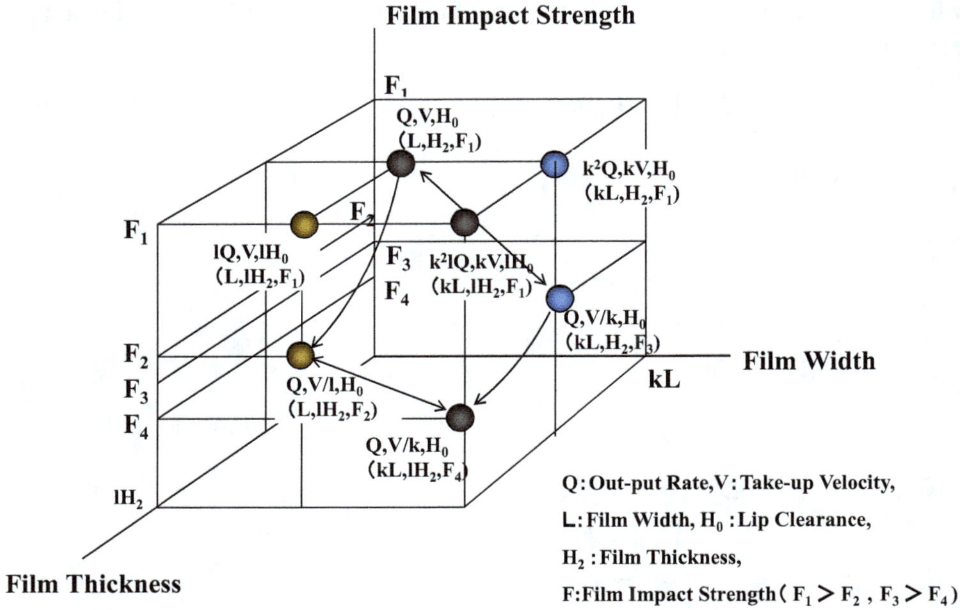

Figure 10.29 Scale-up rule of tubular film

In a study on the scale-up rule for the double bubble tubular process, the conclusion was obtained as follows.

1. The application of the scale-up rule, which was set up theoretically, was evaluated by using both pilot plant and large-scale production, under the conditions of controlling the relationship between output and film width (and so on). According to the scale-up rule in the case of changing bubble radius R and film thickness H into KR and LH, bubble radius R, film thickness H, and output rate Q for the pilot plant correspond to KR, LH, and K^2LQ for the large-scale production plant, respectively.
2. Both pilot plant and large-scale production showed the same stretching bubble stability, the equivalent stretching stress, the equivalent birefringence pattern, and the equivalent crystal structure pattern, when kept at the same deformation rate.
3. In the observation through the polarizing plate of the bubble sample, which shows deformation pattern, the equality of deformation pattern was also confirmed.

It was found that the scale-up rule that was set up theoretically is applicable predicting the physical properties and bubble stability for a large-scale double bubble tubular film process, once the experiment is carried out by using the pilot plant and a small amount of resin.

10.11 Three Different Stretching Processes

Three different stretching processes were studied, namely the double tubular film process, sequential biaxial stretching, and simultaneous biaxial stretching [36].

In terms of the physical properties of biaxially oriented film, the difference between the double bubble tubular film and the tentering process was studied. The simultaneous biaxial stretching film and sequential biaxial film as film samples of the tenter process were used.

Tentering process film has anisotropic tensile properties between MD and TD. This result was influenced by the later stretching process, namely TD stretching. In particular, the tensile modulus and tensile strength have inverse properties.

The double bubble tubular film has good balanced properties, especially thermal shrinkage and impact strength.

Tentering simultaneous stretching film has much larger shrinkage in MD than in TD. The sequential stretching film has larger shrinkage in TD than in MD. The double bubble tubular film has high impact strength between −25°C and 23°C. This result corresponds to the balanced shrinkage properties.

The double bubble tubular film has good balanced properties, particularly the stress-strain curve in tensile property. This property affects high impact strength and good thermoformability. The polarizing fluorescent evaluation result of film orientation in the amorphous portion shows that the double bubble tubular film has uniform orientation in any direction and the tentering process has orthogonal biaxial orientation. This kind of content is also introduced in Chapter 8 of this book.

10.12 Conclusions

There is a strong relationship between the stretching stress and the physical properties of the stretched film. Greater stretching stress and higher strain rate lead to better shrinkability and physical properties. The DBTF has better shrinkability than the LTSF because the DBTF has a higher stretching stress and a higher stretching strain rate than the LTSF. In addition, the DBTF is stretched nonisothermally, whereas the LTSF is stretched isothermally in the biaxial stretcher.

In the DBTF process, the stretching temperature is a very important factor. Bubble stability and film thickness uniformity can be obtained by setting the stretching temperature as low as possible while also keeping the temperature levels low without rupturing the bubble.

The DBTF is not a perfect simultaneous multiaxial stretching because the bubble is stretched slightly in the preheated oven. In order to achieve a good stretching balance between machine direction and transverse direction, selection of the lowest feasible preheater temperature is advantageous.

The most important points for the double bubble tubular film extrusion are the bubble stability and the stretchable temperature range. The bubble stability and stretchable temperature range are very much dependent on the MFR and composition distribution, respectively. The composition distribution, which is measured by temperature rising elution fractionation, is closely related to stretching temperature distribution. The wider the composition distribution the resin has the wider the stretchable process window.

The scale-up rule, which was set up theoretically, relates bubble radius R, film thickness H, and output rate Q for the pilot plant and KR, LH, and K^2LQ for the large-scale production plant, respectively. It is found that the scale-up rule, which was set up theoretically, is applicable in predicting physical properties and bubble stability for the large-scale double bubble tubular film process, once the experiment is carried out by using the pilot plant and a small amount of resin.

References

1. White, J. L., *Film Processing*, Kanai, T., Campbell, G. A. (Eds.) (1999) Hanser Publications, Cincinnati
2. Takashige, M., Kanai, T., *Int. Polym. Process.* (1980) V, p. 287
3. Ree, S., White, J. L., *Int. Polym. Process.* (2001) XVI, p. 272
4. Song, K., White, J. L., *SPE ANTEC Proc.* (1999) 57, p. 1650
5. Goldman, M., U. S. Patent 2,979,777 (1961)
6. Blid, F., Robinson, W., U. S. Patent 3,300,555 (1967)
7. Lindstrom, C. A., William, M. A., Baird, G., Bosse, A. L., U. S. Patent 3,260,776 (1966)
8. Tsuboshima, K., Kanho, T., U. S. Patent 3,510,549 (1970)
9. Nash, J. L., Polich, S. J., U. S. Patent 4,112,034 (1978)
10. Hayashi, K., Morihara, K., Nakamura, K., U. S. Patent 4,279,580 (1981)
11. Schoenberg, J. H., U. S. Patent 4,551,380 (1985)
12. Walter, B. M., U. S. Patent 4,617,241 (1986)
13. Thomas, C. W., U. S. Patent 4,837,084 (1989)
14. David, A. H., U. S. Patent 4,226,905 (1980)
15. Lusing, S., Mack, N. M., Schuetz, J. M., U. S. Patent 5,059,481 (1991)
16. Uehara, H., Sakauchi, K., Kanai, T., Yamada, T., *Int. Polym. Process.* (2004) 19, no. 2, p. 155
17. Uehara, H., Sakauchi, K., Kanai, T., Yamada, T., *Int. Polym. Process.* (2004) 19, no. 2, pp. 163–171

18. Uehara, H., Sakauchi, K., Kanai, T., Yamada, T., *Int. Polym. Process.* (2004) 19, no. 2, pp. 172–179
19. Kanai, T., Uehara, H., Sakauchi, K., Yamada, T., *Int. Polym. Process.* (2006) 21, no. 5, pp. 449–456
20. Takashige, M., Kanai, T., *Int. Polym. Process.* (1990) 5, p. 287
21. Takashige, M., Kanai, T., Yamada, T., *Int. Polym. Process.* (2003) 18, no. 4, pp. 368–375
22. Takashige, M., Kanai, T., Yamada, T., *Int. Polym. Process.* (2004) 19, no. 1, pp. 47–55
23. Takashige, M., Kanai, T., Yamada, T., *Int. Polym. Process.* (2004) 16, no. 1, pp. 56–63
24. Takashige, M., Kanai, T., *Int. Polym. Process.* (2004) 19, no. 2, pp. 147–154
25. Kanai, T., Campbell, G.A., Chapter 3.1 in *Film Processing*, Kanai, T., Campbell, G.A. (Eds.) (1999) Hanser Publications, Cincinnati
26. Matsuzawa, N., Yamada, T., Kanai, T., Sakauchi, K., Uehara, H., *Seikei Kakou '05* (2005) p. 225
27. Matsuzawa, N., Kamatani, Y., Yamada, T., Kanai, T., Uehara, H., Sakauchi, K., *22nd Polym. Process. Soc. Annu. Meet. Proc.* (2006)
28. Kanai, T., Takashige, M., Iwai, A., *Seni Gakkaishi* (1985) 41, p. T-271
29. Kanai, T., Tomikawa, M., White, J.L., Shimizu, J., *Seni Gakkaishi* (1984) 40, no. 12, p. T-465
30. Takashige, M., Kanai, T., *Int. Polym. Process.* (2005) 20, no. 1, pp. 100–105
31. Yamada, T., Matsuo, T., *Proc. CHISA '98*, (1998) Prague, Czech Republic, p. E.6.5
32. Yamada, T., Yoshida, T., Fujii, M., Yoshii, T., *16th Polym. Process. Soc. Annu. Meet. Proc.* (2000) pp. 203–204 and pp. 651–652
33. Nakamura, K., Watanabe, T., Katayama, K., Amano, T., *J. Polym. Sci.* (1972) 16, p. 1077
34. Kanai, T., Uehara, H., Sakauchi, K., Yamada, T., *Int. Polym. Process.* (2006) 13, no. 5, p. 449
35. Kanai, T., Kimura, M., Asano, Y., *J. Plast. Film Sheeting* (1986) 2, pp. 224–241
36. Takashige, M., Kanai, T., *J. Polym. Eng.* (2011) 31, no. 1, pp. 29–35

11 Double Bubble Tubular Film Process and Its Application

Masao Takashige

■ 11.1 Introduction

In recent years, environmental problems have come into question in the packaging industry. These problems, such as dechlorination and waste reduction, have been a major focus in this industry. Waste reduction especially has been a serious problem. To achieve waste reduction, there is a rapidly increasing shift from a disposable bottle to a standing pouch for repackaging use in an effort to utilize resources effectively. Biaxially oriented PA6 films, which are thin and strong, are indispensable for many applications.

Several types of manufacturing processes have been developed to produce biaxially oriented PA6 film that has high strength. Among them, the double bubble tubular process producing biaxially stretched PA6 film is the best process in terms of impact strength. It is highly valued for the safe distribution of many products.

There are several reports published on the double bubble tubular stretching technologies [1–12] on processabilities and structure analyses for resins such as PET (polyethylene terephthalate), PBT (polybutylene terephthalate), PPS (polyphenylene sulfide), PA6-12 (polyamide 6-12), and PA12 (polyamide 12) reported by Kang et al., Song et al., and Rhee et al. [3–12].

There are two different processing routes that produce the biaxially oriented film, namely the tenter process and the double bubble tubular film process [13–22]. Furthermore, there are two tenter processes: the sequential stretching process and the simultaneous biaxial stretching process. However, there are few reports on the double bubble tubular technology and the tenter biaxial stretching process.

In general, biaxially oriented PP film and PET film can be easily produced by the sequential stretching process. By contrast, because PA6 has hydrogen bonds, it is hard to produce biaxially oriented PA6 film by the sequential stretching process. Biaxially oriented PA6 film was first produced by the simultaneous biaxial stretching process developed by Unitika. Hitachi developed the simultaneous biaxial stretching machine using pantograph-type technology, and then Hitachi and Unitika have collaborated and produced the biaxially stretched film commercially [23].

Several years later, Toyobo developed a new type of PA6 resin with reduced hydrogen bonding by using compounding technology [24].

The biaxially oriented PA6 film using the double bubble tubular technology could produce a good balance of physical properties and high impact strength, and thus it is used for applications that take advantage of these film properties [25–32].

There is no report that clearly describes the differences in film physical properties from the three different processes. In this report, the influence of each process on film physical properties will be studied in detail.

PA6 film has inferior oxygen gas barrier performance. K-coated film, which has a coating of polyvinylidene chloride on the surface of biaxially stretched PA6 film in order to raise the oxygen gas barrier performance, was usually used. When this film is incinerated, it generates toxic substances such as chlorine gas and dioxin because the film contains chlorine in its molecular structure. The development of a more environmentally friendly oxygen gas barrier film is desired. Recently, oxygen gas barrier film that satisfied formability and physical property needs was developed using blending technology. As a gas barrier resin, poly(*m*-xylene adipamide) (MXD6) was selected.

As biaxially oriented PA6 film is very strong, the application of PA6 film affixed with an LLDPE sealant film is expanded for toughness and environmental sustainability. However, when the package is opened, it is not easily cut due to the large tearing resistance, and it does not cut in a straight line.

It is desirable that the barrier bag be simple and safe to open, for children and even for elderly people. In the medical field, there are a lot of applications where a pair of scissors cannot be used for safety and hygiene reasons, so the development of a film with easy-open access is desired. The development of barrier film with compatible performance of toughness and openability is the goal. Using polymer blends is a useful method to approach this goal in order to obtain a new material from existing material [33–36]. Research on the blending of PA6 and other resins has been undertaken. It has been explained in a patent that the mixed and stretched PA6 resin and MXD6 resin at a fixed compounding ratio can make a film with easier tearing properties [37]. It was found that blending led to a film that has a straight-line cut as a new property. Blended samples of PA6 and MXD6 are studied by Takeda by changing the melting conditions [38–40]. It is known that in this system reactive blending occurs from research of the past. In particular, Shibayama evaluated the reaction analysis of PA6 and MXD6 [41–42].

There are no reports on double bubble tubular technology of the blended resin. Analyses of deformation behavior of PA6 were reported previously based on stretching stress analyses [2]. The current study was carried out to clarify the relationship between blending composition and biaxial stretching formability structure. Its aim is the development of a film with ease of opening and straight line cuts while keeping the toughness of biaxial stretched PA6 film. The development of straight-line cut film and ease of opening was examined by the combination of tubular biaxial stretching technology and resin blending technology. Detailed studies under various processing conditions, governing the film processability in the double bubble tubular PA6/MXD6 blending film process, are described next.

11.2 Physical Properties of Biaxial Oriented PA6 Film for Simultaneous Stretching and Sequential Processing

11.2.1 Stretching Process (Three Technical Methods)

Figure 11.1 shows a schematic view of the double bubble tubular film process.

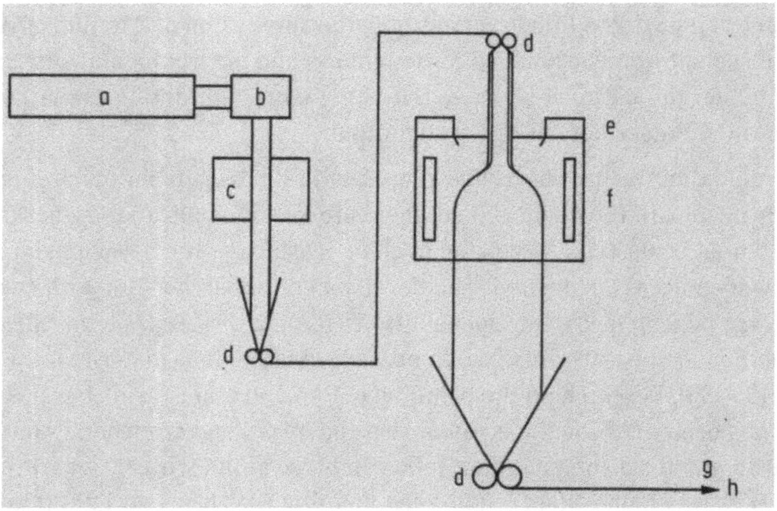

Figure 11.1 Schematic view of double bubble tubular process: (a) extruder, (b) die, (c) cooling bath, (d) take-up roll, (e) air ring, (f) heating furnace, (g) annealing, (h) winding

In addition to the tenter frame process, the double bubble tubular film process is one of the most important polymer processing methods for producing biaxially oriented film.

Particularly for the case of oriented PA6 film, this process is widely used to produce biaxial stretched film, because the tenter frame, which is usually a multistage process, has difficulty in producing a uniform film because of hydrogen bonding generated perpendicular to the machine direction in the first stage of stretching. Thus it is important that we learn how to use the double bubble tubular process more effectively through the choice of resins and process conditions based on processability and film quality.

Biaxial oriented PA6 films are produced by five different Japanese companies using five different processes, and they compete with one another in the same market applications. Figure 11.2 shows three stretching processes.

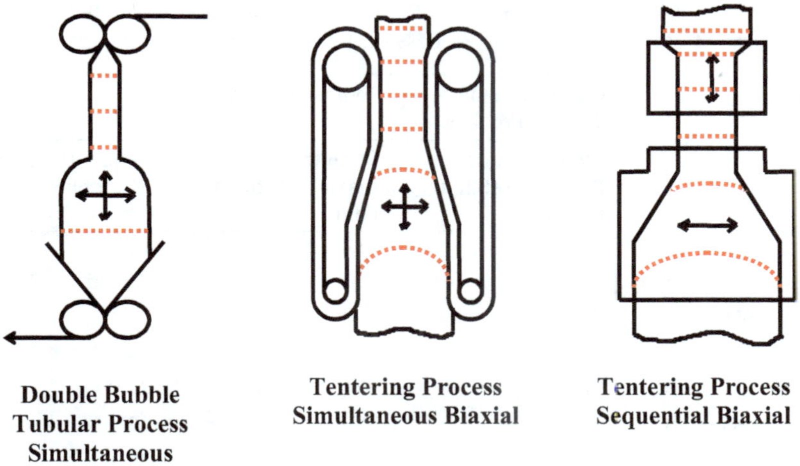

Figure 11.2 Three kinds of biaxial stretching processes for PA6

There are basically three processes employed by these manufacturers: (1) a sequential biaxial stretching in a tenter frame process, (2) a simultaneous biaxial stretching in a tenter frame process, and (3) a simultaneous biaxial stretching version of the double bubble tubular process. These processes have almost the same shares of the market.

In the case of PA6 resin in particular, there is an amide linkage C–N in the molecular structure, and thus hydrogen bonding occurs readily between molecular chains. For this reason, multistage film forming difficulties are encountered in the biaxial stretching process such as film breaks and uneven stretching. If the molecules are oriented in the machine direction (MD) at first, hydrogen bonds always occur; this causes resistance to transverse direction (TD) orientation. On the other hand, the simultaneous biaxial stretching process can give an even force to the surface in both directions without forming these strong hydrogen bonds; this process is widely used for biaxial stretching of PA6.

To compare the physical properties of double bubble tubular film with two of the tenter films, the commercially popular film produced by two different processes were used, namely the sequential process and simultaneous tenter process. Two tentering oriented films (15 µm), produced by the sequential process and the simultaneous tenter process, were obtained.

The resin properties of each sample, such as relative viscosity, molecular weight, melting point, and crystallization temperature, were measured. The quantitative analysis of relative viscosity was measured using a 98% solution of sulfuric acid. The molecular weight was estimated using relative viscosity.

Table 11.1 shows the resin properties of three different types of biaxially oriented film. To keep the bubble stability, high molecular weight PA6 was used. Each film used silica as a filler.

Table 11.1 Resin Properties of Biaxially Oriented PA6; Simultaneous Biaxial Stretching Film vs. Sequential Biaxial Stretching Film; Tubular Process vs. Tenter Process

Properties	Units	Double Bubble Tubular Process	Tenter Process	
		Simultaneous Biaxial Stretching Film	Simultaneous Biaxial Stretching Film	Sequential Biaxial Stretching Film
Resin		PA6	PA6	PA6 (blended with MXD6)
Relative Viscosity	η_r	3.55	3.35	2.97
Molecular Weight	M_n	23,000	19,300	15,800
Melting Point	°C	221	221	222
Crystallization Point	°C	187	187	187
Silica Filler	ppm	1600	1950	3550

11.2.2 Film Physical Properties

11.2.2.1 Impact Strength

Toughness was evaluated by using film impact strength equipment. The film impact strength was evaluated under the different temperatures in the room, which could be controlled as the temperature of atmosphere (−25°C to 23°C).

Figure 11.3 shows the impact strength. The impact strength of double bubble tubular film has the highest value at all temperatures from −25°C to 23°C. This result is

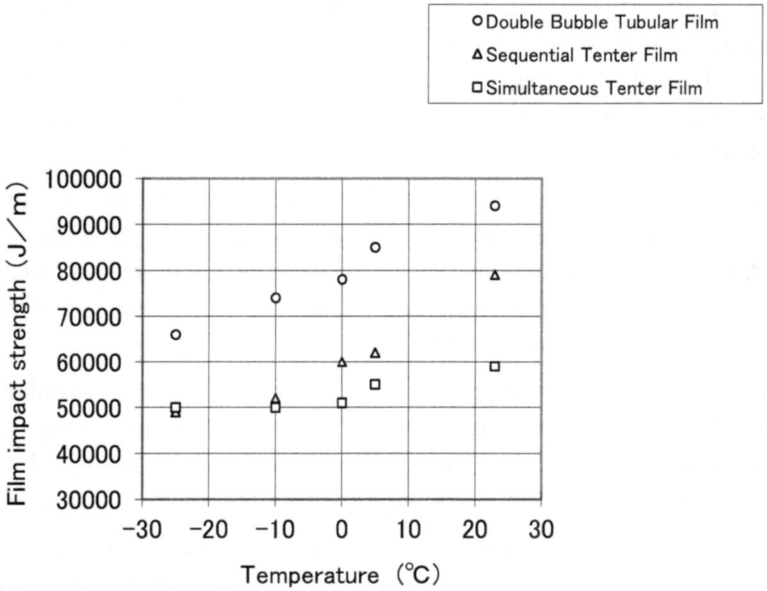

Figure 11.3 Relationship between temperature and film impact strength

owing to the high molecular weight and MD/TD balanced properties, which denotes multiaxis stretching and not biaxial stretching. The film impact strength increases with increasing measured temperature for all samples. Furthermore, the impact strength of the double bubble tubular film at −23°C is stronger than the impact strength of the simultaneous tenter film at 23°C.

The double bubble tubular film has high impact strength between −25°C and 23°C. The result corresponds to the balanced shrinkage properties.

11.2.2.2 Tensile Properties

Table 11.2 shows the physical properties of three different types of biaxially oriented films.

Table 11.2 Physical Properties of Biaxially Oriented PA6; Simultaneous Biaxial Stretching Film vs. Sequential Biaxial Stretching Film; Tubular Process vs. Tenter Process

Properties	Units		Double Bubble Tubular Process	Tenter Process		Testing Method
			Simultaneous Biaxial Stretching Film	Simultaneous Biaxial Stretching Film	Sequential Biaxial Stretching Film	
Tensile Modulus	MPa	MD	2400	2400	2700	ASTM D882
		TD	2100	2200	1700	
Tensile Strength	MPa	MD	260	260	250	ASTM D882
		TD	290	270	320	
Elongation	MPa	MD	130	100	110	ASTM D882
		TD	110	90	80	
Elmendorf Tearing Resistance	N/cm	MD	80	90	80	JIS-Z-1702
		TD	80	80	70	
Film Impact Strength	J/m		91,000	58,000	75,000	IPC Method
Coefficient of Static Friction	-	Inside	1.0	1.0	0.9	IPC Method
		Outside	0.5	1.0	0.4	
Haze	%		1.9	1.7	1.5	ASTM D1003
Gloss	%		140	140	140	ASTM D1003

From these results, the double bubble tubular film has balanced physical properties. By contrast, the tenter sequential stretching film has anisotropic properties between MD and TD. The tensile strength at break in TD produced by the sequential stretching process is higher than in MD. This result is owing to the stretching effect in TD, which remained in effect by the latest stretching. Interestingly, it was found that tensile modulus and tensile strength at break have an opposite trend between MD and TD. In the case of PA6, the tensile modulus is very much higher perpendicular to the TD stretching direction. By contrast, the tensile strength at break is high in the stretching direction. This result is caused by the hydrogen bonding, and it is very much different from PET and PP.

Tentering process film has anisotropic tensile properties between MD and TD. This result was influenced by the later stretching process, namely TD stretching. In particular, the tensile modulus and tensile strength have opposite trends in properties, especially thermal shrinkage and impact strength.

11.2.2.3 Shrinkage Properties in Hot Water (Shrinkage Patterns)

For retort food packaging, PA6 film is soaked in hot water of high temperature for sterilization and processing. Shrinkage in hot water was evaluated by changing cross-direction in two different temperature conditions, namely a hot steam condition of 120°C for 30 min, and hot water at 95°C for 30 min.

Figure 11.4 shows the heat shrinkage pattern. The heat shrinkage pattern is influenced by the biaxial stretching process. The double bubble tubular film has a bal-

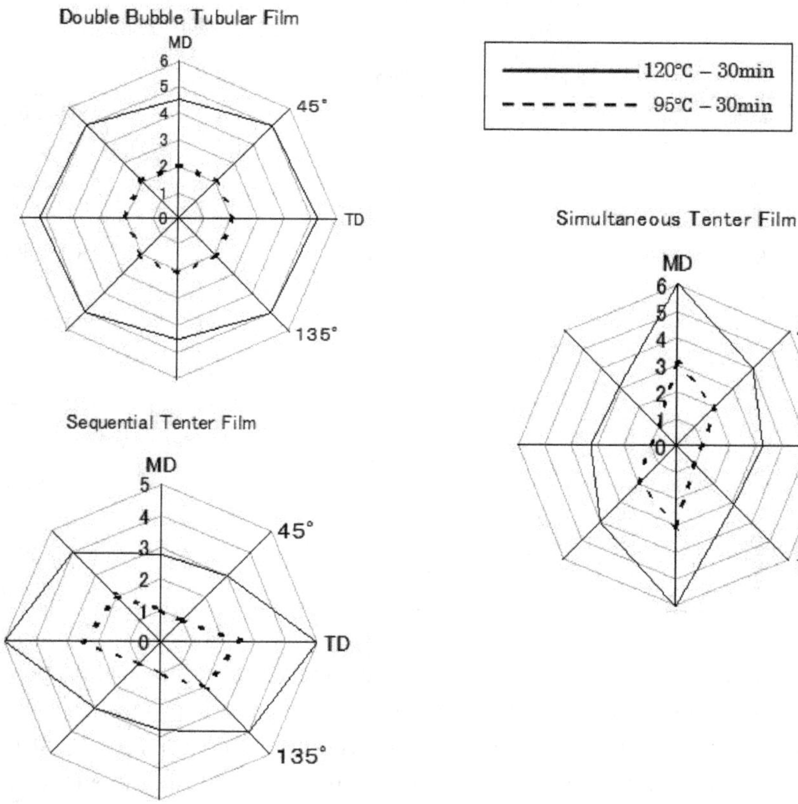

Figure 11.4 Shrinkage percentage pattern in hot water

anced shrinkage pattern. For tentering simultaneous stretching, the film has much larger shrinkage in MD than in TD. By contrast, with the tentering sequential process the stretched film has larger shrinkage in TD than in MD. This result is owing to a stretching effect that is related to the latest stretching. The shrinkage in hot water is related to the amorphous orientation.

11.2.2.4 Stress-Strain Curve Pattern

Tensile properties such as tensile strength at break and elongation at break were measured by changing the cross-direction position (MD, TD, 45°, 135°).

Figure 11.5 shows the stress-strain pattern. The stress-strain pattern is influenced by the biaxial stretching process. The double bubble tubular film has a balanced tensile property pattern. By contrast, the tenter stretched film shows an unbalanced tensile property pattern in the film plane. The result is owing to the stretching effect that is called the bowing phenomenon.

The bowing phenomenon that causes the nonuniformity of film characteristics in the TD is a serious technical problem that needs to be solved in the stretching of film. The bowing phenomenon relates to the discrepancy between the orientation axes at the edge of the film width. The bowing phenomenon causes anisotropy of film characteristics such as heat shrinkage, refractive index, and mechanical properties.

The difference between the direction of the molecular orientation axis and the film line direction becomes large towards the edge of film in the film width. The bowing phenomenon does not occur in the stretching process of the double bubble process owing to the no-edge effect. By contrast, the tentering process has an edge effect. Reducing the bowing phenomenon is one of the most important technical issues in the manufacturing of biaxially oriented film.

The double bubble tubular film has a balanced stress-strain curve in each direction and high elongation at break, which gives good thermoformability. Figure 11.6 shows the tensile strength pattern in-plane and Fig. 11.7 shows the strain pattern (elongation) in-plane.

The double bubble tubular film also has a balanced tensile property in-plane. By contrast, the tenter stretched film shows an unbalanced tensile property pattern in the film plane. In particular, tentering simultaneous stretching film has small strain patterns (elongation) in-plane. This property effect results in low impact strength and bad thermoformability. The double bubble tubular film has good balanced properties, particularly the stress-strain curve in tensile properties.

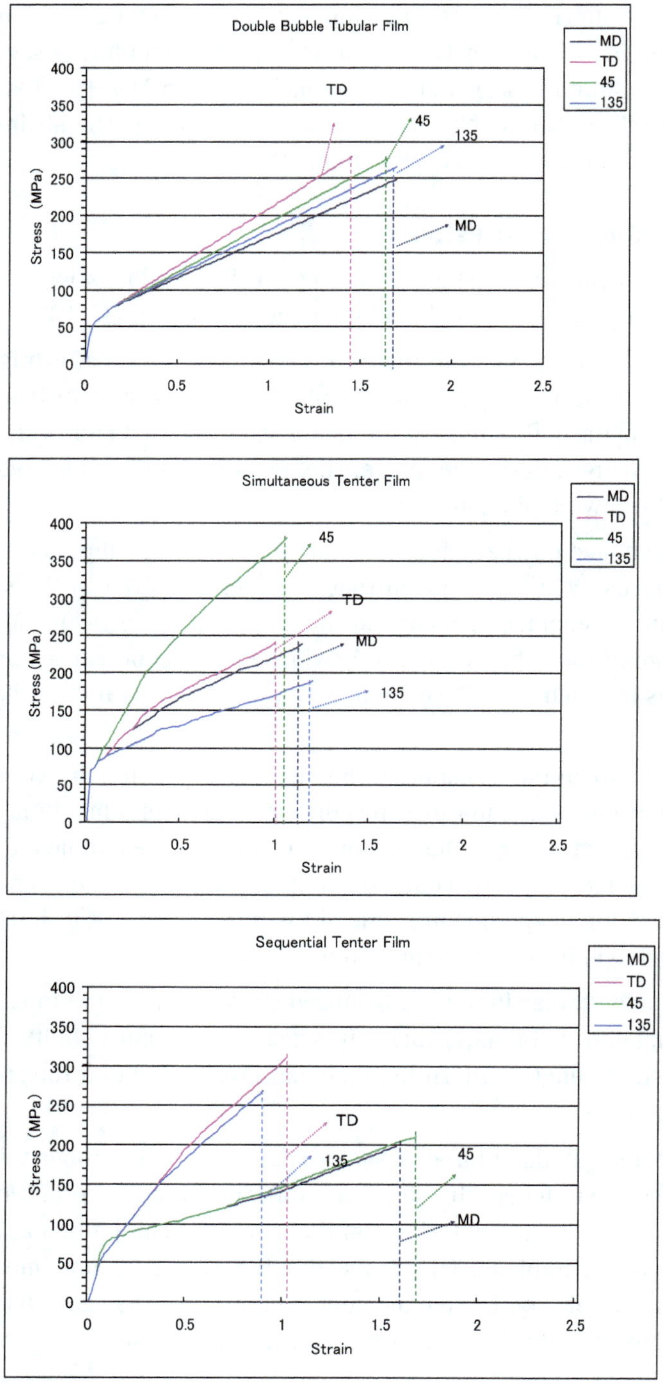

Figure 11.5 Stress-strain pattern in-plane

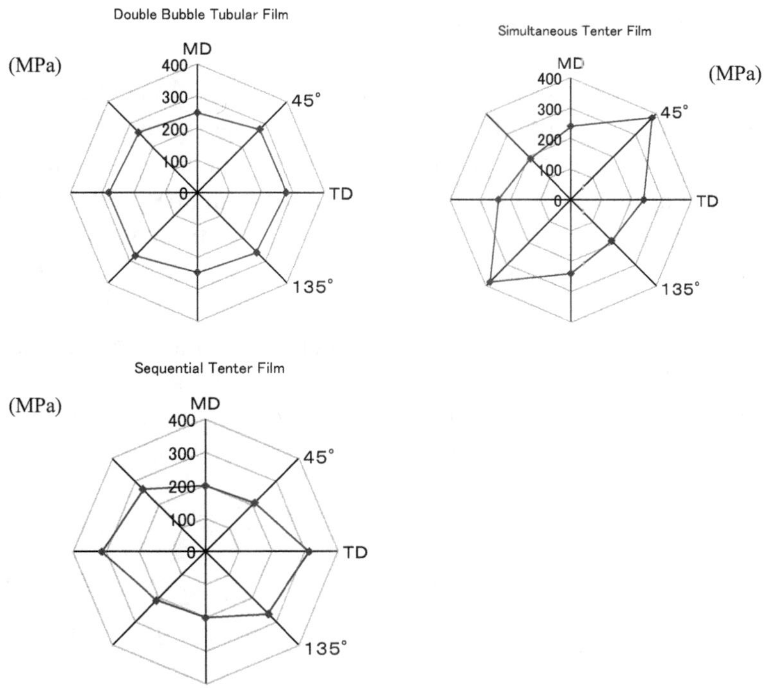

Figure 11.6 Tensile strength pattern in-plane

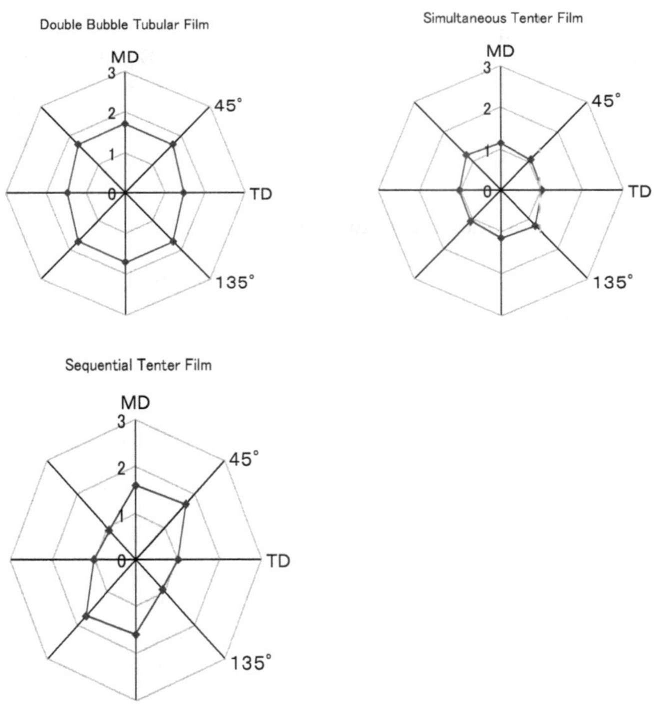

Figure 11.7 Elongation at break pattern in-plane

11.2.3 Wide-Angle X-ray Diffraction Pattern

Film orientation is evaluated by using wide-angle X-ray diffractometry. Figure 11.8 shows the wide-angle X-ray diffraction patterns. These results show the different orientation patterns dependent on the biaxial orientation process. Namely, the double bubble tubular film has balanced orientation between MD and TD, but the tenter stretched films have higher MD orientation than TD. The through-view of the double bubble tubular film shows a uniform diffraction pattern. By contrast, the through-view of the tenter biaxial stretching shows strong diffraction on the meridian direction (MD), and the *b*-axis in the crystalline (molecular chain direction) is predominantly oriented in TD. The simultaneous biaxial stretching has the same trend as the sequential stretching, but the orientation of the simultaneous biaxial stretching is not as strong as the sequential biaxial stretching.

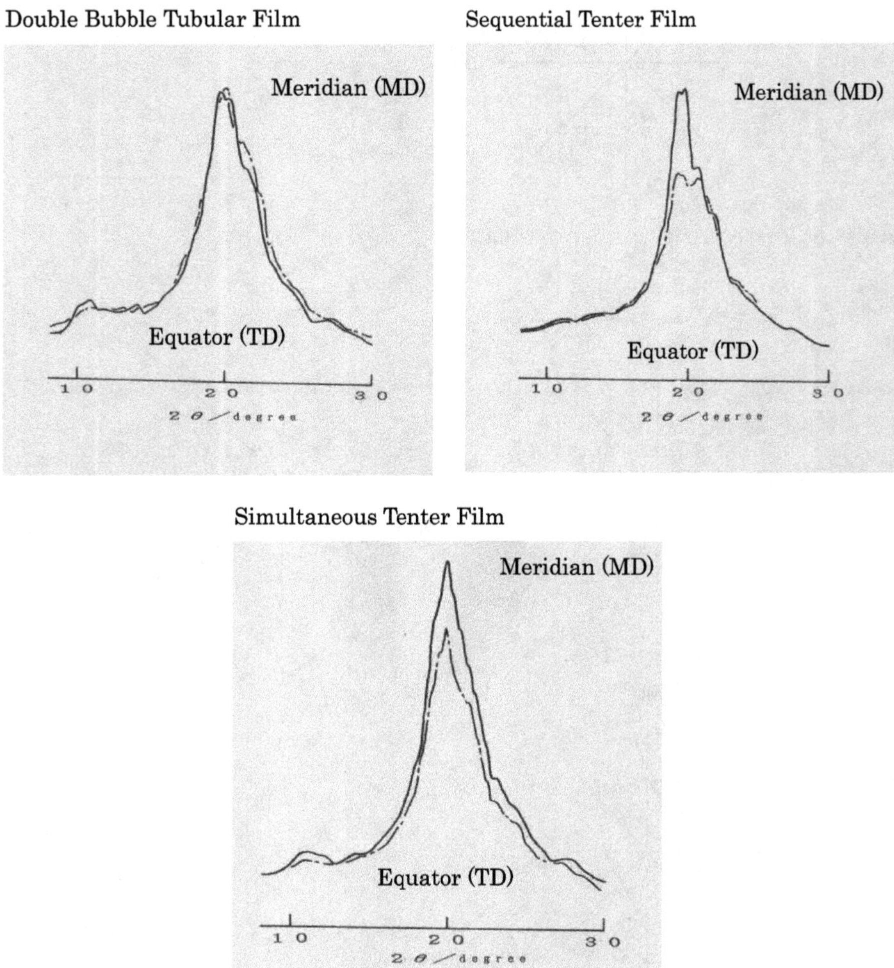

Figure 11.8 Wide-angle X-ray diffraction patterns (WAXD)

11.2.4 Polarized Fluorescence

Molecular orientation in the amorphous region is evaluated by using a polarized fluorescence technique. The film sample was immersed under 23 °C into the 0.2% solution of a fluorescent substance. The excitation light wavelength was 365 nm, and fluorescence intensity in-plane (I// and I⊥), which was measured parallel and perpendicular by rotating a polarizer and an analyzer, was measured as a function of angle. The symbol I⊥ denotes that a polarizer is perpendicular to an analyzer. The excitation light entered from the back side of the sample. Figure 11.9 shows the

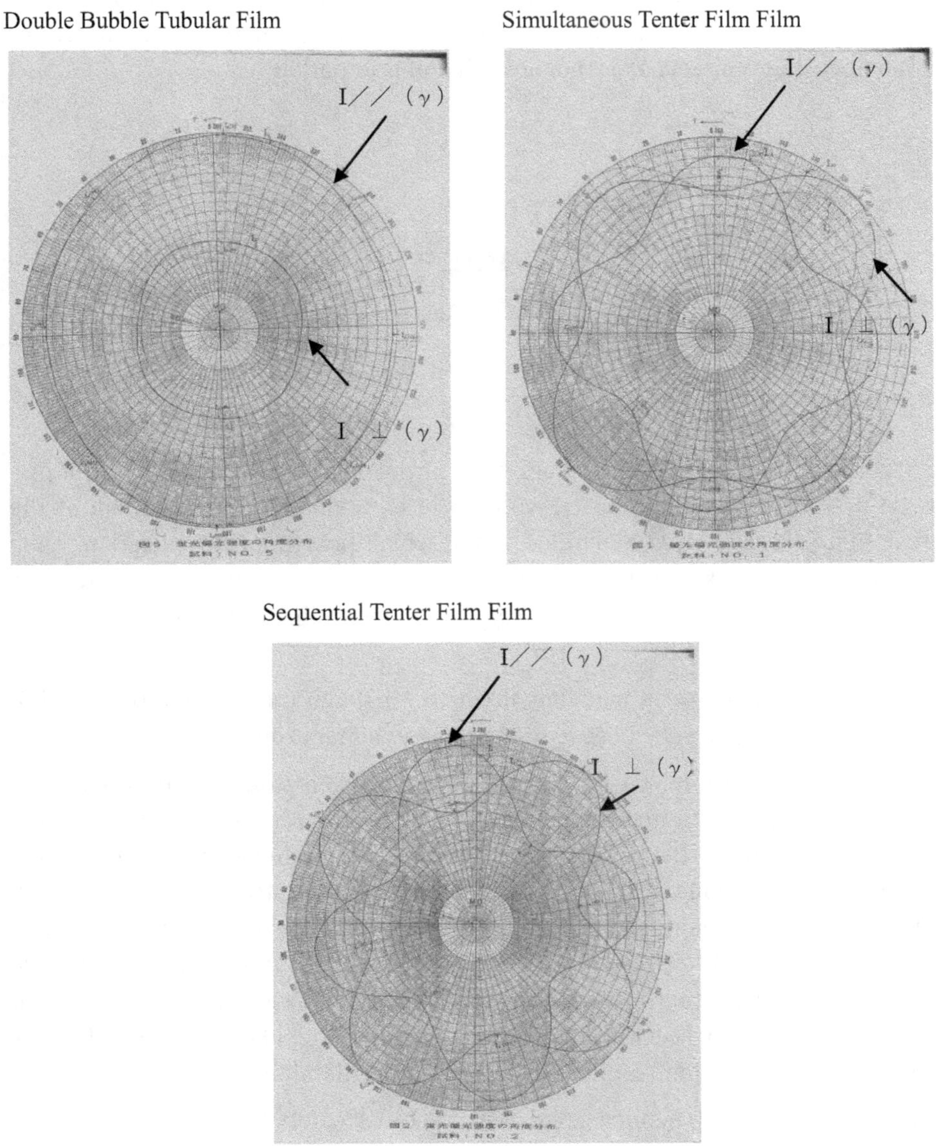

Figure 11.9 Polarized light of fluorescence in-plane

polarized fluorescence observation. This result shows the different orientations among biaxial stretching processes. In particular, the double bubble tubular film shows random orientation, and the tenter stretching process shows the orthogonal biaxial orientation.

From these results, the double bubble tubular film has multiaxis orientation rather than biaxial orientation. The impact strength and shrinkage in hot water are very much influenced by the orientation in the amorphous region. The random orientation gives high impact strength. In addition, the high molecular weight of PA6 for double bubble tubular film contributes to high impact strength.

The polarizing fluorescence results for film orientation in the amorphous portion shows that the double bubble tubular film has uniform orientation in all directions and the tentering process has orthogonal biaxial orientation.

11.3 Easy-Tear Film of Biaxial Oriented PA6/MXD6 Blend by Double Bubble Tubular Film Process

11.3.1 Equipment and Materials

Recall that Fig. 11.1 shows a schematic view of the double bubble tubular film process. At first a PA6 film is produced with low orientation and crystallization by the tubular film process cooled by chilled water, which prevents crystallization. Then the tubular film is biaxially stretched between two pairs of top and bottom nip rolls. Biaxial stretching occurs in the heating apparatus controlled by the heater temperature and cooling air velocity.

The stretching ratio between machine direction (MD) and transverse direction (TD) is controlled by the roll speed between the top and bottom rolls.

While passing through this part of the apparatus, which is composed of two pairs of pinch rolls and a heating furnace (a far-infrared radiation heater is self-contained) for the second blowing, the raw film was stretched simultaneously in the machine and transverse directions by using internal bubble air. The stretched film was heat-set using a heat treatment device (the infrared heater). One material was PA6 with mean molecular weight of 23,000 and relative viscosity of $\eta_r = 3.5$ in 98% sulfuric acid as a solvent. The other material was MXD6 nylon with mean molecular weight of 25,000 and relative viscosity of $\eta_r = 2.7$ in 96% sulfuric acid as a solvent. MXD6 is the polyamide formed from adipic acid and *m*-xylene diamine.

Figure 11.10 shows the chemical structure of both PA6 and MXD6.

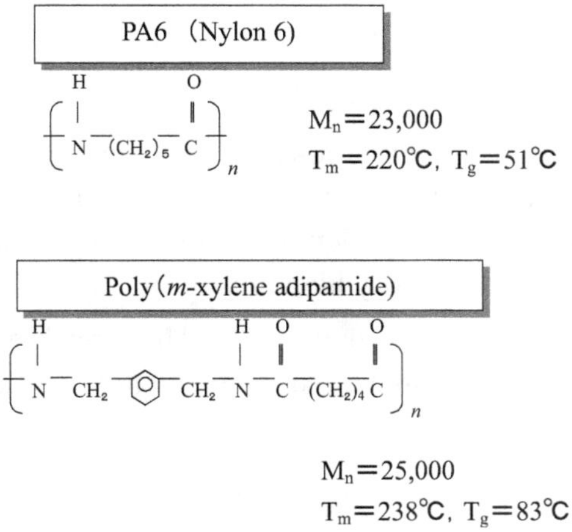

Figure 11.10 Chemical structure of PA6 and MXD6

Figure 11.11 shows the method of measuring stretching stress. Stretching stress was calculated with the help of the tubular theoretical equation reported in [23]. The maximum stress at the end point of stretching may be obtained by using internal bubble pressure.

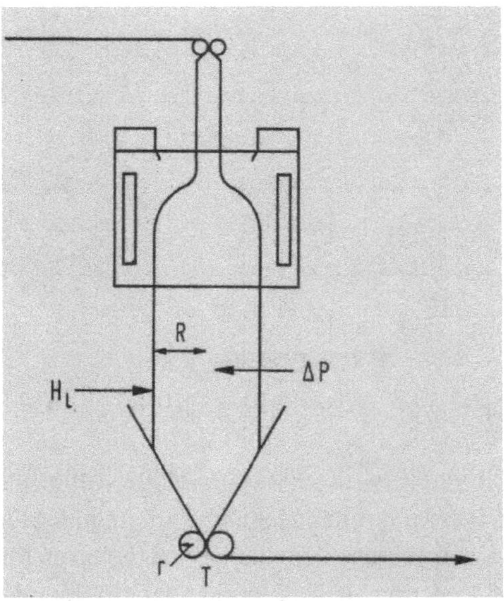

Figure 11.11 Measurement method for stretching stress (the internal bubble pressure was measured using the digital manometer)

11.3.2 Blend Ratio (Dry Blending)

11.3.2.1 Stretchability

The comparative evaluation of stretching stress and formability was done using resins of different blending ratios. Formability and physical properties were evaluated by the change of blending ratio of PA6 and MXD6. The blending ratios of 0%, 10%, 20%, 30%, and 40% in MXD6 was accomplished using dry blending.

The samples of different MXD6 blending ratios were produced, and the stretching stability was evaluated. Figure 11.12 shows the relationship between blending ratio and stretching stress.

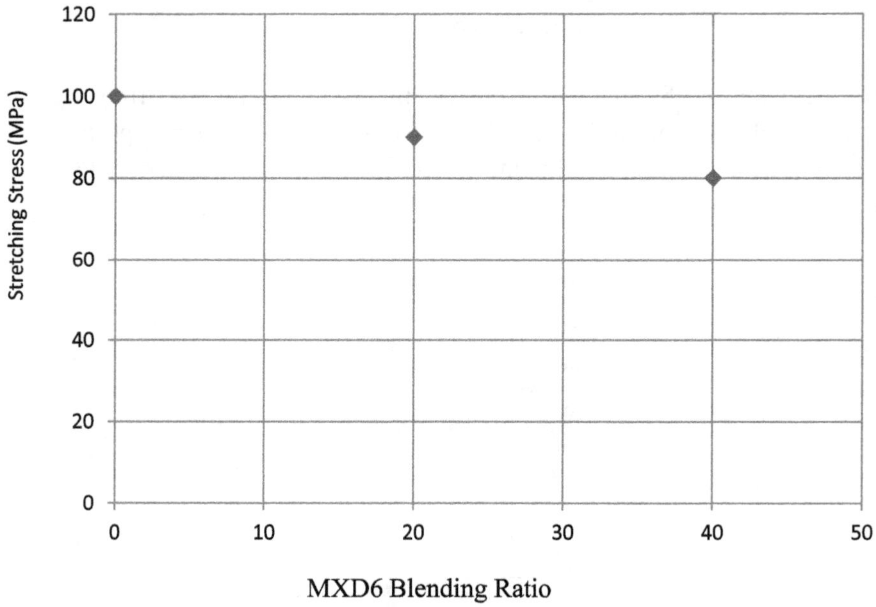

Figure 11.12 Relationship between MXD6 blending ratio and stretching stress

The stretching stress lowered as the MXD6 blending ratio increased. Figure 11.13 shows the relationship between process temperature (temperature of heating furnace) and stretching stress. Temperature dependency of the stretching stress for blended film is larger than for PA6 film. It was found that the blended film had a narrow stretchability region because the temperature dependency of MXD6 resin is large.

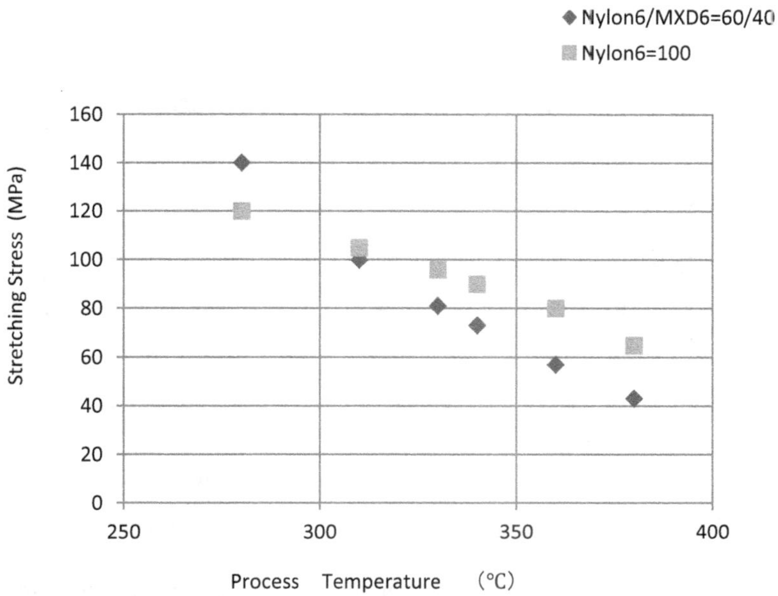

Figure 11.13 Relationship between process temperature and stretching stress

PA6 is also known for its strong hydrogen bonding owing to the molecular structure having polyamide bonding intrinsic to the resin, and therefore it has better simultaneous biaxial stretching than sequential stretching. Hydrogen bonding is suppressed by the steric hindrance of the MXD6 resin. As a result, stretching stress of blended film decreased.

11.3.2.2 Physical Properties

The physical properties were evaluated as a function of the change of blending ratio of PA6 and MXD6. Figure 11.14 shows the relationship between blending ratio and oxygen gas permeability.

Oxygen gas permeability of the blended sample of PA6 and MXD6 was much better than the PA6 film. The oxygen gas permeability was also improved as blending ratio of the MXD6 resin increased. The toughness was also evaluated by using standard film impact strength equipment.

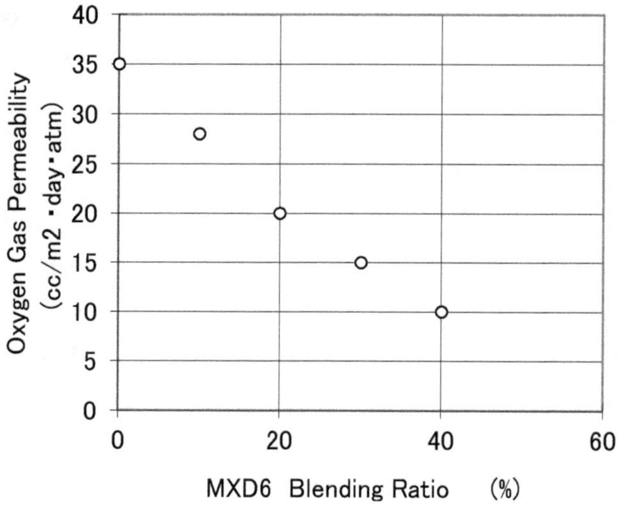

Figure 11.14 Relationship between MXD6 blending ratio and oxygen gas permeability

Figure 11.15 shows the relationship between blending ratio and film impact strength. The film impact strength decreased as the MXD6 blending ratio increased. Because hydrogen bonding is known to be suppressed by the steric hindrance of the MXD6 resin, MXD6 resin is rigid.

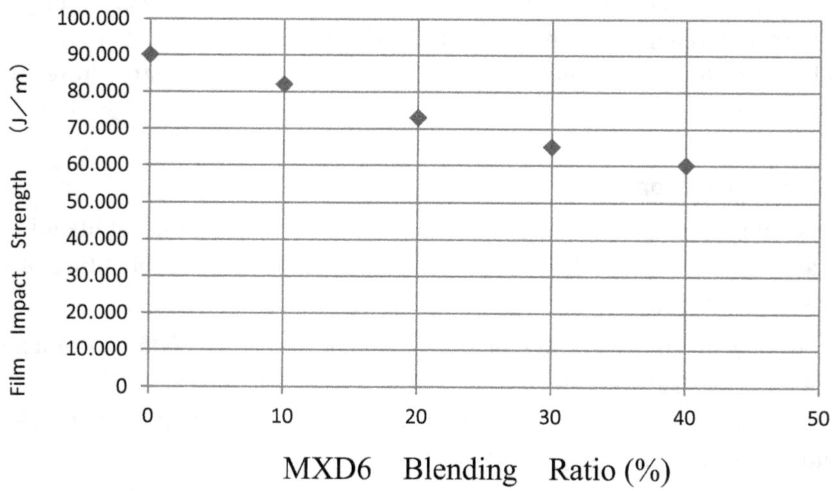

Figure 11.15 Relationship between MXD6 blending ratio and film impact strength

The high gas barrier property of MXD6 is demonstrated in the blended film without hurting PA6 ductility. It is shown that the complementarity of dynamic properties and gas barrier properties of MXD6 blended films has been skillfully carried out. Both PA6 and MXD6 are crystalline resins. Elastic modulus and gas barrier properties are much better in MXD6 than PA6. The ductility of PA6 is higher than that of

MXD6. It is possible to develop material that compensates for the weak point of the other by blending these two components.

11.3.2.3 Easy-Tearing Properties

To evaluate the ease-of-tearing properties, the tearing resistance and straight line cut were measured. The tearing resistance was evaluated by using the Elmendorf tearing test (ASTM D1922) machine.

Figure 11.16 shows the relationship between MXD6 blending ratio and tearing resistance. The tearing resistance in MD and TD were measured. The tearing resistance lowered as shown, as MXD6 blending ratio increased. It showed the same tendency in the MD and TD directions. The MXD6 blended film has smaller tearing resistance than PA6 film.

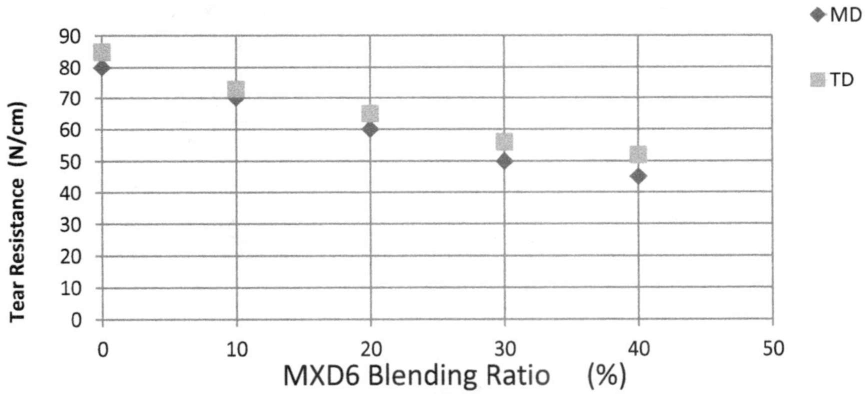

Figure 11.16 Relationship between MXD6 blending ratio and tearing resistance

The straight line cut was evaluated by using the stretched film sample. The film was torn to 200 mm in length, and the deviation length was measured.

Figure 11.17 shows the measurement method of a straight line cut in film. The straight line cut level was expressed using the deviation length. Figure 11.18 shows the relationship between blending ratio and straight line cut. The deviation length of MD drastically decreased with increasing MXD6 blending ratio, but the deviation length of TD was not changed in all MXD6 blending ratios. In PA6 film, the deviation length was about 35 mm, but in 30% MXD6 blended film, the deviation length was almost 0 mm in MD.

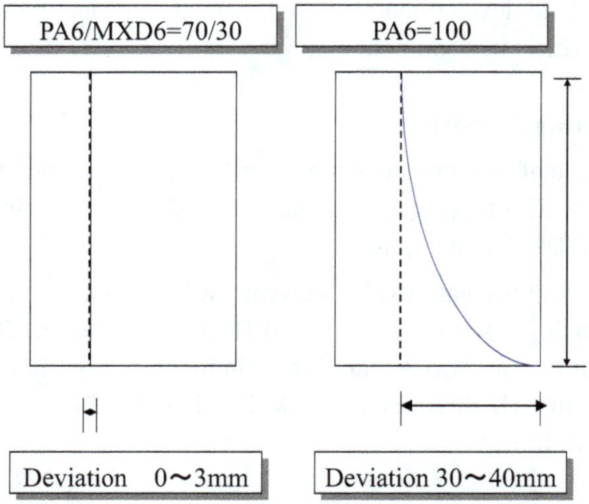

Figure 11.17 Straight line cut measurement method (film)

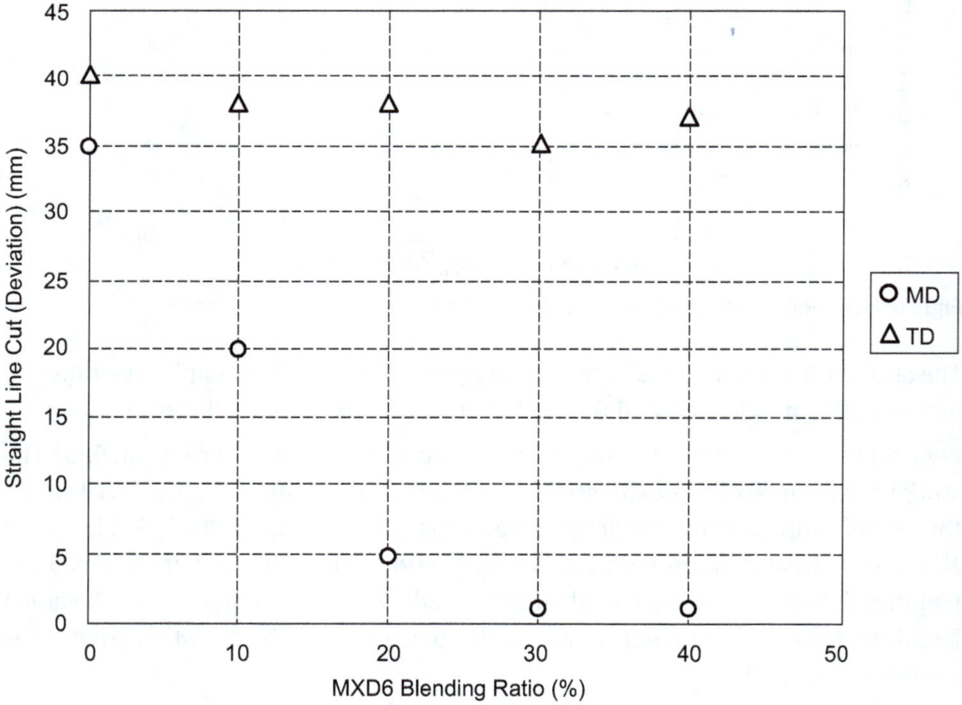

Figure 11.18 Relationship between MXD6 blending ratio and straight line cut

It was found that MXD6/PA6 film blended between 20% and 40% has excellent straight line cut property in the MD direction. The good property of straight line cut disappeared when the MXD6 blend ratio was below 15%. The straight line cut was evaluated by using the package film sample (laminated bag). The film was torn

100 mm in length, and the deviation length was measured. The laminated film (blending film/LLDPE sealant film = 15/50 μm) was measured by the straight line cut too. Figure 11.19 shows the relationship between blending ratio and straight line cut in package film (laminated bag). The straight line cut level was expressed using the deviation length. The deviation length of MD drastically decreased with increasing MXD6 blending ratio. In PA6 film, the deviation length was about 30 mm, but in 30% MXD6 blending film, the deviation length was about 2 mm in MD.

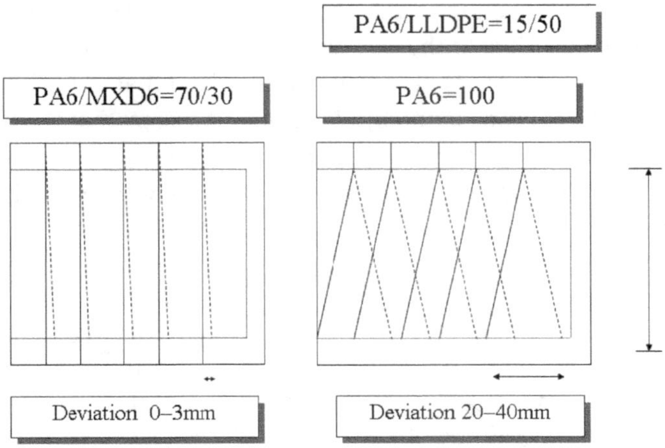

Figure 11.19 Straight line cut measurement method (package)

It was thus verified that MXD6/PA6 film blended between 20% and 40% has very good straight line cut property in the MD direction in package film (laminated bag) too. Figure 11.20 shows the observation of the film fracture plane. Blended film has a sharp cutting plate. This film has no fiber in the fracture plane, and it is easily tearable. By contrast, PA6-only film has large tearing resistance.

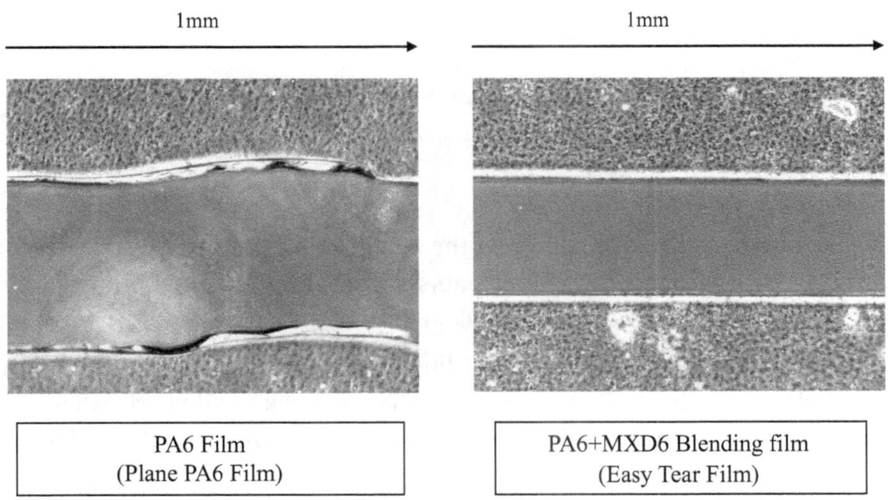

Figure 11.20 Observation of film fracture plane

11.3.2.4 Observation with TEM

Phosphorus wolframic acid was used to dye the film. The thin section was cut down using the microtome. The phase separation structure was evaluated by transmission electron microscopy (TEM). The film in which the blending ratios of the PA6 and MXD6 resin changed was used.

Figure 11.21 shows the measurement direction of TEM. Edge view and end view were selected as measurement directions. Edge view and end view were defined as indicated in Fig. 11.21. Figure 11.22 shows the TEM photographs of biaxially stretched MXD6 blended film at the various blending ratios. The dyeing mechanism by phosphorus wolframic acid is not clear. It was judged that the bright section was a region of MXD6 resin. It was found by electron microscopy that the blending films had a plate-like cylinder structure consisting of MXD6 domains. The MXD6 domain is highly oriented. The difference in tearing anisotropy is partially due to this domain structure.

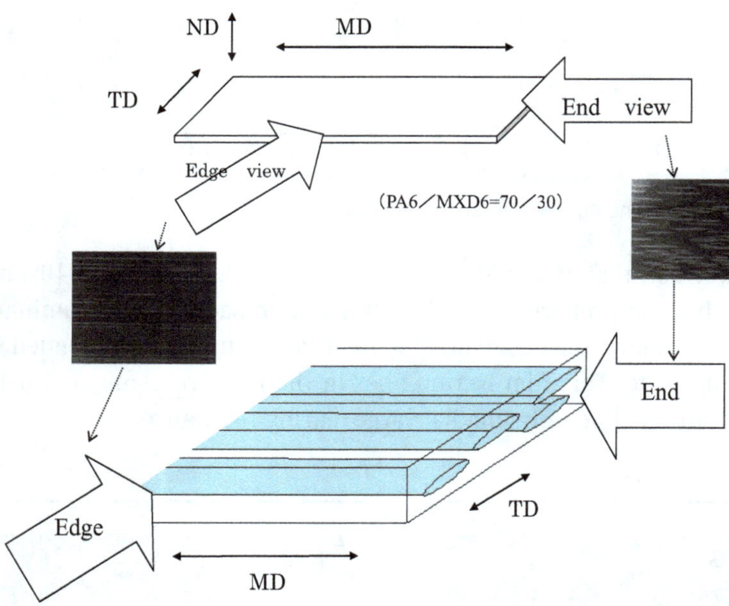

Figure 11.21 Measurement direction of TEM

MXD6 has formed a dispersed phase in the blended film. It was proven by electron microscopy that the slender MXD6 domains are several hundred nm in width and several thousand nm in length, and the domains are lined up in the MD direction.

It was found that the film broke in a rectilinear manner when it was torn along the MD parallel to the slender MXD6 domains. It did not break rectilinearly when it was torn perpendicular to the MD with the slender MXD6 domains at right angles to the tear.

In a PA6 film, the anisotropy of such tearing was not observed. It is estimated that the 100 nm order structure caused the difference in the macroscopic dynamic property of the tearing. It was found that this cylinder structure has produced the straight line cut. Tearing resistance of MD and TD was also decreased by the cylinder structure. ND is normal direction.

As a result of extensive observation of the phase separation structure, it was found that the number of the cylinder structures decreased with decreasing MXD6 blending ratio, and the longitudinal length shortened with decreasing MXD6 blending ratio. The number of cylinder structures increased with increasing MXD6 blending ratio, the longitudinal length increased with increasing MXD6 blending ratio, and the property of straight line cut became excellent.

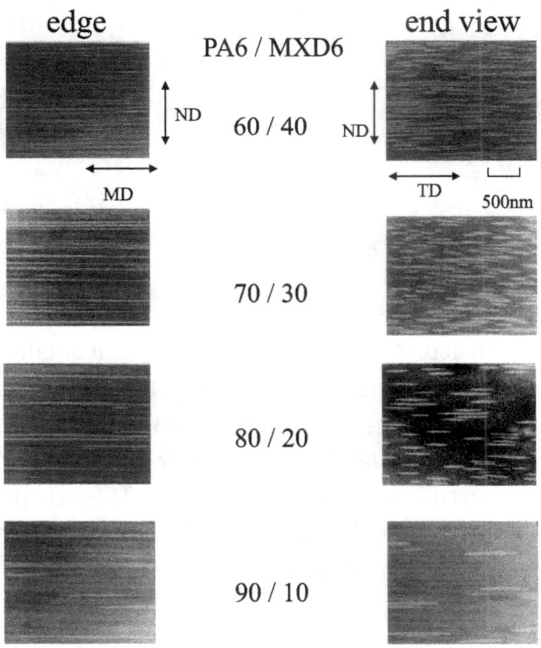

Figure 11.22 Observation of TEM (biaxially stretched film)

Figure 11.23 shows the TEM photographs of nonstretched blended film at the various compounding ratios. The continuity of the MXD6 layer in MD was confirmed in samples that had not been stretched. It was thought that the continuity of the MXD6 layer in the nonstretched film occurs as a result of the extrusion processing. ND in the figure means normal direction.

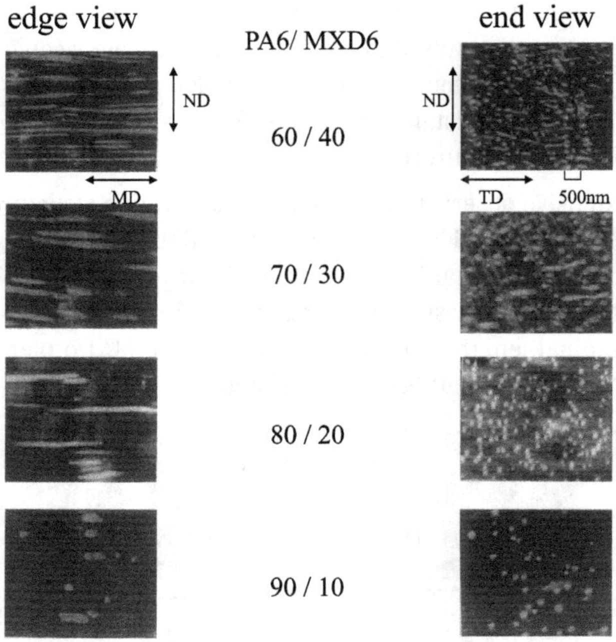

Figure 11.23 Observation of TEM (nonstretched film)

11.3.2.5 Observation with SALS

The phase separation structure was evaluated by using a small-angle light scattering device (SALS). Figure 11.24 shows the photograph of a typical light scattering image. The light scattering intensity increased with increasing MXD6 blending ratio. It was confirmed that the MXD6 phase separation structure had lined up in the MD, because it appeared in the TD tear resistance. This streak reflects the anisotropic shape, which was extended in MD of the MXD6 layer, while only an isotropic diffuse scattering was observed in the PA6 film.

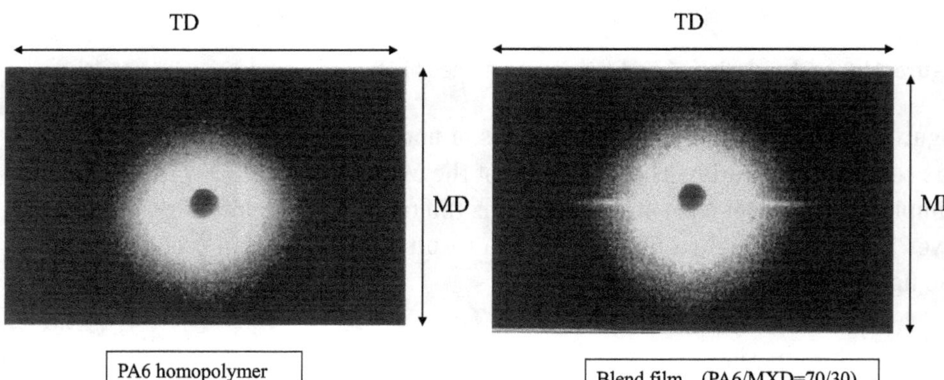

Figure 11.24 Observation of light scattering image of PA6 and PA6/MXD

11.3.2.6 Mechanism of Developed Property

Figure 11.25 shows the mechanism for properties developed by blending MXD6. It was found that each phase controlled many properties. The plate-like cylinder structure was formed in MD.

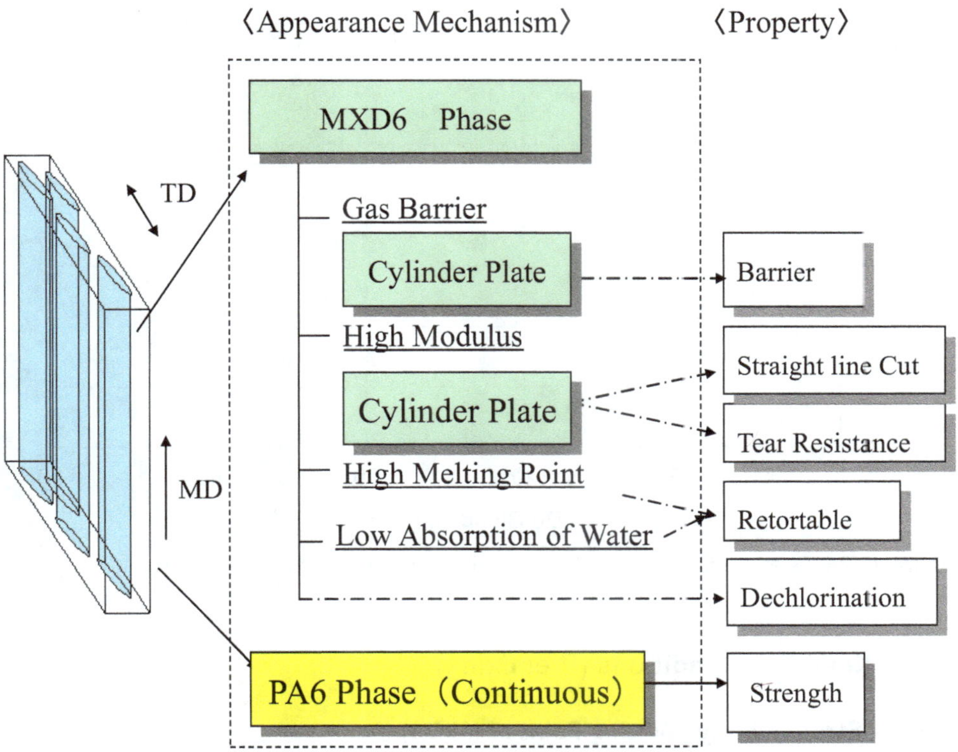

Figure 11.25 Mechanism of property appearance

The MXD6 phase controls the straight line cut, tearing resistance, and oxygen gas barrier performance. The PA6 phase is continuous, and the PA6 phase controls the strength. A high-strength film was achieved because it was oriented biaxially using the double bubble tubular process. The blended film keeps the compatible performance of strength and easy tear property.

11.3.2.7 Thickness Uniformity

The thickness uniformities of nonstretched film and heat-treated stretched film were measured.

Figure 11.26 shows the relationship between MXD6 blending ratio and thickness uniformity. The thickness uniformity deteriorates with increasing MXD6 blending ratio. In particular, the thickness uniformity deteriorates when the MXD6 blending ratio exceeds 40%. All properties were satisfied at about 30% blended MXD6 in the case of dry blending.

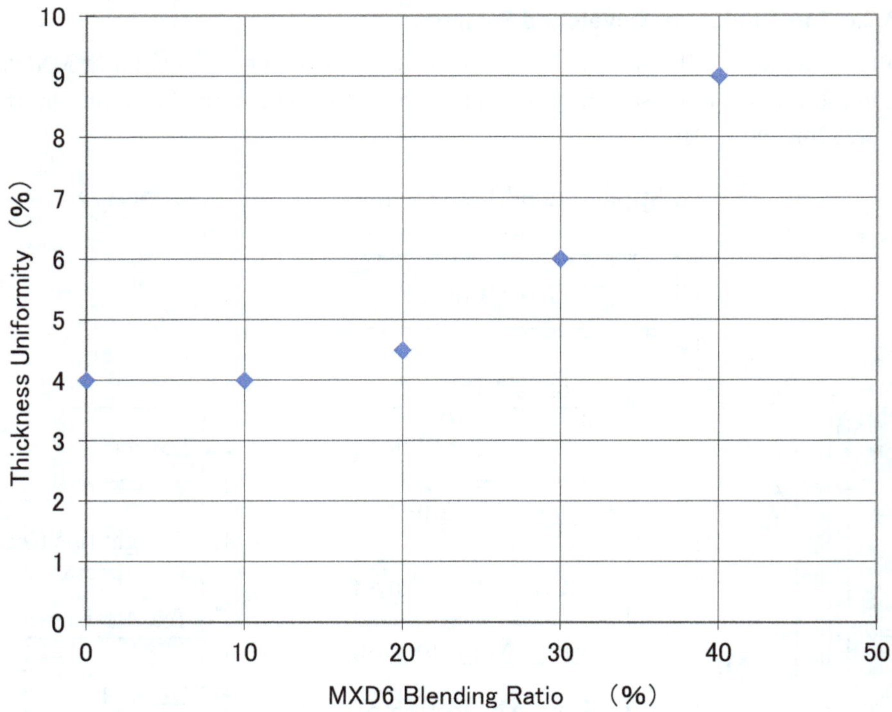

Figure 11.26 Relationship between MXD6 blending ratio and thickness uniformity

11.3.3 Kneading Conditions (Premixing)

11.3.3.1 Stretchability (Melting Point of MXD6)

The improvement of thickness uniformity by a change in the kneading level was studied. It was examined by selecting 40% as the MXD6 blending ratio. The kneading was carried out using the twin extruder with corotation screws. Fig. 11.27 shows the relationship between premixing and thickness uniformity. It was confirmed that the thickness uniformity improved when the MXD6 melting point went down.

Figure 11.28 shows the relationship between melting point of MXD6 and stretching stress.

The MXD6 melting point was used as a scale of the kneading level. The stretching stress drastically lowered as the MXD6 melting point lowered. For the low melting point of MXD6, the stretching bubble was not stable.

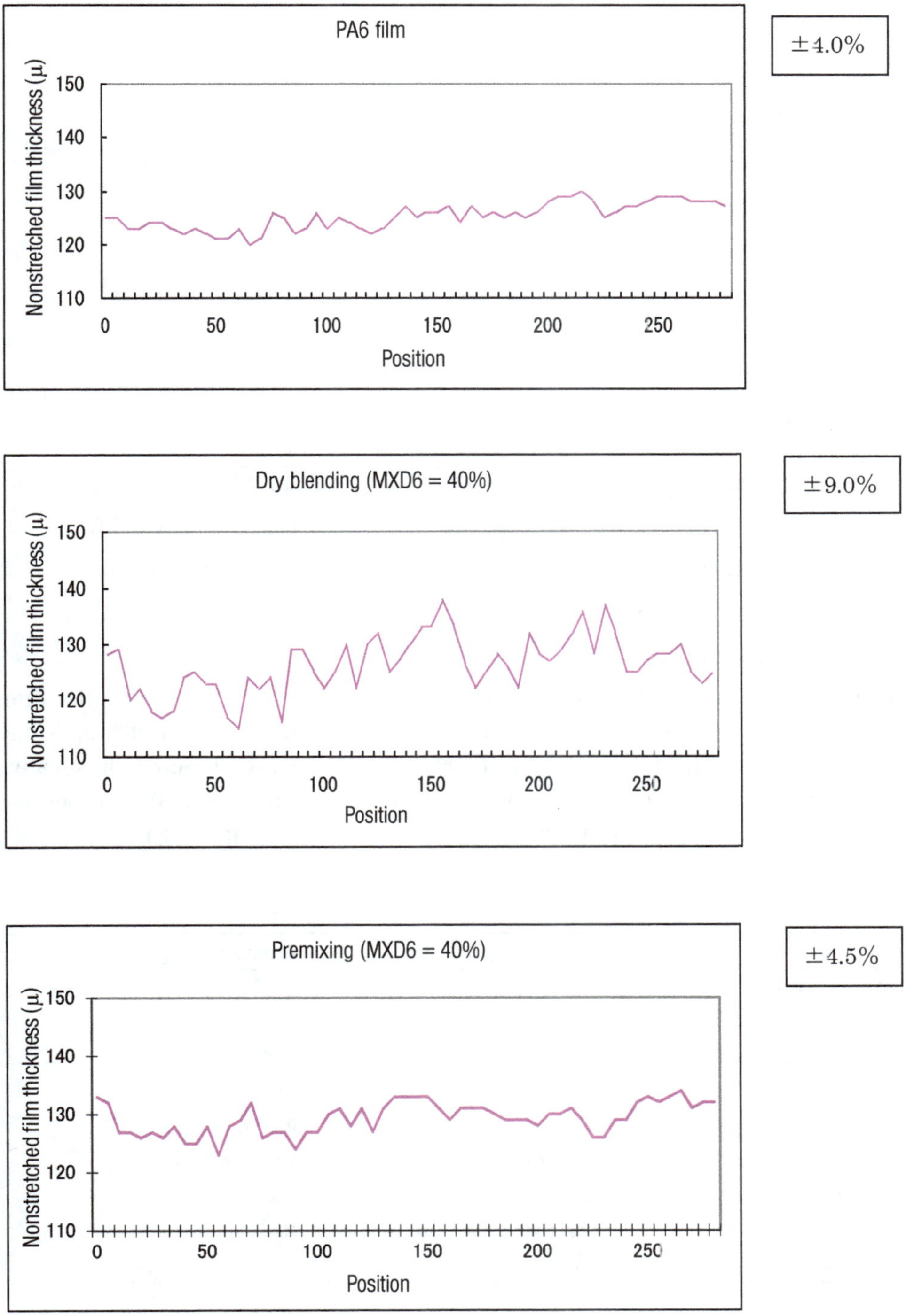

Figure 11.27 Relationship between premixing condition and thickness uniformity

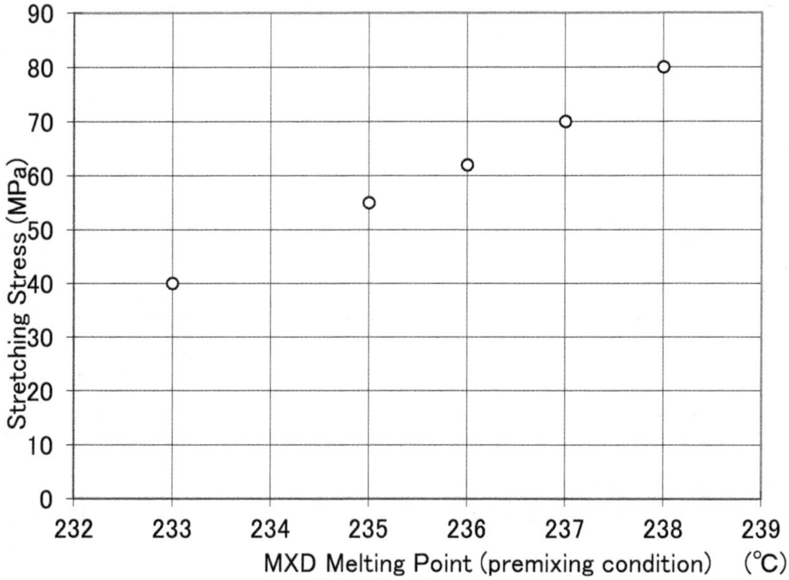

Figure 11.28 Relationship between MXD melting point (premixing condition) and thickness uniformity

11.3.3.2 Physical Properties

Table 11.3 shows the relationship between melting point of MXD6 and the film physical properties. When the MXD6 melting point is lowered, the stretching formability deteriorated, and the physical properties were also lowered drastically. Regarding the physical properties, the film impact strength lowered, the gas permeability deteriorated, and the straight line cut performance disappeared.

Table 11.3 Melting Point of MXD6 vs. Physical Properties

	Premixing (Low Level)	Premixing (High Level)
Resin Temperature	286°C	300°C
Melting Point of MXD6	236.6°C	232.3°C
Stretching Formability	Good	Bad
Stretching Stress	70 MPa	36 MPa
Film Impact Strength	60,000 J/m	43,000 J/m
Oxygen Gas Permeability	15 cc/m^2 × 24 h	24 cc/m^2 × 24 h
Straight-Line Cut	Good	Bad
Observation of TEM	Phase separation structure	Nothing
Thickness Uniformity	±6.0%	±4.5%

11.3.3.3 Structure Analysis

The melting point was evaluated by using differential scanning calorimetry (DSC). The melting point is influenced by the degree of copolymerization of the two kinds of resins.

The difference in phase separation structure between the kneading conditions was observed using electron microscopy too.

The phase separation structure completely disappeared at resin temperatures over 300°C in the kneading process (Fig. 11.29(top)). It was shown by DSC that the two peaks of the melting point became one (210°C); in other words, the copolymerization occurred completely (Fig. 11.29(bottom)).

In addition, the stretching was not stabilized, so the film sample could not obtained.

It is known from past research that this system undergoes reactive blending. Cutting and recombination of amide groups in the molecular chain are generated under the higher temperature over the melting point on both components, and the copolymerization of the sample advances. The excessive reaction of the blending resins is not desirable. In the case of premixing by dry blending, the MXD6 melting point remains over 236°C. It was confirmed that this is the condition that satisfies all properties.

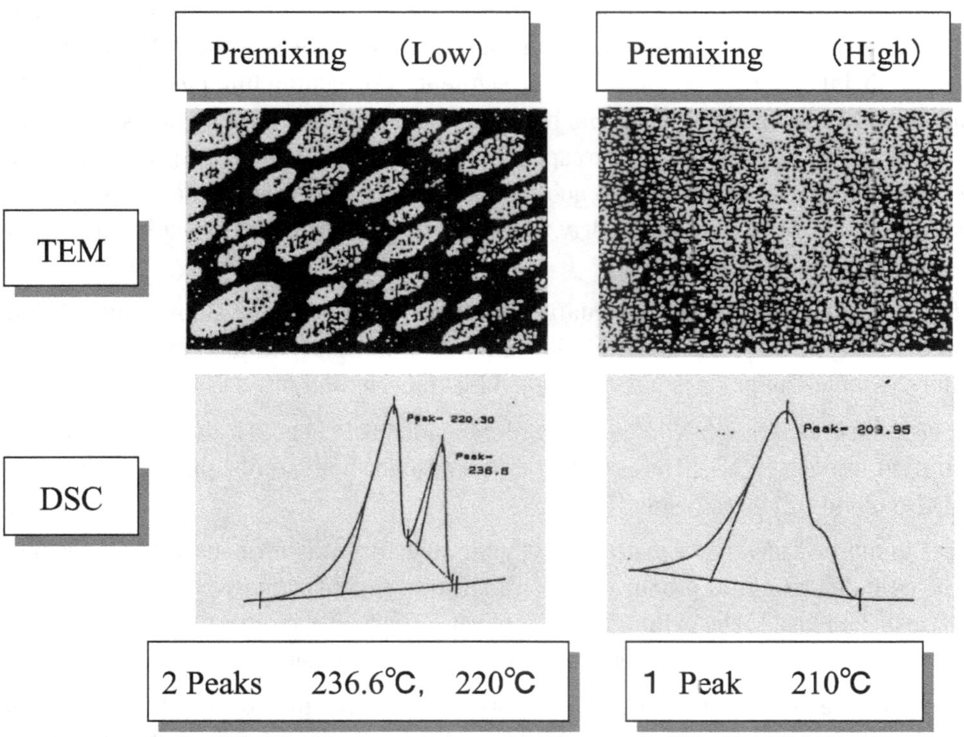

Figure 11.29 (top) Observation of TEM (premixing); (bottom) observation of DSC

11.4 Summary

As these examples have shown, there are advantages with the double bubble tubular process (simultaneous biaxial stretching process) as well as with tentering (sequential biaxial stretching process), and thus it is impossible to say that one is better than the other. It is clear, however, that these processes will play an important role in enhancing the quality and function of goods made if they are matched to the film property needs. It is necessary to match the method to appropriate applications and choose the methods with a clear understanding of their capabilities.

The tentering process gives high productivity because the process can scaled to a large machine, so the tentering process has cost merit. The tentering process film has high tensile modulus, so its film has good printing properties. Furthermore, the tentering process film has good ability to laminate because the film has good thickness uniformity. In particular, PA6 film has irregular stretching properties because of its hydrogen-bonding property.

Tentering processed film has anisotropic tensile properties between MD and TD. This result was influenced by the later stretching process, namely TD stretching. In particular, the tensile modulus and tensile strength have opposite properties.

The double bubble tubular film has good balanced properties, especially thermal shrinkage and impact strength. Tentering simultaneous stretching film has much larger shrinkage in MD than in TD. The sequential stretching film has larger shrinkage in TD than in MD. The double bubble tubular film has high impact strength at -25°C and 23°C. This result corresponds to the balanced shrinkage properties. The double bubble tubular film has good balanced properties, particularly the stress-strain curve in the tensile property. This property affects high impact strength and good thermoformability.

The polarizing fluorescence evaluation of film orientation in the amorphous portion shows that the double bubble tubular film has uniform orientation in any direction and the tentering process has orthogonal biaxial orientation.

It was confirmed that MXD6 blending film blended between 20% and 40% provides a straight line cut in MD. The property of the straight line cut disappeared when the MXD6 blend ratio was below 15%.

It is found by TEM that a plate-like cylinder structure was formed for these blends. The cylinder structure produced the straight line cut. Tearing resistance of MD and TD also decreased. The cylinder structure was observed even in a nonstretched film. The thickness uniformity deteriorates when the MXD6 blend ratio exceeds 40%.

The improvement of thickness uniformity by a change in the kneading level was studied. In the condition in which MXD6 melting point was lowered, the stretching formability deteriorated, and the mechanical properties also drastically lowered.

The phase separation structure completely disappeared at resin temperatures over 300 °C in the kneading process. It was shown by DSC that the peaks of the melting point became one, that is, that the copolymerization occurred completely. The excess reaction of the blending resins is not desirable.

The PA6 phase controls the strength, and the MXD6 phase controls the straight line cut property. A high-strength film was achieved because it was stretched biaxially using the double bubble tubular process.

The stretchability, easy-tearing property, and oxygen gas barrier property are maintained by the following process conditions. In the case of dry blending, MXD6 at 30% is the proper content. And in the case of premixing, the MXD6 melting point stays over 236 °C. It is confirmed that this is the condition for satisfying all properties.

The physical blended film keeps the compatible performance of strength and easy-tearing property. The material and production technology of such a film could solve the environmental and barrier-free problems. The excess reaction of the blending resins is not desirable. The PA6 phase controls the strength, and the MXD6 phase controls the straight cut property.

A high-strength film was achieved because it was stretched biaxially using the double bubble tubular process.

This is an interesting process that is based on retaining stretching technique and providing an oriented film with easy tearability in a straight line by controlling PA6 and MXD6 in the form of a special, phase-separated structure.

Generally speaking, this easy-tearing property is provided by a uniaxial stretching process (including roll stretching). However, the product with these traits has good breakability but very poor service strength such as impact strength, and it is not useful in most practical cases unless it is covered with another base material (including PA6 film). Therefore, it has been shown that the double bubble tubular biaxial stretching technology when used with blends of PA6 and MXD6 has led to film that combines good service strength and good breakability (easy to tear in a straight line).

In structure terms, it can be said that MXD6 becomes phase separated as small islands in a sea of PA6. This phase-separated state makes it possible to obtain this unique morphology through simultaneous biaxial stretching in the vertical and transverse directions. Here, in connection with multilayer and blend stretching, it is likely that the double bubble tubular stretching technology will be applied and promoted in the future.

References

1. Kanai, T., Takashige, M., *Seni Gakkaishi* (1985) 41, p. 272
2. Takashige, M., Kanai, T., *Int. Polym. Process.* (1990) 5, p. 287
3. Kang, H.J., White, J.L., *Polym. Eng. Sci.* (1990) 30, p. 1228
4. Kang, H.J., White, J.L., Cakmak, M., *Int. Polym. Process.* (1990) 1, p. 62
5. Kang, H.J., White, J.L., *Int. Polym. Process.* (1990) 5, p. 38
6. Rhee, S., White, J.L., *Int. Polym. Process.* (2001) 16, p. 272
7. Rhee, S., White, J.L., *Polym. Eng. Sci.* (1999) 39, p. 1260
8. Song, K., White, J.L., *Polym. Eng. Sci.* (2000) 40, p. 902
9. Song, K., White, J.L., *Int. Polym. Process.* (2000) 15, p. 157
10. Song, K., White, J.L., *Polym. Eng. Sci.* (2000) 40, p. 1122
11. Rhee, S., White, J.L., *SPE ANTEC Tech. Papers* (2001) 59, p. 1446
12. Rhee, S., White, J.L., *SPE ANTEC Tech. Papers* (2001) 59, p. 1451
13. Tobita, K., Miki, T., Takeuchi, N., *Film Processing*, Kanai, T., Campbell, G. (Eds.) (1999) Hanser, Munich
14. Cakmak, M., *Film Processing*, Kanai, T., Campbell, G. (Eds.) (1999) Hanser, Munich
15. Tsunashima, K., Toyoda, K., Yoshii, T., *Film Processing*, Kanai, T., Campbell, G. (Eds.) (1999) Hanser, Munich
16. Takashige, M., *Film Processing*, Kanai, T., Campbell, G. (Eds.) (1999) Hanser, Munich
17. Yamada, T., Nonomura, C., Kase, S., *Film Processing*, Kanai, T., Campbell, G. (Eds.) (1999) Hanser, Munich
18. Takashige, M., Kanai, T., Yamada, T., *Int. Polym. Process.* (2004) 19, p. 47
19. Takashige, M., Kanai, T., Yamada, T., *Int. Polym. Process.* (2004) 19, p. 147
20. Takashige, M., Kanai, T., *Int. Polym. Process.* (2005) 20, p. 100
21. Takashige, M., Kanai, T., *Int. Polym. Process.* (2006) 21, p. 86
22. Takashige, M., Kanai, T., *J. Polym. Eng.* (2008) 28, p. 179
23. Unitika, Japan Patent 44-2598
24. Toyobo, Japan Patent 48-80672
25. Takashige, M., Kanai, T., Yamada, T., *Int. Polym. Process.* (2003) 18, p. 368
26. Takashige, M., Kanai, T., Yamada, T., *Int. Polym. Process.* (2004) 19, p. 56
27. Takashige, M., Hayashi, T., U.S. Patent 5,541,011 (1996)
28. Takashige, M., Hayashi, T., Iwamoto, T., U.S. Patent 5,716,696 (1998)
29. Takashige, M., Hayashi, T., Iwamoto, T., U.S. Patent 6,221,502 (2000)
30. Uehara, H., Sakauchi, K., Kanai, T., Yamada, T., *Int. Polym. Process.* (2004) 19, p. 155
31. Uehara, H., Sakauchi, K., Kanai, T., Yamada, T., *Int. Polym. Process.* (2004) 19, p. 163
32. Uehara, H., Sakauchi, K., Kanai, T., Yamada, T., *Int. Polym. Process.* (2004) 19, p. 172
33. Nishi, T., Wang, T.T., *Macromolecules* (1975) 8, p. 909

34. Chuang, H.K., Han, C.D., *J. Appl. Polym. Sci.* (1985) 30, p. 165
35. Greco, R., Malincoico, M., *Polymer* (1988) 29, p. 1418
36. Jo, W.H., Kim, G., Chae, S.H., *Polym. J.* (1993) 25, p. 1023
37. Takashige, M., Hayashi, T. (Idemitsu Petrochemical), U.S. Patent 5,541,011 (1996)
38. Takeda, Y., Paul, D.R., *Polymer* (1992) 33, p. 899
39. Takeda, Y., Keskkula, H., Paul, D.R., *Polymer* (1992) 33, p. 3394
40. Takeda, Y., Paul, D.R., *Polymer* (1991) 32, p. 2771
41. Shibayama, M., Uenoyama, K., Oura, J., Iwamoto, T., *Polymer* (1995) 36, p. 4811
42. Shibayama, M., Oura, J., Iwamoto, T., *Kobunshi Ronbunshu* (1996) 53, p. 453

12 Highly Transparent Polypropylene Sheets

Akira Funaki

12.1 Introduction

Polypropylene (PP) is an excellent resin from the viewpoint of it having a reasonable price, good physical properties, good recycling features, and so on. However, because it is a crystalline resin, it is rather difficult to obtain a highly transparent sheet, and it has had limited development with regards to usage when a high transparency is required. The primary factors in obtaining a highly transparent PP sheet are how to reduce internal haze and external haze and also how to make an excellent glossy surface. To achieve these features, a random copolymer PP as a base material and various kinds of nucleating agents were used in the polymer modification technology. Also, a quick quenching system and a rolling process were adopted for the sheeting process technology. However, the rigidity and the heat resistance of PP sheets decrease with use of a random copolymer PP. Using a nucleating agent is not enough to obtain a highly transparent sheet, and the sheet has a problem: a narrow thermoforming process window.

There has been a lot of research in the past exploring the unstable flow phenomenon of melted polymer that obstructed the transparency [1–12]. In the latest research, it was reported that "shark skin" was a phenomenon in which only the surface of melted resin film got rough. The cause of it is a rapid change from the shear stress at the wall vicinity in the die land to extension stress in the die exit section [13]. However, even in the state where the critical range of shear stress is not exceeded, the surface roughness of melted web depends on the screw geometry of the extruder. Therefore, the transparency should be discussed from another viewpoint.

There has also been a lot of research in the past addressing the higher order of quenched isotactic PP sheet [14–23]. It is well known that a highly transparent sheet of a crystalline PP resin can be obtained by quick quenching a molten resin sheet in the laboratories. There are mainly two theories regarding the higher order structure that is formed during its solidification process. One is the crystallization theory (by Hoffman and Lauritzen [24]) of bulk polymers with chain folding based on the nucleation of spherulites and then growth through the molecular diffusion of spherulites. Another is a solidification process based on the thermodynamic balance theory of Fischer [25].

There is also another theory based on the so-called mesomorphic phase generated as a stable intermediate state during the induction period of crystallization in the glassy or semicrystalline states [26–33].

In one published piece of the research on spherulite nucleation and growth of spherulites, it was reported that PP crystallization is suppressed by rapid cooling, and the volume fraction of small monoclinic structure spherulite in smectic crystalline matrix was found by Shibayama et al. [34] to be about 20 vol%.

Meanwhile, concerning the solidification process, Fischer, Nitta, and Takayanagi [35, 36] proposed the lamellar clustering model with thickness being comparable with the end-to-end chain distance.

On the other hand, on the generation theory of the mesomorphic phase, Kaji et al. [37–40] reported that the molten isotactic PP transforms into the mesomorphic phase through a spinodal decomposition-type separation when it is quenched below 0 °C with a rate faster than 80 °C/sec. By the heat treatment of this quenched sheet, it was also reported that a novel morphology with "bamboo-leaf-like" shape with predominantly α-monoclinic high crystallinity is formed.

However, there are few articles on flow-induced crystallization that focused on obtaining highly transparent PP sheet by the industrial extrusion process under the influence of stress rather than the ideal cooling conditions [41–53].

In this chapter, we focus on the shear stress history in an extruder, which has a major influence on the nucleation of spherulites. In addition, cooling conditions and resin characteristics that affect the growth of spherulites will be addressed. To produce highly transparent PP sheets by an industrial process, we discuss various factors contributing to the transparency, especially (a) shear stress control of the melted web before cooling solidification, (b) quick quenching and heat treatment conditions in the subsequent process of extrusion and melted web section, and also (c) resin characteristics.

12.2 Influence of Screw Geometry on External Haze of Melted Web

The evaluation process was conducted as follows. In the first step, to clarify factors in achieving low external haze, the preliminary extrusion tests were conducted using a typical and simple geometry screw. Based on these results and also taking a wide applicability for a large-size extruder into consideration, the screw geometry was optimized to satisfy low external haze and extrusion stability under higher throughput conditions [54].

12.2.1 Preliminary Extrusion Tests Using Typical and Simple Geometry Screw

Concerning the results of extrusion experiments with the different screw geometries, Table 12.1 shows the summary of extrusion performance such as throughput, extrusion stability, specific energy consumption, melt temperature, and transparency. A commercial cast film grade of homo-PP with a melt flow rate (MFR) of 2.0 g/10 min (2.16 kg, 230°C) and molecular weight distribution of 6.6 was used as the material.

Table 12.1 Summary of Extrusion Performance

Screw No.	Screw geometry	Throughput (kg/h)	Pressure fluctuation range at screw tip (MPa)	Specific energy consumption (kWh/kg)	Melt temperature (°C)	Transparency
1	Low compression, metering depth (4 mm)	54	± 18.5	0.242	248	4
2	Low compression, metering depth (1.79 mm)	39	±0.3	0.314	252	2
3	Low compression, metering with mixer	43	±0.3	0.292	261	1
4	Straight channel depth (4 mm)	43	±5.3	0.169	237	5
5	Straight channel depth (5 mm)	57	±9.5	0.170	228	3
6	Straight channel with torpedo (δ = 1 mm, L = 50 mm)	43	±0.3	0.178	245	5
7	Straight channel with torpedo (δ = 1 mm, L = 150 mm)	43	±0.3	0.241	256	1
8	Straight channel with torpedo (δ = 0.5 mm, L = 50 mm)	45	±0.4	0.217	254	1

Definition of transparency: Excellent 5, Very Good 4, Good 3, Fair 2, Poor 1

At first, screws that are the most general and gently tapered compression type (screws 1 to 3) were examined. The evaluation of transparency was done by using a five-point scale based on visibility. A larger number of transparency using this definition means better transparency.

As a result, screw 1, which has the deep channel depth in the metering section, showed better transparency. Screw 2, which has the shallow channel depth in the metering zone, showed worse transparency than screw 1. And it was confirmed that the gently tapered compression-type screw with the strong mixer parts in the metering section (screw 3) had poor transparency of the melted web.

Next, the straight channel screw 4 with the same metering channel depth as screw 1 and, in comparison with it, screw 5 with a deeper channel depth were examined. The transparency of the melted web using screw 4 was greatly improved. Aside from being the same kind of screw, in screw 5 of the lower shear stress type, the unmelted resin particles were observed in the melted web and the transparency had deteriorated.

Based on these results, the screw geometry was selected to enable the melt plasticization in the early stage of the extruder under even low shear stress conditions. Screws 6 to 8 of straight channel depth screw with torpedo-type barrier section at the middle part of the screw were examined. Depending on the torpedo geometry, the shear amount, which is the multiplication of shear rate and residence time at the torpedo section, was different. The contribution of the shear amount at the torpedo section to the transparency of melted web was suggested from these experiments.

To clarify factors in achieving low external haze, the transparency order of melted web as a function of average melt temperature at the center of the die exit and specific energy consumption at screw speed 150 rpm were plotted in Fig. 12.1. It was found from the results obtained by experiments of the screw geometry change that high transparency was achieved under the small specific energy consumption and the low melted temperature. In the case of screw 5, the transparency was not as good owing to the unmelted resin particles in the melted web even though it showed the low specific energy consumption and the low melt temperature of the resin.

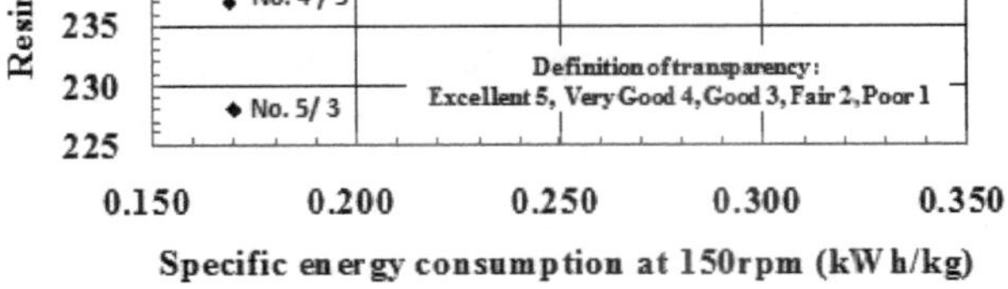

Figure 12.1 Transparency as a function of average melt temperature and specific energy consumption

Based on the aforementioned experimental results, to obtain highly transparent melted web, it is very important not to generate excessive heat owing to the shear stress or the shear rate on the assumption that the unmelted resin particles are not left. The highest transparency of melted resin sheet was obtained by a screw geometry such that the specific energy consumption was small and the melted temperature was low.

This transparent film mechanism is fundamentally different from the shark skin generated at the die exit. That is, the shark skin is generated above a certain limit of shear stress. It means that the lower the melt temperature, the easier the shark skin is generated. On the other hand, the lower melt temperature causes a more transparent appearance of melted web depending on the screw geometry. And even in the state that the critical range of shear stress is not exceeded, the surface roughness of melted web depends on the screw geometry of the extruder.

12.2.2 Optimization of Screw Geometry

Based on these preliminary test results and also taking a wide applicability for a large-size extruder of the production machine into consideration, a screw geometry to satisfy a low external haze and extrusion stability was designed. The results of optimization of the screw geometry are shown in Fig. 12.2 and Table 12.2.

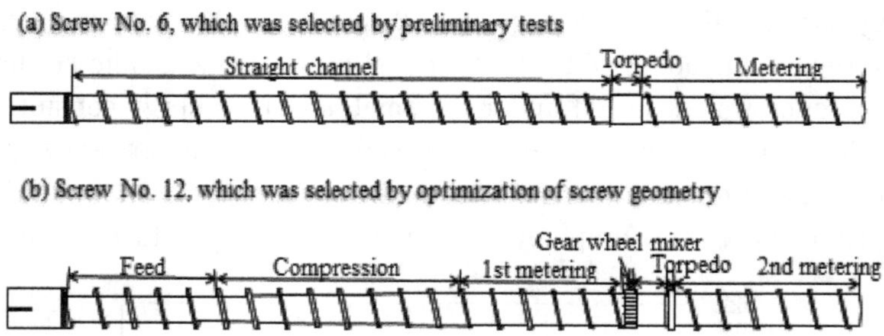

Figure 12.2 Schematic representation of screw geometries that were selected by (a) preliminary tests and (b) optimization of screw geometry

Table 12.2 Results of Optimization of Screw Geometry

Screw No.	Limit screw speed (rpm)	Throughput (kg/h)	Melt temperature (°C)	Pressure fluctuation range at screw tip (MPa)	Solid ratio before torpedo section (%)	Transparency
6	150	43	245	±0.3	-	5
12	210	78	228	±0.1	less than 20	5

Definition of transparency: Excellent 5, Very Good 4, Good 3, Fair 2, Poor 1

The optimization was conducted according to the basic idea that the specific energy consumption becomes small and melted temperature becomes low. As a result, higher throughput with stable extrusion and also lower melt temperature were achieved with screw 12 than with the straight channel depth screw with torpedo-type barrier (screw 6), which was selected by preliminary tests. Excellent transparency of the melted web was obtained by using this optimized screw.

12.3 Influence of Shear Stress in Die on Internal Haze

Transparency has a strong correlation with the number of spherulites [55]; to improve the transparency of PP sheets, it is necessary to decrease the number of spherulites in the sheets. In an extrusion process, the shear stress in an extruder and a die is known to strongly contribute to the nucleation of spherulites. To decrease the number of spherulites, it is necessary to decrease the shear stress in a die in addition to controlling specific energy consumption and heat generation by shear in an extruder, as described in the previous section.

The transparency improves by enlarging the lip openings of a lip section with the largest shear stress in a die, or by raising the skin temperature of a melted web using lip heaters. Here, an example is introduced in which the shear stress was controlled by a multilayer process, and as a result the internal haze was improved. Multilayer extrusion was conducted using a resin with lower melt viscosity (MFR 20 g/10 min) for the surface layers than one (MFR 3 g/10 min) for the core layer (Table 12.3) [56].

Table 12.3 Characterization of PP Sheet Samples

Sheet structure	Total thickness (μm)	Each layer thickness (μm)	MFR (g/10 min)	Meso pentad fraction (mol%)
A	344	344	3	92.5
B/A/B	335	21/293/21	7/3/7	92/92.5/92
C/A/C	347	19/309/19	20/3/20	92/92.5/92

By using multilayer extrusion, the number of spherulites near the surface decreased dramatically and the internal haze of the sheet was improved (Figs. 12.3 and 12.4 and Table 12.4). The number of spherulites in the sheet cross section and the internal haze were closely correlated to each other (Fig. 12.5).

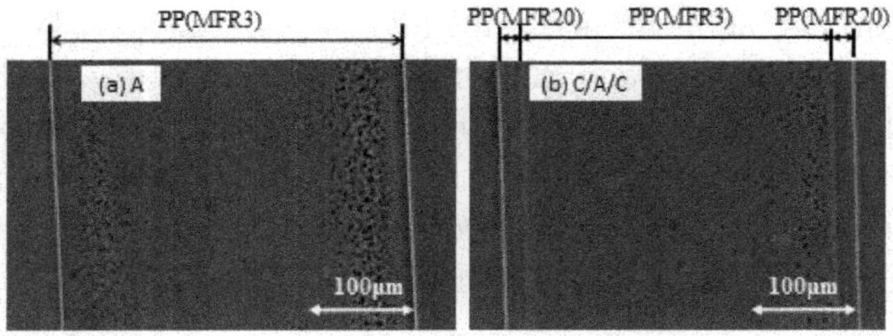

Figure 12.3 Micrographs obtained by phase-contrast microscopy of cross sections of (a) A and (b) C/A/C structure sheets

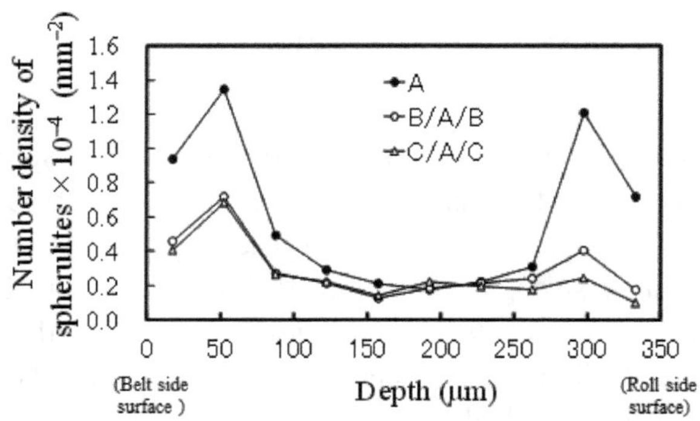

Figure 12.4 Depth profiles of the number density of spherulites for A, B/A/B, and C/A/C sheets

Table 12.4 Transparency of Various PP Sheet Samples Normalized to 350 μm Thickness

Sheet structure	Total haze (%)	Internal haze (%)	Outer haze (%)
A	14.0	13.5	0.5
B/A/B	9.1	8.6	0.5
C/A/C	8.3	7.6	0.7

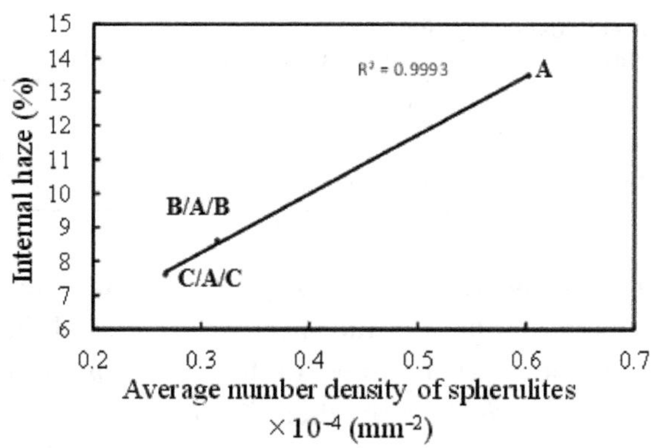

Figure 12.5 Relationship between average number density of spherulites and internal haze for A, B/A/B, and C/A/C sheets

Each distribution on the flow velocity, shear rate, and shear stress in a die lip section was calculated using a finite element method. When the resin with lower melt viscosity was laminated on both surfaces, the multilayer flow velocity distribution in a die lip section was different from monolayer flow, even if the sheeting was carried out under the same throughput conditions (Fig. 12.6).

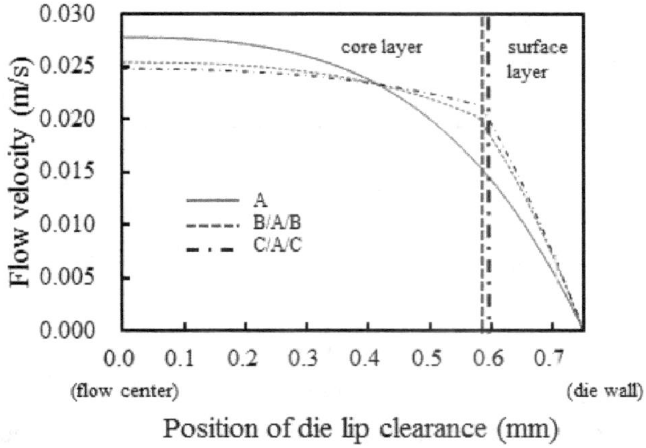

Figure 12.6 Flow velocity distribution at the die lip section in the case of A, B/A/B, and C/A/C flow

As a result, compared with monolayer flow, the shear stress distribution across the sheet thickness direction was markedly reduced even under higher shear rate conditions (Figs. 12.7 and 12.8). When it progressed to a cooling process with this aforementioned state, the sheet surfaces cooled rapidly relative to the interior, and growth of spherulites was restrained. Although the inside of a sheet cooled relatively slowly, because the stress level was low, the stress-induced crystallization was considerably restrained, and the number of spherulites decreased as a result (Fig. 12.4).

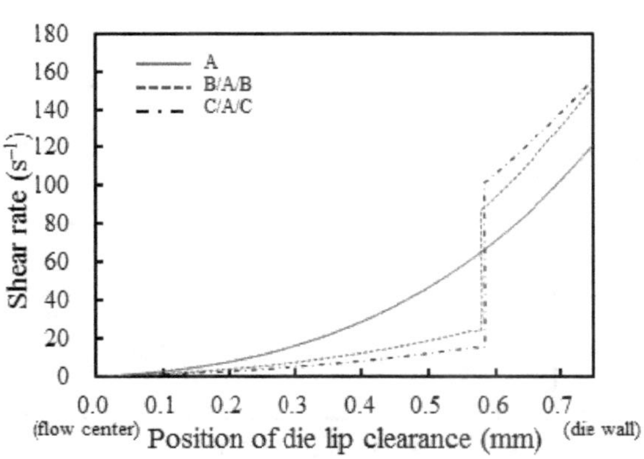

Figure 12.7 Shear rate distribution at the die lip section in the case of A, B/A/B, and C/A/C flow

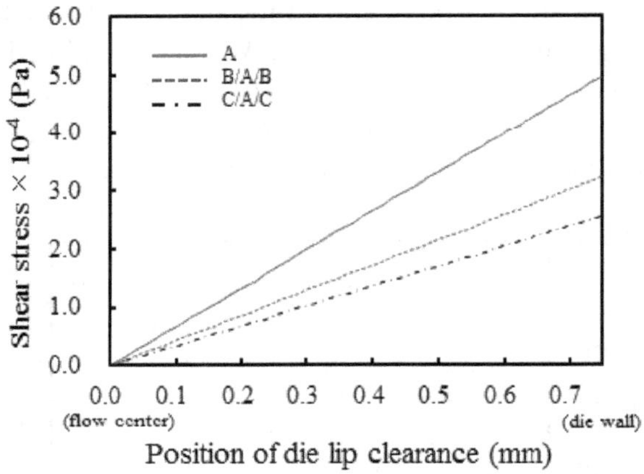

Figure 12.8 Shear stress distribution at the die lip section in the case of A, B/A/B, and C/A/C flow

Furthermore, the cooling simulation in a belt quenching process (Fig. 12.9), which is also discussed in Section 12.4, was conducted. Figure 12.10 shows the temperature distribution in each position that divided the cross section of the sheet into ten layers for cooling time.

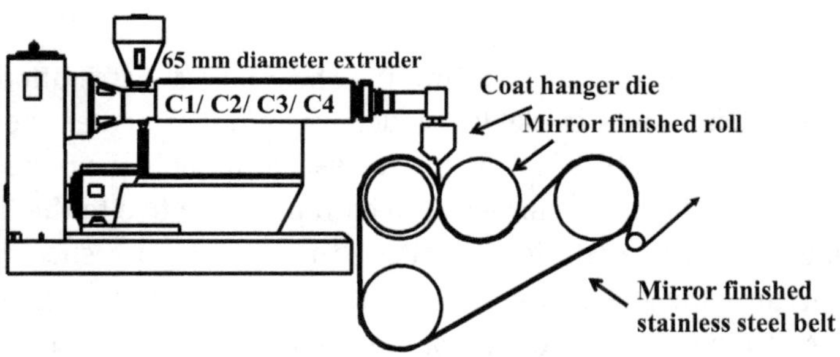

Figure 12.9 Schematic diagram of extrusion sheeting system (single belt quick quenching process)

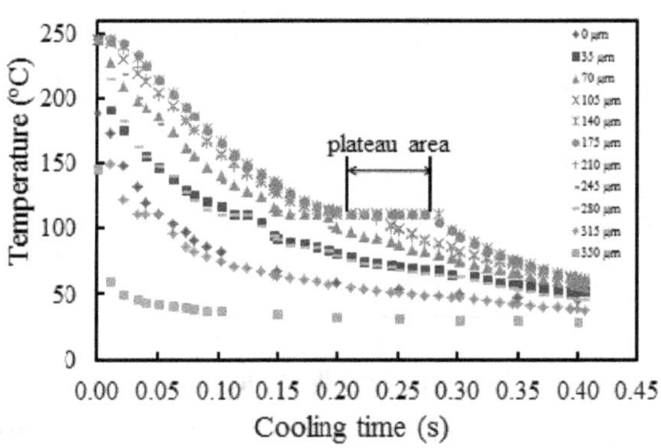

Figure 12.10 Temperature change in the sheet thickness direction in the cooling single belt process

As a result of examining various effects of each independent variable, the prediction formula (Equation 12.1) was derived by multiple regression analysis, using the number density of spherulites as a response variable and both the plateau area time of the temperature in each thickness direction and shear stress as independent variables, respectively:

$$Y = -1.12 \times 10^{4} + 2.96 \times 10^{-1} X_1 - 1.61 \times 10 X_1 X_2 + 9.42 \times 10^{-3} (X_1 X_2)^2 + 1.22 \times 10^{6} X_2 - 2.05 \times 10^{7} X_2^2$$

$$R_f^2 = 0.90 \tag{12.1}$$

where Y (mm^{-2}) is the number density of spherulites, X_1 (Pa) is shear stress at the die lip section, X_2 (s) is the plateau area time of the temperature in each thickness direction, and R_f^2 (–) is the adjusted coefficient of determination.

The number density of spherulites was predicted using the obtained regression formula (Equation 12.1) in each sheet layer composition. The predicted value could express the increasing tendency of the number density distribution of spherulites near the surface and the decreasing effect of the number density of spherulites by lamination of lower melt viscosity resin for the surface layers.

12.4 Analysis of Contributing Factors to Production of Highly Transparent PP Extrusion Sheets

To obtain highly transparent isotactic polypropylene sheet by an industrial process, various factors contributing to transparency were analyzed.

A 65 mm diameter single screw extruder ($L/D = 28$) and a coat-hanger die with 900 mm width and 2 mm lip openings were used. A direct belt system was used for the cooling system (Fig. 12.9). This take-up device was composed of a mirror-finished stainless steel endless belt with 0.65 mm thickness and a mirror-finished roll. The illustrated metal rolls were cooled by a chiller unit. The molten resin sheet, which was extruded from the coat-hanger die, was nipped and quenched by the mirror-finished belt and roll at 18 °C. Because in a direct belt system, quick quenching of both sheet sides is carried out, the number of spherulites decreases markedly and then the transparency improves, compared with the conventional chill roll process.

12.4.1 Influence of Isotacticity on Transparency

In order to consider the influence on the transparency of isotacticity, PP with 92.5 mol% of meso pentad fraction was compared with the high-isotacticity PP with 97.9 mol% [57]. Figure 12.11 shows micrographs obtained by phase-contrast microscopy of a cross section of sheet samples.

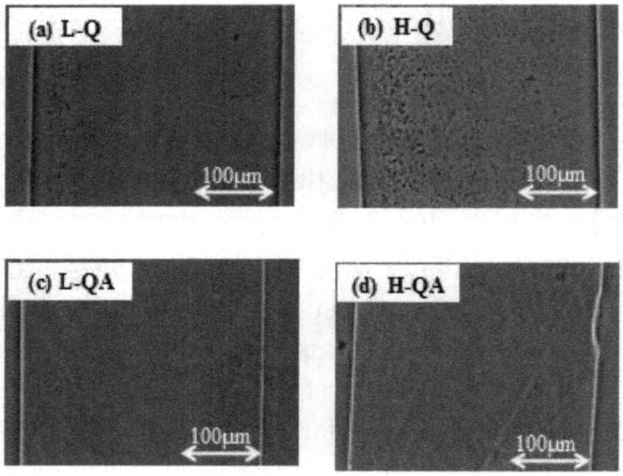

Figure 12.11 Micrographs obtained by phase-contrast microscopy of cross sections of (a) L-Q, (b) H-Q, (c) L-QA, and (d) H-QA sheets

The first characters "H" and "L" of the sample code express the difference in isotacticity. The higher one is set to H and the lower one is set to L. The code "Q" means a quick quenching process, and the code "A" means annealing after cooling.

For both L-Q and H-Q sheet, a lot of small spherulites were observed, especially near the sheet surfaces. Compared with the higher isotacticity PP, even if the sheeting was carried out under the same conditions, the lower isotacticity PP generated fewer spherulites. As a result of heat treating these sheets at 140°C, the image of the spherulites in both sheets was obviously weakened. This shows that the difference in the refractive index of matrix and spherulites was made very small by the heat treatment.

Figure 12.12 shows micrographs of the same samples obtained by polarizing microscopy. In both sheet samples, the maltese cross pattern used detect spherulites could be observed, especially near the sheet surfaces, same as in Fig. 12.11. In the cases of quick quenching samples for which big differences were observed by phase-contrast microscopy, at first glance there were no differences (especially regarding the size and number of spherulites) between quenching sheets and annealed samples observed by polarizing microscopy. It means that there was no big change from the point of view of a birefringence or an optical anisotropy difference between a matrix and spherulites even by the heat treatment of the quick quenched sheet. This phenomenon was not a result of their isotacticities.

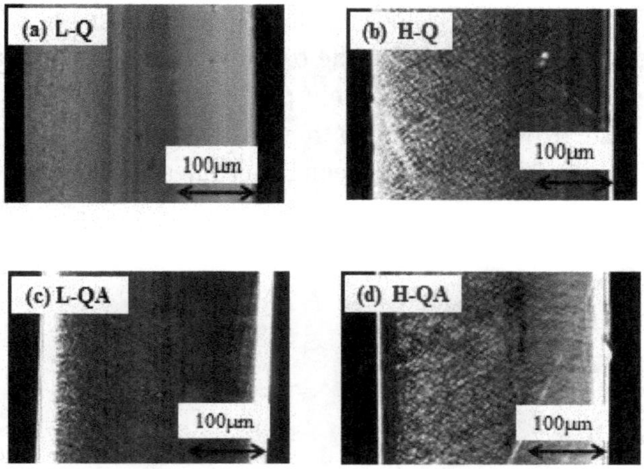

Figure 12.12 Micrographs obtained by polarizing microscopy of cross sections of (a) L-Q, (b) H-Q, (c) L-QA, and (d) H-QA sheets

The micrographs obtained by polarizing microscopy and phase-contrast microscopy of essentially the same cross sections of the same sheet sample were observed. As a result, the images of spherical substances in the phase-contrast micrograph were extremely coincident with that of the maltese cross pattern in a polarizing micrograph.

Figure 12.13 shows the variation of WAXD profiles of the sliced H-Q sheet in the thickness direction as 75, 150, and 75 μm, respectively. The core 150 μm thickness sample and top surface 75 μm thickness sample showed two broad peaks at around 2θ = 15° and 21°, which are known as characteristic features of the mesomorphic phase of isotactic PP [16, 58]. On the other hand, in the case of a bottom surface sample that contained a large fraction of spherulites, the peak position of the first peak was shifted slightly to a lower scattering angle. On the basis of this data, it is proposed that the matrix is almost the mesomorphic phase and the spherulites consisted of α-crystallites.

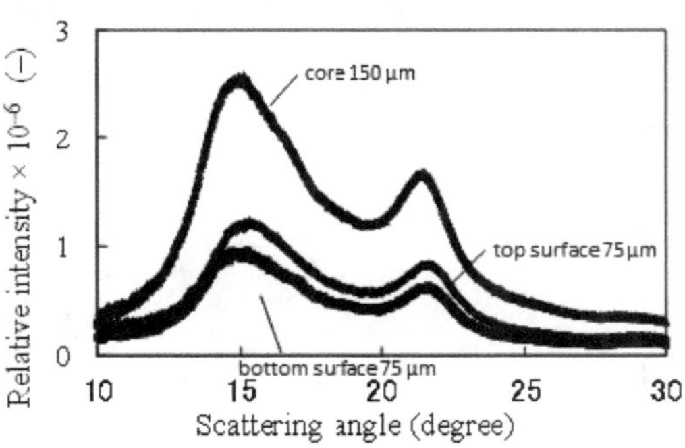

Figure 12.13 Variation of WAXD profiles of sliced H-Q sheet in the thickness direction as 75, 150, and 75 μm, respectively

In Figure 12.14, the scattering patterns of WAXD of (a) H-Q sheet, (b) H-Q sheet annealed at 110°C, and (c) 140°C were measured. In the case of (a), only a mesomorphic phase type peak was detected. At 110°C, the α-crystalline Bragg peaks have already appeared, and at 140°C, the α-crystalline peaks increased further in intensity. It means that the mesomorphic phase was transformed into α-crystal. Thanks to the meso-α transition due to the heat treatment, the density difference and the refractive index difference between spherulites and matrix were decreased.

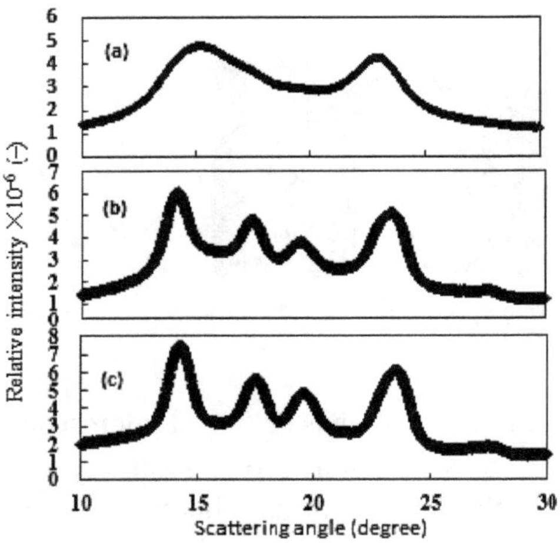

Figure 12.14 Variation of WAXD profiles of (a) H-Q sheet, (b) H-Q sheet annealed at 110°C, and (c) annealed at 140°C

Figure 12.15 shows the variation of V_v light-scattering intensity distribution curves for L-Q, L-QA, H-Q, and H-QA. The V_v light-scattering intensity consists of an optical anisotropy ingredient and a density fluctuation ingredient contribution. These data were measured in the direction of the equator where the optical anisotropy ingredient contribution was comparatively small. For both quick quenching samples, it turned out that these relative intensities were quite small for the heat treatment samples. In particular, the H-Q sheet sample showed a marked downward tendency of relative intensity and had a large effect on transparency by the heat treatment. On the other hand, the relative intensity itself of L-QA sheet is smaller, and it means that the absolute density fluctuation is smaller than that of H-QA sheet. Anyway, by carrying out quenching and heat treatment, density fluctuation was decreased, and it is thought that transparency was improved as a result.

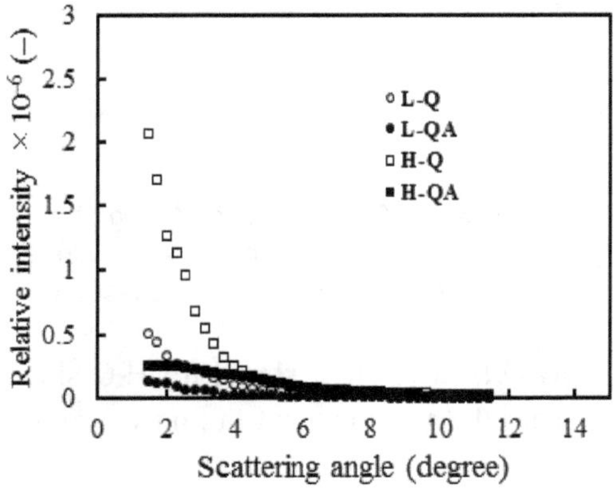

Figure 12.15 Variation of V_v light-scattering intensity distribution curves for L-Q, L-QA, H-Q, and H-QA

Figure 12.16 shows the variation of H_v light-scattering intensity distribution curves for L-Q, L-QA, H-Q, and H-QA sheets. The peak angle of H_v scattering denotes the size of spherical substances, and relative intensity denotes the number of them. Because intensity peaks were clearly observed for both L-Q sheet and H-Q sheet, the existence of spherulites was suggested. Also, this figure shows that the size and number of spherulites did not change substantially, even if both quick quenched sheets were heat treated. As observed in Figs. 12.11 and 12.12, it turns out clearly that the L-Q sheet has markedly smaller numbers and average diameter of spherulites.

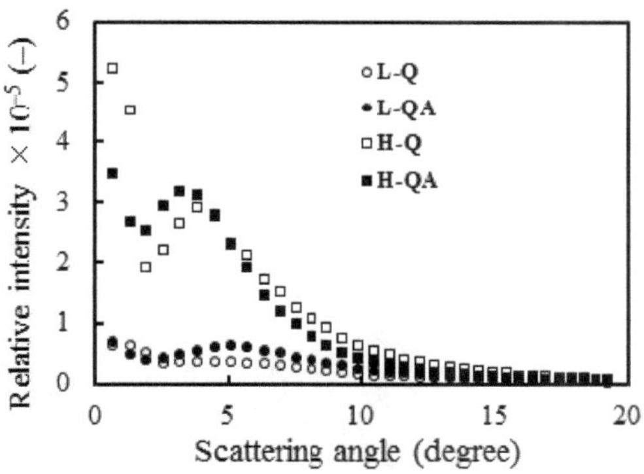

Figure 12.16 Variation of H_v light-scattering intensity distribution curves for L-Q, L-QA, H-Q, and H-QA

Figure 12.17 shows the variation in SAXS intensity distribution curves for L-Q, L-QA, H-Q, and H-QA. When both quick quenched sheets were heated at 140°C, the peak position of relative intensity shifted to the lower scattering angle and increased. It means that in each the average long period is increased.

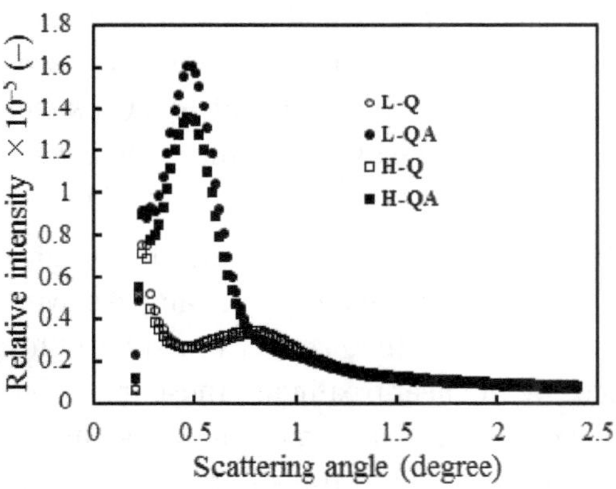

Figure 12.17 Variation of SAXS intensity distribution curves for L-Q, L-QA, H-Q, and H-QA

Table 12.5 Summary of Transparency and Higher Order Structure of PP Sheets

Sample code	H-Q	H-QA	L-Q	L-QA
Meso pentad fraction (mol%)	97.9	97.9	92.5	92.5
Sheeting process	Quick quenching (single belt process)	Quick quenching & annealing at 140°C	Quick quenching (single belt process)	Quick quenching & annealing at 140°C
ΔH (J/g)	82.6	108.8	76.0	102.5
Crystallinity (%)	50.4	66.3	46.3	62.5
Total haze (%)	19.6	8.5	9.4	5.3
Internal haze (%)	18.3	7.7	6.6	3.6
Outer haze (%)	1.4	0.8	2.8	1.7
Average diameter of spherulites (µm)	6.2	7.5	4.7	4.7
Peak value of H_v scattering intensity $\times 10^{-5}$ (-)	2.94	3.15	0.38	0.64
Average correlation length (µm)	1.74	0.51	1.60	0.64
Average long period (nm)	11.0	19.2	11.6	18.4

The transparency and higher-order structure of L-Q, L-QA, H-Q, and H-QA are summarized in Table 12.5. As shown in Table 12.5, it turns out that L-Q sheet, which has lower isotacticity, has markedly smaller numbers and average diameter of spherulites. As a result, the obtained sheet has better transparency with low internal haze.

Although the crystallinity of L-QA sheet calculated from the melting enthalpy data of differential scanning calorimetry (DSC) was increased by heat treatment, the size and number of spherulites were almost unchanged. The internal haze of L-QA sheet was markedly improved by the heat treatment. These results show the same behaviors as the higher-isotacticity PP.

It was found that the average correlation length of both sheets calculated by V_v light-scattering intensity data was decreased by the heat treatment. Each sample's average long period was increased almost twice. Because the sizes of the spherulites themselves do not result in large differences by the heat treatment, these increases of average long period suggest that small crystallites of mesomorphic phase in the matrix merge into some larger structure during the meso-α transition process [59].

From the above-mentioned points, it is considered that the lower-isotacticity PP, with which it is relatively difficult to make chain-folded structures, has a slower nucleation and growth rate of spherulites. As a result, the internal haze of the lower-isotacticity PP became better than higher-isotacticity PP.

Figure 12.18 shows the transparence improvement model of a quick quenching PP sheet by heat treatment.

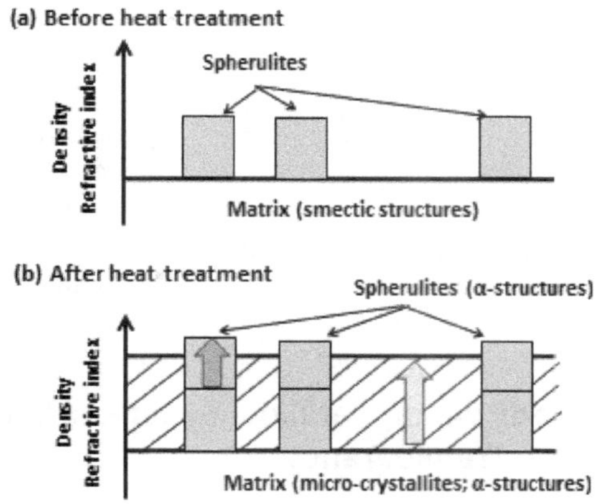

Figure 12.18 Transparence improvement model of quick quenched PP sheets by heat treatment

Before heat treatment, the density difference of spherulites and the matrix domain, which was observed by phase-contrast microscopy, infringed on the transparency of the PP sheet. After the heat treatment, it is thought that the matrix transformed from smectic structure to α-monoclinic crystal phase, and both the density and refractive index of spherulites and matrix were increased. The density and refractive index difference between spherulite and matrix were decreased, and, as a result, the transparency was drastically improved. On this occasion, because the density difference of spherulite and the matrix became small by the use of the heat treatment, it is thought that apparent correlation length also became small.

12.4.2 Influence of Molecular Weight Distribution on Transparency

Two PP grades used for the experiment have the same meso pentad fraction and MFR but different molecular weight distributions (PP-1: Mw/Mn = 4.26; PP-2: Mw/Mn = 4.66). The optical characteristics of the sheets quenched by the belt process are shown in Fig. 12.19. It was found that PP-1 with a narrow molecular weight distribution has relatively better transparency than PP-2. Specifically, the external haze of PP-1 sheet became better because PP-1 with a narrow molecular weight distribution had shorter relaxation time.

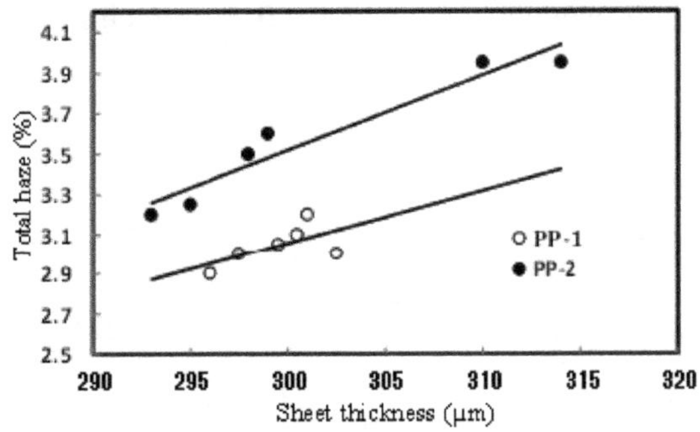

Figure 12.19
Effect of molecular weight distribution on sheet transparency:
(a) PP-1 (Mw/Mn = 4.26);
(b) PP-2 (Mw/Mn = 4.66)

12.4.3 Influence of Addition of Metallocene Linear Low Density Polyethylene on Transparency

The contribution to the transparency by 10 wt% addition of various metallocene LLDPEs, which has narrow composition distribution with wide range density, was investigated [57].

Figure 12.20 shows the relationship between the density of blended LLDPE and total haze. These total haze data, which were obtained by quenching and heat treatment at 120, 130, and 140°C, are indicated. In the case of the quenched 100% PP sheet, the total haze was 9.4% under the same quenching condition as Fig. 12.20, and it was 5.3% after heat treatment. It was clear that, after heat treatment, the transparency of the blended sheets markedly depended on the density of the LLDPE grade. The LLDPE grades that showed excellent transparency of 2% or less of the total haze had a density range of 903 to 905 kg/m³. The LLDPE grades with other density ranges could not achieve the transparency of 2% or less of the total haze. The suitable heat treatment temperature condition that showed the best transparency was changed according to the density of added LLDPE.

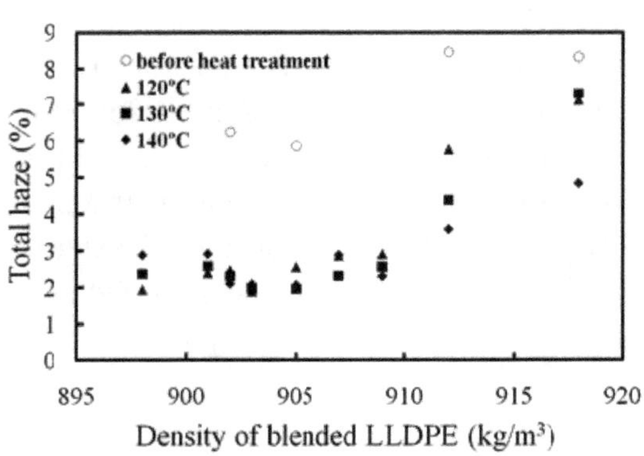

Figure 12.20
Effect of density of blended metallocene LLDPE into PP on sheet transparency for different heat treatment temperatures

According to the observation results of the H_v light scattering patterns and the transmission electron microscape TEM micrographs, in the case of L-Q with 10 wt% LLDPE, the number and size of spherulites remarkably decreased. However, the appearance of the quick quenching sheet became rather clouded. From this transparency behavior, Fig. 12.21 focuses on the sheet density and the refractive index. Figure 12.21 shows the relationship between sheet density and refractive index for metallocene LLDPE (898, 905, and 912 kg/m³) blended PP sheets. These sheets were obtained by quick quenching and then heat treatment at 120, 130, and 140°C. It was found that the refractive index increased linearly with the sheet density by the various heat treatment conditions.

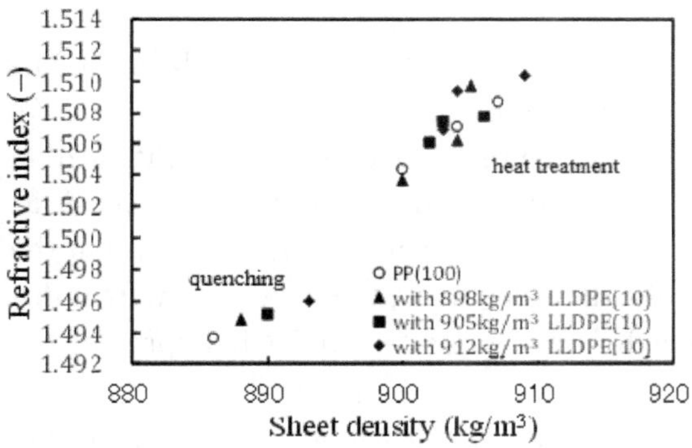

Figure 12.21
Relationship between sheet density and refractive index for various metallocene LLDPE blended PP sheets

Figure 12.22 shows the relationship between "absolute value of refractive index difference" at each heat treatment temperature and the total haze based on the data obtained from Fig. 12.21. Figure 12.22 shows that the total haze became small, regardless of the density of added LLDPE, when the absolute value of refractive index difference became small.

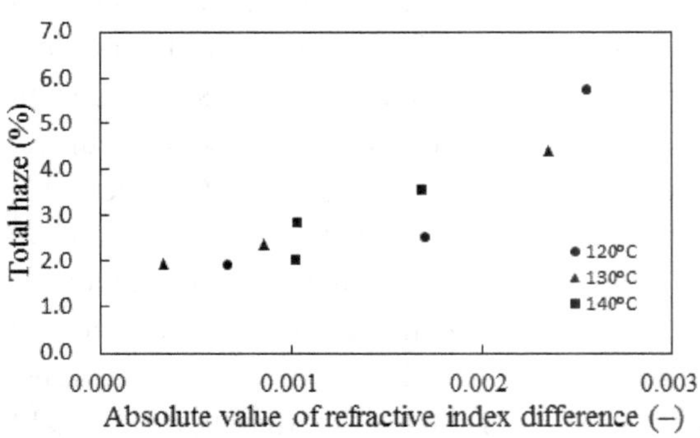

Figure 12.22
Relationship between transparence and absolute value of refractive index difference of PP sheet and LLDPE blended sheet as a function of heat treatment temperatures

From these results, the following can be considered. In the case of 100% PP sheet, the spherulites increased the internal haze. In the case of the LLDPE-added sheet, although the number and size of spherulites remarkably decreased, the refractive index difference of LLDPE particles and PP matrix increased the internal haze. That is, in the case of a blend of PP and LLDPE, because the LLDPE particles have different refractive index from PP material, the transmitted light waves are scattered and the sheet becomes cloudy. When the refractive indexes of LLDPE and PP are approached by the suitable heat treatment temperature, the cloudy sheet appearance changes to a highly transparent appearance because the degree of light scattering decreases. However, because the sensitivity of the refractive index to the heat treatment temperature differs in accordance with density of added LLDPE, the optimum temperature conditions to obtain the best transparency differ.

12.5 Conclusion

To improve further the transparency of an isotactic PP sheet obtained by an industrial extrusion quenching process, the important factors were investigated. We focused on an external haze control of PP melted resin sheet before its solidification and the transparency improvement behavior under a quick quenching process, as the subsequent process of a melt web section. Specifically, it was clarified that the screw geometry, the shear stress in a die, the isotacticity, the molecular weight distribution, and the mixing of metallocene LLDPE were very effective in improving the transparency of PP sheet.

The transparency of melted resin sheet was obtained by screw geometry such that the specific energy consumption was small and the melted temperature was low. When the shear stress in a die lip section was reduced, the number of spherulites near the sheet surface remarkably decreased. In particular, by using multilayer extrusion using a resin with lower melt viscosity for the surface layers than for the core layer, the internal haze of the sheet was improved.

When quick quenching of both sheet sides was carried out, the number of spherulites decreased markedly and then the transparency improved, compared with the conventional chill roll process. Even if the sheeting was carried out under the same quenching conditions, the lower-tacticity PP generated fewer and smaller-sized spherulites, and most transparent sheet was obtained. It is thought that the lower-isotacticity PP, with which it is relatively difficult to make chain-folded structures, has slower nucleation and growth rate of spherulites, and as a result the internal haze became better than in higher-isotacticity PP. Before the heat treatment, the density difference of spherulites and the matrix domain, which was observed by phase-contrast microscopy, infringed on the transparency of PP sheet. After the heat treatment, it is thought that the matrix transformed from a smectic structure to an α-monoclinic crystal phase, and both the density and refractive index of spherulites and matrix were increased. The density and refractive index difference between spherulite and matrix were decreased, and as a result the transparency was drastically improved.

Concerning the influence of the molecular weight distribution on the transparency, the quick quenched PP sheet with a narrow molecular weight distribution and shorter relaxation time resulted in relatively better external haze and transparency.

Moreover, when metallocene LLDPE was added into PP, the number of generated spherulites and the growth rate of them were restrained by the fine distribution particles of LLDPE in the PP matrix. In the case of addition of LLDPE with the specific density to PP, the transparency was markedly improved by the heat treatment of quick quenching sheets.

References

1. Ramamurthy, V., *J. Rheol.* (1986) 30, p. 337
2. Ghanta, V.G., Ruese, B.L., Denn, M.M., *J. Rheol.* (1999) 43, p. 435
3. Kurtz, S.J., Adv. Rheol. Proceedings of 9th International Congress on Rheology, Acapulco, Mexico (1984) p. 399
4. El Kissi, N., Piau, J.M., Toussaint, F., *J. Non-Newtonian Fluid Mech.* (1997) 68, p. 271
5. Hatzikiriakos, S.G., Stewart, C.W., Dealy, J.M., *Int. Polym. Process.* (1993) 8, p. 30
6. Hatzikiriakos, S.G., Dealy, J.M., *Int. Polym. Process.* (1993) 8, p. 36
7. Piau, J.M., El Kissi, N., Toussaint, F., Mezghani, A., *Rheol. Acta* (1995) 34, p. 40
8. Cogswell, F.N., *J. Non-Newtonian Fluid Mech.* (1977) 2, p. 37
9. Wang, S.Q., Drda, P.A., Inn, Y.W., *J. Rheol.* (1996) 40, p. 875
10. Barone, J.R., Plucktaveesak, N., Wang, S.Q., *J. Rheol.* (1998) 42, p. 813
11. Mackley, M.R., Rutgers, R.P.G., Gilbert, D.G., *J. Non-Newtonian Fluid Mech.* (1998) 76, p. 281
12. Rutgers, R.P.G., Mackley, M.R., *J. Rheol.* (2000) 44, p. 1319
13. Kometani, H., Kitajima, H., Matsumura, T., Suga, T., Kanai, T., *Seikei Kakou* (2007) 19, p. 118
14. Slichter, W.P., Mandell, E.R., *J. Appl. Phys.* (1958) 29, p. 1438
15. Natta, G., *SPE J.* (1959) 15, p. 373
16. Miller, R.L., *Polymer* (1960) 1, p. 135
17. Wyckoff, H.W., *J. Polym. Sci.* (1962) 62, p. 83
18. Gletin, M., Rahalkar, R.R., Hendra, P.J., Cudby, M.E.A., *Polymer* (1981) 22, p. 731
19. Hendra, P.J., Vile, J., Willis, H.A., Cudby, M.E.A., *Polymer* (1984) 25, p. 785
20. Corradini, P., Petraccone, V., DeRosa, C., Guerra, G., *Macromol.* (1986) 19, p. 2699
21. Gomez, M.A., Tanaka, H., Tonelli, A.E., *Polymer* (1987) 28, p. 2227
22. Corradini, P., DeRosa, C., Guerra, G., Petraccone, V., *Polym. Commun.* (1989) 30, p. 281
23. Wang, Z-G., Hsiao, B.S., Srinivas, S., Brown, G.M., Tsou, A.H., Cheng, S.Z.D., Stein, R.S., *Polymer* (2001) 42, p. 7561
24. Hoffman, J.D., Lauritzen, J.I., *J. Res. Natl. Bur. Stand.* (1961) 64, p. 297
25. Fischer, E.W., *Polym. J.* (1985) 17, p. 307
26. Imai, M., Mori, K., Mizukami, T., Kaji, K., Kanaya, T., *Polymer* (1992) 33, p. 4451
27. Imai, M., Mori, K., Mizukami, T., Kaji, K., Kanaya, T., *Polymer* (1992) 33, p. 4457
28. Kaji, K., Nishida, K., Kanaya, T., Matsuba, G., Konishi, T., Imai, M., *Adv. Polym. Sci.* (2005) 191, p. 187

29. Konishi, T., Nishida, K., Kanaya, T., Kaji, K., *Macromol.* (2005) 38, p. 8749
30. Kaji, K., Nishida, K., Kanaya, T., Matsuba, G., Konishi, T., Imai, M., *Seni Gakkaishi* (2006) 62, p. 88
31. Coccorullo, I., Pantani, R., Titomanlio, G., *Polymer* (2003) 44, p. 307
32. Gezovich, D. M., Geil, P. H., *Polym. Eng. Sci.* (1968) 8, p. 202
33. Grubb, D. T., Yoon, D. Y., *Polym. Commun.* (1986) 27, p. 84
34. Shibayama, M., Katoh, K., Iwamoto, T., Takahashi, D., Nomura, S., *Polym. J.* (1991) 23, p. 837
35. Nitta, K., Takayanagi, M., *J. Macromol. Sci. Phys.* (2003) B42, p. 107
36. Nitta, K., Takayanagi, M., *Polym. J.* (2006) 38, p. 757
37. Hsu, C. C., Geil, P. H., Miyaji, H., Asai, K., *J. Polym. Phys. Ed.* (1986) 24, p. 2379
38. Ogawa, T., Miyaji, H., Asai, K., *J. Phys. Soc. Jpn.* (1985) 54, p. 3668
39. Nishida, K., Konishi, T., Kanaya, T., Kaji, K., *Polymer* (2004) 45, p. 1433
40. Konishi, T., Nishida, K., Kanaya, T., *Macromol.* (2006) 39, p. 8035
41. Swartjes, F. H. M., Peters, G. W. M., Rastogi, S., Meijer, H. E. H., *Int. Polym. Process.* (2003) 18, p. 53
42. Schrauwen, B. A. G., van Breemen, L. C. A., Spoelstra, A. B., Govaert, L. E., Peters, G. W. M., Meijer, H. E. H., *Macromol.* (2004) 37, p. 8618
43. van der Beek, M. H. E., Peters, G. W. M., Meijer, H. E. H., *Macromol.* (2006) 39, p. 1805
44. Balzano, L., Rastogi, S., Peters, G. W. M., *Macromol.* (2009) 42, p. 2088
45. Housmans, J-W., Steenbakkers, R. J. A., Roozemond, P. C., Peters, G. W. M., Meijer, H. E. H., *Macromol.* (2009) 42, p. 5728
46. Vega, J. F., Hristova, D. G., Peters, G. W. M., *J. Therm. Anal. Calorim.* (2009) 98, p. 655
47. Shen, C.-Y., Zhou, Y.-G., Zheng, G.-Q., Liu, C.-T., Chen, J.-B., Li, Q., *Polym. Eng. Sci.* (2008) 48, p. 2454
48. Wang, R. S. K., Zhao, P., Zhang, Q., Du, R., Fu, Q., Li, L., Li, L., *Polymer* (2007) 48, p. 4529
49. Zhang, J., Shen, K., Na, S., Fu, Q., *J. Polym. Sci., Part B* (2004) 42, p. 2385
50. Wang, Y., Na, B., Fu, Q., Men, Y., *Polymer* (2004) 45, p. 207
51. Li, L., de Jeu, W.H., *Faraday Discuss.* (2005) 128, p. 299
52. Zhang, C. G., Hu, H. Q., Wang, D. J., Yan, S. K., Han, C. C., *Polymer* (2005) 46, p. 8157
53. Elmoumni, A., Winter, H. H., *Rheol. Acta* (2006) 45, p. 793
54. Funaki, A., Takubo, T., Kanai, T., *Polym. Eng. Sci.* (2010) 50, p. 420
55. Matsuzawa, N., Kamatani, Y., Kanai, T., Yamada, T., Sakauchi, K., Uehara, H., *22nd Annu. Meet. Polym. Process. Soc.* (2006) p. 157
56. Funaki, A., Kuratani, S., Yamada, T., Kanai, T., *Seikei Kakou* (2011) 23, p. 299
57. Funaki, A., Kondo, K., Kanai, T., *Polym. Eng. Sci.* (2011) 51, p. 1068
58. Natta, G., Corradini, P., *Nuovo Cimento* (1960) 15, p. 40
59. Funaki, A., Kanai, T., Saito, Y., Yamada, T., *Polym. Eng. Sci.* (2010) 50, p. 2356

Author Index

I

J

K

L

M

S

T

U

V

W

X

Y

Z

Subject Index

I

K

L

M

N

O

P

Q

R

S

X

Z

www.ingramcontent.com/pod-product-compliance
Lightning Source LLC
LaVergne TN
LVHW081927200926
848484LV00009B/2387